SAS® and SPSS® Program Solutions

for use with

Applied Linear Regression Models

Fourth Edition

Michael H. Kutner
Emory University

Christopher J. Nachtsheim
University of Minnesota

John Neter
University of Georgia

Prepared by
William Replogle
University of Mississippi Medical Center

William Johnson
University of Mississippi Medical Center

**McGraw-Hill
Irwin**

Boston   Burr Ridge, IL   Dubuque, IA   Madison, WI   New York   San Francisco   St. Louis
Bangkok   Bogotá   Caracas   Kuala Lumpur   Lisbon   London   Madrid   Mexico City
Milan   Montreal   New Delhi   Santiago   Seoul   Singapore   Sydney   Taipei   Toronto

SAS® and SPSS® Program Solutions for use with
APPLIED LINEAR REGRESSION MODELS
Kutner, Nachtsheim, and Neter

Published by McGraw-Hill/Irwin, an imprint of The McGraw-Hill Companies, Inc., 1221 Avenue of the
Americas, New York, NY 10020. Copyright © 2004 by The McGraw-Hill Companies, Inc. All rights reserved.

1 2 3 4 5 6 7 8 9 0 2QSR/2QSR 0 9 8 7 6 5 4

ISBN 0-07-298958-0

www.mhhe.com

The McGraw-Hill Companies

Preface

The first edition of *Applied Linear Regression Models* was published in 1983 and immediately became a popular text with both instructors and students. Each edition has brought many improvements while the clarity in presentation and usefulness of examples and exercises have remained consistent. We have used the book as a required text for many years; however, the students in our courses typically have limited statistical programming skills. As a result, many students spend a disproportionate amount of time mastering programming skills rather than statistical concepts. In an effort to minimize the amount of assignment time students spend learning SAS® or SPSS® computer code, we present source programming code and sample output for the most salient chapter examples given in *Applied Linear Regression Models*, *4th Edition* by Kutner, Nachtsheim and Neter (2003). The resulting text is intended to be a supplement to *Applied Linear Regression Models*, although it also may be a helpful reference for the practicing analyst/statistician.

This book is divided into two sections: SAS® Solutions and SPSS® Solutions. Chapters 1-14 of each section parallel the 14 chapters of *Applied Linear Regression Models*. In addition, we have added a Chapter 0 that explains how to get started using the SAS and SPSS statistical packages. This chapter includes a brief discussion of how to access SAS or SPSS, enter data, perform computations, and store and retrieve results. Each of the remaining chapters gives specific SAS and SPSS instructions and resulting output for solving the examples presented in *Applied Linear Regression Models*.

We display illustrative SAS and SPSS syntax almost entirely in *Insert* boxes. Each *Insert* is intended to represent a fully functional program that is not dependent on syntax from preceding *Inserts*. We have taken this approach to simplify the steps for new users. As you become familiar with SAS and SPSS programming, you will be able to combine some *Inserts* to decrease the number of steps needed to complete a specific analysis. Additionally, most programming tasks can be accomplished through many different choices of computer syntax. Some of the more esoteric programming techniques, however, can be difficult for beginners. We have taken a pragmatic approach and presented syntax using strategies that we have found to be fairly easy for first year graduate students to grasp. These strategies can be easily modified to analyze problems similar to the illustrative examples. As you gain experience, you may choose to use more sophisticated and perhaps more efficient programming solutions.

The modules required to perform the analyses in the SAS chapters are SAS/STAT™, SAS/IML™, and SAS/GRAPH™. The modules used in the SPSS chapters are SPSS Base™, SPSS Advanced Models™, and SPSS Regression Models™. Both SAS and SPSS have a variety of other products that may be useful for some applications presented in *Applied Linear Regression Models*. The products we used are commonly available to students enrolled in statistics courses and can be used to solve almost all of the examples and exercises in *Applied Linear Regression Models*.

William H. Replogle

William D. Johnson

CONTENTS

SAS Program Solutions

Chapter 3 SAS®

Diagnostics and Remedial Measures

Chapter 4 SAS®

Simultaneous Inferences and Other Topics in Regression Analysis

Chapter 5 SAS®

Matrix Approach to Simple Linear Regression Analysis

SPSS Program Solutions

Chapter 3 SPSS®

Diagnostics and Remedial Measures

Chapter 4 SPSS®

Simultaneous Inferences and Other Topics in Regression Analysis

Chapter 5 SPSS®

Matrix Approach to Simple Linear Regression Analysis

SAS® Program Solutions

Chapter 0 SAS®

Working with SAS®

The purpose of this chapter is to introduce the requisite methods for entering, creating, and opening the data files provided with *Applied Linear Regression Models*, *5th Edition* by Kutner, Nachtsheim and Neter (2003). Although we will work with many of the SAS programs as we progress through *Applied Linear Regression Models,* the present chapter is not intended to be a comprehensive review of SAS capabilities.

SAS is a modular, integrated, product line for the analytical process. The SAS products can be used for:

- Planning
- Data collection
- Data access and management
- Report writing
- Statistical analysis
- File handling

SAS is available for most mainframe and personal computers running under the Windows® or Macintosh® operating systems. The modules required to perform the analyses in the forthcoming chapters are SAS/STAT, SAS/IML, and SAS/GRAPH.

Starting SAS: Most computer systems have a SAS icon on the desktop. We double click the icon to start SAS. When we first start SAS software, the five main SAS windows open: the Explorer, Results, Editor, Log, and Output windows. In this book, we will be working with the Editor, Output, and Log windows. The Editor window allows the user to enter, edit, and submit SAS programs.

The first step in executing a SAS program is to enter the program on the Program Editor screen. If we are in the Program Editor, we can enter a program by clicking the File tab at the top of the screen and then clicking Open on the drop-down menu. We can then identify the folder and name of the program to be opened. Alternatively, we can type a program directly onto the Program Editor. Once we have entered the program onto the Program Editor screen, we can save it to a folder for future use. There is a *run* icon at the top right portion of the screen. When we are ready to execute a program that is entered on the Program Editor screen, we select Run > Submit from the dropdown menu or click the symbol.

When the program is executed, the Output screen will appear on the monitor. The Output screen contains the results that are printed when the submitted program has executed. The Program

Editor, Log, and Output tab bars appear at the bottom of the monitor. If there is a graph in the output, there may also be a Graph tab bar at the bottom of the monitor. We can click on these tab bars to go to the Editor, Log, Output, and Graph screens. Alternatively, we can click on the View tab bar at the top of the monitor and then click on one of the tab bars on the pop-down menu. *Caution*: If nothing happens when we execute the program, we probably have an error in the program. We can click on the Log tab to look for error messages that may be printed by the SAS System. We can scroll down the Log screen to search for errors. Once we discover the errors, we must go back to the Program Editor screen, make corrections, and re-execute the program. When writing a new program, it is not unusual to go through this process several times. We can scroll down the Output screen to see different portions of the output. If applicable, we can click on the Graph bar to inspect graphical output.

At the top of the screen, there is a tab labeled Help. We can click on the Help tab and search for answers to questions about the SAS System. As discussed above, when we submit a SAS program from within the Editor window, SAS creates the Output window, which automatically moves to the front of other windows. The Log window displays messages including error messages regarding the SAS syntax in the submitted program.

The SAS Program: The typical SAS program is made up of two or more steps that are usually specified in the program in the following order:

- Data Step
- Procedure Step

The Data step defines the data file. The Procedure Step identifies the specific SAS programs (Procedures) you want to implement. After we define the data file in a Data Step, we can specify Procedures to be executed. However, once we have specified a Data Step and then one or more Procedures, you cannot alter the data without going through a new Data Step. Thus, a typical organization of a SAS program might be:

- Data a
- Procedure
- Procedure
- Procedure
- Data b
- Procedure
- Procedure
- Data c
- Procedure

Data b and Data c may specify entirely new data sets or they may specify data sets created from results of procedures executed in previous data steps.

The SAS System is made up of many computer programs (software) designed to perform the five functions listed above. As users of the SAS System, we must use SAS instructions to call on the "canned" programs within the SAS System to execute a specific application of a task.

We shall refer to a single SAS instruction as a *statement*. It is essential that all SAS statements end with a semicolon, and the most common SAS programming error is to omit the semicolon. We refer to a set of SAS statements used in an execution of a task as *SAS code* or a *SAS program*. We begin with a basic program to enter data and then to print the data. We then show how to write SAS statements to modify the data using algebraic expressions.

A Basic SAS Program to Input/Output Data: The simple SAS program in Insert 0.1 serves as a starting point for learning SAS syntax:

```
                              INSERT 0.1

options ls = 72 ps = 65 nodate nonumber;
        title 'Chapter 0 Example 1';
        data ch00ex01;
          input firstname $ age gender $ glucose;
        datalines;
        eric    41  male    102
        karen   38  female  93
        ann     43  female  121
        eugene  35  male    109
        frank   47  male    134
        gary    39  male    117
        gloria  42  female  95
        gwen    46  female  99
        robert  37  male    103
        mary    40  female  110
      ;
   proc print;
    var firstname glucose;
   run;
```

In this SAS program the first statement begins with the SAS keyword *options*. This option statement instructs the SAS System to print output according to the specified instructions. The *ls = 72* option specifies that output will have line size with a maximum of 72 printed characters per line. The *ps = 65* option specifies that output will have page size with a maximum of 65 printed lines per page. The *nodate* option specifies that the date is not to be printed on output pages. The *nonumber* option specifies that the page numbers are not to be printed on the output. *ls*, *ps*, *nodate*, and *nonumber* are all SAS code or keywords.

The second statement in the program begins with the SAS keyword *title*. This statement instructs the SAS System to print the words that appear in single quotes at the top of all pages of printed output; i.e., *Chapter 0 Example 1* is to be printed at the top of all output pages.

The third statement in the program begins with the SAS keyword *data*. This statement initiates a data step in the program. We have chosen the name *ch00ex01* for this data step. Thus, *ch00ex01* is not a SAS keyword but rather a name we have assigned for convenience. The data defined in this step will be stored under the data file name *ch00ex01* and can be accessed in a future step using this name.

The fourth statement in the program begins with the SAS keyword *input*. This statement assigns variable names to data fields and specifies the type of data that is in the field. We have chosen "firstname" as a name for the first variable, "age" as a name for the second variable, "gender" as a name for the third variable, and "glucose" as a name for the fourth variable. The dollar sign ($) after *firstname* and *gender* specifies that these variables may contain alphanumeric data; if nothing appears after a variable name in an *input* statement, the corresponding data field must contain only numeric data. The variable names must be arranged in order corresponding to the order of the data fields in the input data set.

The fifth statement in the program begins with the SAS keyword *datalines*. This statement indicates that the typed data set follows. The data fields in the typed data set are separated by one or more blank spaces. Each line of data contains data for a single study unit. A semicolon appears after the last line of data. There are many allowable variations in the way the data are entered in a SAS program, but the way described here is probably the simplest.

The sixth statement in the program begins with the SAS keyword *proc*. This keyword initiates a procedure step in the program. Many *procedures* are available in the SAS system. For illustration, we have chosen the *print* procedure. The SAS syntax *proc print* requests printed output.

The seventh statement in the program begins with the SAS keyword *var*. This keyword identifies the variables to be used in executing the *proc print* statement. Here, we have chosen *firstname* and *glucose* as the variables to be printed. Thus, the first name and glucose value will be printed for all data lines in the data set. Output from *proc print* and output that may be automatically printed by other SAS procedures is printed to the Output window described above. We can obtain a paper copy of the Output window by using the File>Print drop down menu.

The last statement in the program begins with the SAS keyword *run*. This is an important statement because it instructs SAS to run all procedures specified since the end of the data file or since the previous *run* statement. If we omit the *run* statement in the program shown in Insert 0.1, the *proc print* statement will not be executed.

Reading the ALRM data files: The ALRM text comes with a disk containing numerous data sets. We recommend that you transfer these data from the supplied disk to a permanent storage device. For instance, we created a directory named *ALRM* on the C: drive and transferred all the ALRM data sets to this directory. All SAS illustrations in our book specify a path to this directory. As you write SAS code to perform statistical analyses of other data, you will need to modify the path specified within the illustrative SAS syntax to identify the storage device and directory for the data sets relevant to each application.

In text Chapter 1, the authors present the Toluca Company data set in Table 1.1. To read this data set, we start SAS and in the Editor window type the syntax as shown in Insert 0.2 – Method 1. (In Insert 0.2 and throughout this book, an * in an insert box represents a comment that explains some of the syntax.) The *data* statement tells SAS that we want to create a data set named Toluca. The *infile* statement identifies the path to the Toluca Company data set. Note that the directories for chapters 1 – 9 have 2 spaces between the word "chapter" and the corresponding chapter number. Thus, the *infile* statement must also contain 2 spaces as in "chapter 1." The input statement tells SAS the variable names to be assigned to the 2 variables in the data set. Variable names are determined by the user but must be consistent with naming conventions within SAS. In general, the variable name can consist of no more than 10 characters, cannot begin with a number, and cannot contain special characters. Execution of the syntax in Insert 0.1 – Method 1 will read the data for the Toluca Company data set into the variables *lotsize* and *workhrs*. *Lotsize* and *workhrs* will then be available for use with other SAS syntax. We execute the syntax in Insert 0.1 by selecting Run > Submit from the dropdown menu or by clicking the 🏃 symbol.

We can enter data directly into the SAS Editor (syntax file) as shown in Insert 0.2 – Method 2. On text page 3, the authors present a data set containing the number of units sold (*X*) and the dollar sales amount (*Y*) for 3 time periods. As with Method 1, the *data* statement tells SAS that we want to create a data set named *example* and the *input* statement tells SAS the variable names to be assigned to the 3 variables in the imbedded data set. We name the three variables *period*, *units*, and *sales*, respectively, as shown in Insert 0.2. These names are arbitrary but follow conventions for variable names within SAS. The *datalines* statement indicates to SAS that the subsequent lines will contain data until a semicolon ";" is encountered. The syntax in Insert 0.2 – Method 2 will read the data for the example given on text page 3 into the variables *period*, *units*, and *sales*. *Period*, *units*, and *sales* will then be available for use with other SAS syntax.

The user must define by the *input* statement the type of variable to be read. That is, we can define a variable as a numeric, string (alphanumeric), or various other types of variables. For instance, each of the 3 variables defined on text page 3 is a numeric variable and can have only numeric values. If we were to enter the name of a city, such as "Memphis," we would have to define the variable as a string variable so that alpha-numeric characters could be entered to represent the city name. A string variable is defined by placing a $ character after the variable name, for example:

input city $

In Insert 0.2 – Method 2, we can define *units* as a string variable. But, once a variable is defined as a string, mathematical operations cannot be performed on the variable. So, we can take the square root of *units* if it is defined as a numeric variable but not if it is defined as a string variable.

INSERT 0.2	INSERT 0.2 continued
* Method 1; data Toluca; infile 'C:\alrm\chapter 1 data sets\ch01ta01.txt'; input lotsize workhrs;	* Method 2; data example; input period units sales; datalines; 1 75 150 2 25 50 3 130 260 ;

Creating a new variable: Suppose in the Toluca Company example above we wish to find the square of *lotsize* for use in polynomial regression. In Insert 0.3 we open the Toluca Company example as before and then compute a new variable, *lotsqr*, the square of *lotsize*. We can now use SAS *proc* statements to invoke predefined statistical routines. For instance, we can instruct SAS to use the predefined *means* routine to find the mean of *lotsize* and *lotsqr*. Since there will often be more than one data set created and available for use in a SAS session, we can optionally include on the *proc* statement a *date=dataset* statement (e.g., *data=toluca*) to identify the data set for use when executing the *proc* statement. Execution of Insert 0.3 will place the results as shown in SAS Output 0.1 in the Output window.

INSERT 0.3

```
data Toluca;
infile 'C:\alrm\chapter  1 data sets\ch01ta01.txt';
input lotsize workhrs;
lotsqr=lotsize**2;
proc means data=toluca;
 var lotsize lotsqr;
run;
```

SAS Output 0.1

Variable	N	Mean	Std Dev	Minimum	Maximum
lotsize	25	70.0000000	28.7228132	20.0000000	120.0000000
lotsqr	25	5692.00	4015.48	400.0000000	14400.00

In Insert 0.4, we read the Toluca Company data set as before and name the variables *lotsize* and *workhrs*. We then perform a regression analysis (*proc reg*) where the model statement specifies the dependent and independent variables to be used in the *model*, i.e.,

dependent variable = list of one or more independent variables.

In Insert 0.4, the dependent variable is *workhrs* and the single independent variable is *lotsize*. We can optionally request that residuals from the fitted model be saved by the *r* option on the *model* statement. We then tell SAS in which data set (*out=step2*) to store the results and to store the residuals (*r*) in a variable named *residual*. Now, the data set named *toluca* will contain *lotsize* and *workhrs*, and the data set named *step2* will contain *lotsize*, *workhrs*, and *residual*. Thus, to find the square of *residual*, we use *data step3* (the data file name *step3* is arbitrary) to issue a new data statement and then use *set step2* to identify *step2* as the data set to be used to find *ressqr=residual**2;*. Thus, we point SAS to the new data file (*step2*) to find the variable *residual* to be squared. We can then print (*proc print*) the values of *ressqr*.

7

```
                          INSERT 0.4

data Toluca;
  infile 'c:\alrm\chapter  1 data sets\ch01ta01.txt';
  input lotsize workhrs;
proc reg;
  model workhrs = lotsize / r;
  output out=step2 r=residual;
run;
data step3;
 set step2;
 ressqr=residual**2;
 run;
proc print;
 var ressqr;
 run;
```

SAS provides an extensive help system available from the Help dropdown menu. Additional help can be obtained at www.sas.com using SAS Notes® and other technical support options. SAS and all other SAS Institute Inc. product or service names are registered trademarks or trademarks of SAS Institute Inc. in the USA and other countries. We are grateful to SAS Technical Support for their assistance. SAS® software is licensed by:

SAS Institute Inc.
100 SAS Campus Drive
Cary, NC 27513-2414
USA
www.SAS.com

Linear Regression with One Predictor Variable

This chapter of the text introduces the simple linear regression model. The idea is to use statistical methods to find a model that allows us to predict the value of a dependent variable Y given the value of an independent variable X. We are given Y_i and X_i for a sample of n subjects and we want to find a "best fitting" model for predicting Y given X. The usual approach is to begin by assuming the model is of the form:

$$Y_i = \beta_0 + \beta_1 X_i + \varepsilon_i$$

for the ith subject, $i = 1, \ldots, n$. This is often referred to as a *simple linear regression model*. The first step is to find estimates of β_0 and β_1 so that the "fitted" model is best, as defined in the text, among all models of this form. The statistical analysis that involves estimating, constructing confidence intervals, and testing hypotheses about β_0 and β_1 is called *regression analysis*. The model is generalized in subsequent chapters to include more than one independent variable. The text introduces the basic concepts and gives easy to follow numerical examples that illustrate the ideas. Also, the text presents excerpts of a large data set in Table 1.1 followed by a description of the problem and details of a regression analysis. The text refers to this example as the Toluca Company example. A step-by-step set of SAS solutions is given below.

The Toluca Company Example

Data File: ch01ta01.txt

The Toluca Company manufactures refrigeration equipment and replacement parts (text page 19). The parts are manufactured in "runs" and groups of runs comprise lots of varying sizes. The data disk provided with the text contains data on 25 manufacturing runs under the file name ch01ta01.txt. For each run, this file contains data on lot size and work hours, defined as the number of labor hours required to produce the lot. The aim of the example is to investigate the relationship between work hours and lot size. We wish to fit a simple linear regression model that allows us to predict work hours from lot size. We let the variable names *workhrs* denote work hours and *lotsize* denote lot size, and take these to be the dependent variable Y and the independent variable X, respectively, in the simple linear regression model. See Chapter 0, "Working with SAS," for instructions on opening the Toluca Company data file. Note that the directories for Chapters 1-9 on the data disk provided with your text contain 2 spaces between the word "chapter" and the corresponding chapter number. Thus, the *infile* statement must also contain 2 spaces to identify the correct path. After opening and printing ch01ta01.txt, the first 5 lines of output should look like SAS Output 1.1.

SAS Output 1.1

Obs	lotsize	workhrs
1	80	399
2	30	121
3	50	221
4	90	376
5	70	361

Scatter plots: The text begins a regression analysis of the Toluca Company data with a scatter plot showing work hours plotted against lot size in Figure 1.10 (a & b), text page 20. We provide SAS syntax for constructing this scatter plot in Insert 1.1 and display the results in SAS Output 1.2 (a & b) below. To replicate the axes of text Figure 1.10 (a & b), we set the values of the horizontal axis (*haxis*) to range from 0 to 150 while incrementing by 50 units, and we set the values of the vertical axis (*vaxis*) to be range from 0 to 600 while incrementing by 100 units. Additionally, we could use the *proc gplot* procedure to generate the scatter plot with the regression line (text Figure1.10 b) by changing the *interpol=none* statement to *interpol=rl* (regression line) in Insert 1.1. The SAS *proc plot* procedure will also generate scatter plots although this procedure lacks some features available in *proc gplot*.

```
INSERT 1.1

*Produce Output 1.2a;
data Toluca;
infile 'C:\alrm\chapter  1 data sets\ch01ta01.txt';
input lotsize workhrs;
proc gplot;
  symbol1 value=dot interpol=none;
  plot workhrs*lotsize / haxis=0 to 150 by 50 vaxis=0 to 600 by 100;
run;

*Produce Output 1.2b;
data Toluca;
infile 'C:\alrm\chapter  1 data sets\ch01ta01.txt';
input lotsize workhrs;
proc reg;
  model workhrs = lotsize;
   plot workhrs*lotsize / haxis=0 to 150 by 50 vaxis=0 to 600 by 100;
run;
```

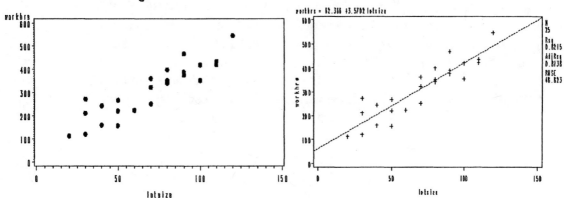

Estimation of β_0 and β_1: To find the least squares estimates b_0 and b_1 of β_0 and β_1, respectively, we follow the overview of the required computations that are outlined in Table 1.1, text page 19. In Insert 1.2, we show SAS syntax that performs these computations. We first find the means for *workhrs* ($\overline{X} = 70$) and *lotsize* ($\overline{Y} = 312.28$). Next, we find $X_i - \overline{X}$ (*xdev*), $Y_i - \overline{Y}$ (*ydev*), $(X_i - \overline{X})^2$ (*xdev2*), $(Y_i - \overline{Y})^2$ (*ydev2*), $(X_i - \overline{X})(Y_i - \overline{Y})$ (*xdev * ydev = cross*). The last step in Insert 1.2 is to find the sum of $(X_i - \overline{X})(Y_i - \overline{Y})$ ($\sum$ *cross*=70,690) and $(X_i - \overline{X})^2$ ($\sum$ *xdev2*=19,800). Using equation 1.10, text page 17, we calculate:

$$b_1 = \frac{\sum (X_i - \overline{X})(Y_i - \overline{Y})}{\sum (X_i - \overline{X})} = \frac{70,690}{19,800} = 3.5702$$

and

$$b_0 = \overline{Y} - b_1 \overline{X} = 312.28 - 3.5702 * 70 = 62.365$$

INSERT 1.2

```
* Find mean of lotsize and workhrs;
data Toluca;
infile 'C:\alrm\chapter 1 data sets\ch01ta01.txt';
input lotsize workhrs;
proc means;
  var lotsize workhrs;
run;

* Calculate least squares estimates;
data Toluca;
infile 'C:\alrm\chapter 1 data sets\ch01ta01.txt';
input lotsize workhrs;
  xdev=(lotsize-70);
  ydev=workhrs-312.28;
  xdev2=(lotsize-70)**2;
  ydev2=(workhrs-312.28)**2;
  cross=xdev*ydev;
proc means sum;
  var cross xdev2;
run;
```

Estimated Regression Function: We used the SAS syntax shown in Insert 1.2 to demonstrate how SAS code can be written to provide intermediate results when we compute the parameter estimates b_0 and b_1. We now use the SAS Regression procedure *proc reg* to find b_0 and b_1. This procedure is shorter and easier to use, but it does not provide the results of all the intermediate computations shown on text page 19. In Insert 1.3, we request a regression analysis via the SAS statement *proc reg*. Next, we identify the dependent and independent variables via the *model* statement. Here, the statement: *model workhrs = lotsize* specifies the simple linear regression model with *workhrs* taken to be the Y variable (the dependent variable) and *lotsize* taken to be the X variable (the independent or predictor variable). Partial results are shown in SAS Output 1.3. The "Unstandardized Coefficients" shown under column heading "Parameter Estimate" give the results $b_0 = 62.36586$ and $b_1 = 3.57020$. Here, b_0 is given in the row labeled "Intercept" and b_1 is given in the row labeled "lotsize". These results are consistent with those shown on text page 20 and in the calculations of Insert 1.2 above. From the above regression analysis, we estimate the regression function to be: $\hat{Y} = 62.366 + 3.570(X)$.

INSERT 1.3

```
data Toluca;
infile 'C:\alrm\chapter 1 data sets\ch01ta01.txt';
input lotsize workhrs;
proc reg;
  model workhrs = lotsize;
run;
```

SAS Output 1.3

Analysis of Variance					
Source	**DF**	**Sum of Squares**	**Mean Square**	**F Value**	**Pr > F**
Model	1	252378	252378	105.88	<.0001
Error	23	54825	2383.71562		
Corrected Total	24	307203			

Parameter Estimates					
Variable	**DF**	**Parameter Estimate**	**Standard Error**	**t Value**	**Pr > |t|**
Intercept	1	62.36586	26.17743	2.38	0.0259
lotsize	1	3.57020	0.34697	10.29	<.0001

In Insert 1.4, we request the predicted value and residual for each lot size specified in the data, as in column 3 of text Table 1.2, in the *model* statement of *proc reg*. We put a slash after the predictor variable(s) listed in the model statement and then write *p* for predicted values and *r* for residuals. Thus, the statement: *model workhrs = lotsize / p r* instructs SAS to compute

12

unstandardized predicted values and residuals for the specified regression model (Insert 1.4). Although not shown here, when we write *model workhrs = lotsize / p r*, SAS prints six columns of output: (1.) the dependent variable, (2.) the unstandardized predicted values, (3.) the standard errors of the predicted values, (4.) the residuals, (5.) the standard error of the residuals, and (6.) the Studentized residuals. Some of these results are not discussed until later in the text.

To get output results into the data file so we can perform calculations involving these results, we must use the *output* statement. The *output* statement instructs SAS to create a new data file that contains the original data and specified output. The *out = results* specification instructs SAS to store the new data file under the name we chose to call *results*. The code *p = yhat* specifies that the predicted values will be stored in the *results* data file as a new variable that we chose to call *yhat*. The code *r = residual* specifies that the residuals will be stored in the *results* data file as a new variable that we chose to call *residual*. That is, we store the predicted and residual values from the regression analysis and the original variables in a new data file that now comprises *lotsize, workhrs, run, yhat,* and *residual*. It is necessary to go through this procedure to get the calculated predicted values and residuals out of the output file and into a data file so we can perform additional calculations that involve these results.

At this point, we must specify a new *data step*. We do this by writing *data step2* where *step2* is the arbitrary name we have chosen for the new data set. We then write *set results* to identify *results* as the old data set to be used in the current instructions. Initially, the data set *step2* will be identical to data set *results*. Next, we compute the squared residual value using the statement *sqres=residual*residual*. The SAS system appends the data for *sqres* to the data set *step2*. Finally, we use *proc print* to print the first 5 lines of the data file. Because we do not specify in the *proc print* statement the data set to be printed, the most recent data set, *step2*, is printed. The printed results shown in SAS Output 1.4 are consistent with those shown in Table 1.2, text page 22. Note that the statement *run+1* that appears after the *input* statement was used to number the data lines in the data file. These numbers are stored in the original data file under the name we chose to call *run*. Once the data lines are numbered, the *where* statement can be used after the *print* statement to specify specific lines to be printed. Thus, the statement *where run le 5* used after the *print* statement instructs SAS to print the data lines whose *run* number is less than or equal to 5. The *var* statement that appears after the *print* statement identifies the variable names whose data values are to be printed.

INSERT 1.4	**INSERT 1.4 continued**
| ```
data Toluca;
infile 'C:\alrm\chapter  1 data sets\ch01ta01.txt';
input lotsize workhrs;
run+1;
proc reg;
  model workhrs = lotsize/ p r;
    output out=results p=yhat r=residual;
run;
``` | ```
data step2;
  set results;
sqres=residual*residual;
proc print;
  where run le 5;
  var lotsize workhrs yhat residual sqres;
run;
``` |

SAS Output 1.4

Obs	lotsize	workhrs	yhat	residual	sqres
1	80	399	347.982	51.018	2602.83
2	30	121	169.472	-48.472	2349.53
3	50	221	240.876	-19.876	395.05
4	90	376	383.684	-7.684	59.04
5	70	361	312.280	48.720	2373.64

Chapter 2 SAS®

Inferences in Regression and Correlation Analysis

In this chapter, the authors introduce the fundamentals for making statistical inferences about the parameters of the regression model. The model of interest is:

$$Y_i = \beta_0 + \beta_1 X_i + \varepsilon_i$$

where:

β_0 and β_1 are unknown parameters
X_i are known constants for $i = 1, 2, \ldots, n$.
ε_i are unknown errors associated with the true model.

We assume the ε_i are independent $N(0, \sigma^2)$ for $i = 1, 2, \ldots, n$. The assumptions made here about the errors differ from those made in Chapter 1. The normality assumption is required by theoretical considerations used to develop confidence intervals and test statistics for making inferences involving the model parameters. Inferences were not considered in Chapter 1 and, therefore, normally distributed error terms were not required. The inferences of specific interest pertain to confidence intervals and tests of hypotheses for model parameters and functions of model parameters. The statistical analysis associated with estimation and inference involving regression models is called *regression analysis*.

Chapter 1 of the text presents formulas for the estimators b_0 and b_1 of the parameters β_0 and β_1, respectively. These estimators yield estimates that vary from sample to sample and, therefore, have a (sampling) distribution of their own. It is desirable then to make inferences about the true values of the parameters based on their sample estimates. Let β_j represent either β_0 or β_1 depending on whether $j = 0$ or 1 and let b_j denote the corresponding estimator. We can estimate the variance of b_j from a single sample of Y's and X's and the square root of this variance estimate is an estimate of the standard error of b_j. The text denotes the variance and standard error estimators by $s^2\{b_j\}$ and $s\{b_j\}$, respectively, and gives their computational formulas (see equation 2.9 page 43 and equation 2.23, text page 49).

Consider the null hypothesis:

$$H_0: \beta_j = 0$$

A statistic for testing this hypothesis against either a one or two-sided alternative is:

$$t = \frac{b_j}{s\{b_j\}}$$

Under the assumptions of the model this statistic is distributed as Student's t with $n-2$ degrees of freedom.

Let $t(1-\alpha/2; n-2)$ denote the $(\alpha/2)100$ percentile of the t distribution with $n-2$ degrees of freedom. The $1-\alpha$ confidence limits for β_j are:

$$b_j \pm t(1-\alpha/2; n-2)\, s\{b_j\}$$

Chapter 2 of the text presents these formulas and others of inferential interest in the simple linear regression model.

Further Aspects of the Toluca Company Example

Data File: ch01ta01.txt

We return to the Toluca Company example first presented in Chapter 1 (text page 19). We let the variable names *workhrs* denote work hours and *lotsize* denote lot size, and take these to be the dependent variable Y and the independent variable X, respectively, in the simple linear regression model. See Chapter 0, "Working with SAS," for instructions on opening the Toluca Company data file. After opening and printing ch01ta01.txt, the first 5 lines of output should look like SAS Output 1.1.

SAS Output 1.1

Obs	lotsize	workhrs
1	80	399
2	30	121
3	50	221
4	90	376
5	70	361

We wish to perform a regression analysis using the model:

$$\text{workhrs} = \beta_0 + \beta_1(\text{lotsize}) + \varepsilon$$

Confidence Interval for β_1: On text page 46 the management team wants an estimate of β_1 with a 95% confidence interval. The $1-\alpha$ confidence limits for β_1 are (equation 2.15, text page 45):

$$b_1 \pm t(1-\alpha/2; n-2)\, s\{b_1\}$$

We request confidence intervals for the regression coefficients using the *clb* option in the *model* statement given in Insert 2.1. From SAS Output 2.2 we see that the 95% confidence interval is $2.852 \le \beta_1 \le 4.288$, consistent with text page 46.

```
data Toluca Company;
  infile 'c:/neter/ch01ta01.dat';
  input lotsize workhrs;
proc reg;
  model workhrs = lotsize / clb;
run;

* Find tvalue by quantile function;
data tvalues;
tvalue=tinv(0.95,23);
proc print;
run;
```

SAS Output 2.2

Analysis of Variance					
Source	**DF**	**Sum of Squares**	**Mean Square**	**F Value**	**Pr > F**
Model	1	252378	252378	105.88	<.0001
Error	23	54825	2383.71562		
Corrected Total	24	307203			

Root MSE	48.82331	**R-Square**	0.8215	
Dependent Mean	312.28000	**Adj R-Sq**	0.8138	
Coeff Var	15.63447			

Parameter Estimates							
Variable	**DF**	**Parameter Estimate**	**Standard Error**	**t Value**	**Pr > |t|**	**95% Confidence Limits**	
Intercept	1	62.36586	26.17743	2.38	0.0259	8.21371	116.51801
lotsize	1	3.57020	0.34697	10.29	<.0001	2.85244	4.28797

Tests of Hypothesis Concerning β_1: The t-test for H_0: $\beta_1 = 0$ is:

$$t^* = \frac{b_1}{s\{b_1\}}$$

with $n - 2$ degrees of freedom (equation 2.17, text page 47). We get the numerical value for this test statistic as part of the output from the SAS instructions given in Insert 2.1. In SAS Output

2.2 we see (t^*) = 10.290 for *lotsize* and the associated *p*-value = 0.000 (Pr>|t|). Since $p < 0.05$ we conclude $\beta_1 \neq 0$.

To test that $\beta_1 > 0$ (a one-sided test), we require that $|t^*| > t(1-\alpha; n-2)$. The standard output in SAS displays the *p*-value for a two-sided test. To get the *p*-value for a one-sided test, first determine whether the numerical value of the *t*-test is negative or positive and hence whether it tends to favor the specified one-sided alternative. A negative *t*-test favors H_a: $\beta_1 < 0$ whereas a positive *t*-test favors H_a: $\beta_1 > 0$. If the *t*-test favors the specified one-sided alternative, just divide the *p*-value displayed in the SAS output by 2. In the example, (t^*) = 10.290 is positive and this favors H_a: $\beta_1 > 0$ with *p*-value = 0.000/2 = 0.000. If the numerical value of the *t*-test does not favor the specified alternative, it is probably sufficient to quote $p > 0.50$ as the *p*-value. More precisely, the *p*-value for the one-sided *t*-test in this situation is 1 minus the *p*-value displayed in the SAS output divided by 2. In the present example, if we had specified H_a: $\beta_1 < 0$ and found (t^*) = 10.290, the *p*-value would be $1 - 0.000/2 = 1.000$. In this circumstance, we would not reject H_0.

Sometimes it is desirable to find the critical value $t(1-\alpha; n-2)$ corresponding to a *t*-test. To find $t(1-\alpha; n-2)$, use the SAS *"tinv"* quantile function (Insert 2.1 above) to request a new variable named *tvalue*. Let T denote a random variable that has a *t*-distribution with degrees of freedom = *df*. Suppose we want to find the value t such that $P(T > t) = \alpha$ for a value of α that we specify. Equivalently, suppose we want to find the value t such that $P(T \leq t) = 1-\alpha$. We can use the SAS *quantile function tinv*($1-\alpha$, *df*) for this purpose. When we specify $1-\alpha$ and *df*, this function returns the associated *t*-value. To find the critical value for testing H_a: $\beta_1 > 0$, controlling the level of significance at $\alpha = 0.05$, we must find *tvalue* = $t(1-\alpha; n-2)$ or *tvalue* = $t(0.95, 23)$. This critical value ($t = 1.71$) is obtained from the *quantile function* in Insert 2.1 and is printed to the output file by the *proc print* statement. Since the obtained *t*-value ($t = 10.29$) is greater than the critical value $t = 1.71$, we conclude $\beta_1 > 0$. If we want to obtain the critical value for a two-sided test where $\alpha = 0.05$, we find $t(1-\alpha/2; n-2)$ or $t(0.975, 23)$ using the above quantile function.

Confidence Interval for β_0: To obtain the 90% confidence interval for β_0 (text page 49), we specify *alpha* = 0.10 in the *model* statement and the 90% (rather than 95%) confidence intervals for the model parameters are printed. Notice that if we do not specify *alpha* in the *model* statement, SAS uses *alpha* = 0.05 by default. The complete model statement is

$$model\ workhrs = lotsize\ /\ clb\ alpha = \textbf{\textit{0.10}};$$

Alternatively, we can find the 90% confidence interval through direct calculations as follows. We must find $t(1-\alpha/2; n-2)$. Note that SAS Output 2.2 presents the 95% confidence interval. Since we desire a 90% confidence interval, $\alpha = 0.10$ and $t(0.95, 23) = 1.71$. From SAS Output 2.2, we find the standard deviation of b_0 (Intercept) is $s\{b_0\} = 26.17$. From text equation 2.25 (equation 2.25, text page 49) the 90% confidence interval is:

$$b_0 \pm\ t(1-\alpha/2;\ n-2)\ s\{b_0\}$$

or

$$62.366 \pm 1.71 * 26.17$$

Simplifying this expression, we find the 90% confidence interval:

$$17.5 \le \beta_0 \le 107.2$$

Confidence Interval for $E\{Y_h\}$: Once specific sample data have been used to produce parameter estimates b_0 and b_1, we denote the fitted model as:

$$\hat{Y}_i = b_0 + b_1 X_i \qquad i = 1, 2, \ldots, n$$

We call $\hat{Y}_i$ the *predicted value* of Y when the predictor variable is X_i. Let X_h denote a specific value of the independent variable in the simple linear regression model. As noted by the authors, "X_h may be a value which occurred in the sample, or it may be some other value of the predictor variable within the scope of the model." The corresponding predicted value is denoted $\hat{Y}_h$. We can select many random samples of data and different samples may produce different estimates b_0 and b_1 and hence different $\hat{Y}_h$. Hence, for any such X_h there is a distribution corresponding to the random variable $\hat{Y}_h$. This random variable has a mean and variance denoted $E\{\hat{Y}_h\} = E\{Y_h\}$ and $\sigma^2\{\hat{Y}_h\}$, respectively. Estimators of the mean and variance are $\hat{Y}_h$ and $s^2\{\hat{Y}_h\}$ (see equation 2.30, text page 53 for the formula for $s^2\{\hat{Y}_h\}$).

On text page 54, the authors use Toluca Company Example to find the 90% confidence interval for $E\{Y_h\}$ when lot size (X_h) = 65 units. The $1 - \alpha$ confidence limits are given in equation 2.33, text page 54:

$$\hat{Y}_h \pm t(1 - \alpha/2; n - 2) s\{\hat{Y}_h\}$$

To find $s^2\{\hat{Y}_h\}$ (equation 2.30, text page 53) we need the mean squared error (*MSE*), which is given in SAS Output 2.2 above (*MSE* = 2383.716). We use *proc means* (Insert 2.2 – Step 1) to find $\bar{X}$ = *lotsize mean* = 70. To find $(X_i - \bar{X})^2$ we first compute a new variable, *xdevsq* (Step 2). To find the sum of $\sum(X_i - \bar{X})^2$ = 19,800 we use *proc means* (Step 2) with the *sum* option (SAS Output 2.3). We can now find:

$$s^2\{\hat{Y}_h\} = MSE\left[\frac{1}{n} + \frac{(X_h - \bar{X})^2}{\sum(X_i - \bar{X})^2}\right]$$

Thus,

$$s^2\{\hat{Y}_h\} = 2383.7\left[\frac{1}{25} + \frac{(65 - 70)^2}{19,800}\right] = 98.37$$

$$s\{\hat{Y}_h\} = \sqrt{s^2\{\hat{Y}_h\}} = 9.918$$

To find the 90 percent confidence interval, we also need $t(1-\alpha/2; n-2)$. Using the *quantile function* introduced above where $t(1-\alpha/2; n-2) = t(0.95; 23)$, *tinv*(0.95, 23) returns $t(0.95, 23) =$ 1.71. Using $X_h = 65$ units and the parameter estimates in SAS Output 2.2 above, we find

$$\hat{Y}_h = b_0 + b_1 X_h$$

or

$$\hat{Y}_h = 62.366 + 3.570(65) = 294.4$$

Substituting these numerical values into equation 2.33, text page 54, we obtain SAS Output 2.4 and

$$294.4 \pm 1.71\big(9.918\big)$$

or

$$277.44 \le E\{Y_h\} \le 311.36$$

SAS Output 2.3

Analysis Variable : xdevsq
Sum
19800.00

Obs	yhat65	sdyhat	ci65hi	ci65lo
1	294.429	9.91758	311.428	277.430

Confidence Band for Regression Line: To find the confidence band for the regression line as shown in equations 2.40 and 2.40a, text page 62, we use the Working-Hotelling $1-\alpha$ confidence band to establish boundary values for each *lotsize* value:

$$\hat{Y}_h \pm W\, s\{\hat{Y}_h\}$$

where:

$$W^2 = 2\, F(1-\alpha;\ 2, n-2)$$

We use text equations 2.28, 2.30, 2.40, and 2.40a to find the confidence band for the Toluca Company regression example. In Insert 2.3 we use the SAS *finv* quantile function of the form *finv(p,df1,df2)* to find F (*fvalue = finv*(0.9, 2, 23) = 2.549). We then find the upper (*bandhi*) and lower (*bandlo*) confidence bands and request the plot (*proc plot*) shown in SAS Output 2.7. This plot has an appearance similar to text Figure 2.6, text page 63.

```
                          INSERT 2.3

data Toluca Company;
  infile 'c:\alrm\chapter  1 data sets\ch01ta01.txt';
  input lotsize workhrs;
fvalue=finv(.9,2,23);
w2=sqrt(2*(fvalue));                       * Text eq 2.40a;
yhat=62.366+3.57*lotsize;                  * Text eq 228;
sdyhat=sqrt(2383.716*((1/25)+(((lotsize-70)**2)/19800)));   * Text eq 2.30;
bandhi=yhat+(w2*sdyhat);                   * Text eq 2.40;
bandlo=yhat-(w2*sdyhat);                   * Text eq 2.40;
proc plot;
 plot bandhi*lotsize bandlo*lotsize yhat*lotsize / overlay;
run;
```

SAS Output 2.7

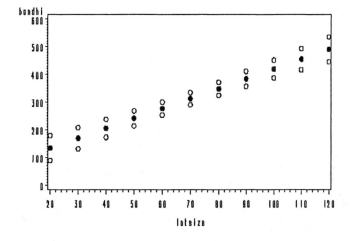

21

To replicate the plot on text page 63, we use *proc gplot* in Insert 2.4. We do not want to plot the predicted values of *Y* as in Insert 2.3, so we do not use *yhat*lotsize* in the *plot* statement in Insert 2.4. Additionally, we use *interpol=join* to connect the data points with straight lines. The data points will be connected in the order they occur in the input data set. Therefore, the data should be sorted by the horizontal axis variable prior to the *proc gplot* statement. We then specify that no symbol be used to plot the data points (*value=none*). Results of Insert 2.4 replicate text Figure 2.6 (SAS Output 2.8).

```
                              INSERT 2.4

data Toluca Company;
  infile 'c:\alrm\chapter  1 data sets\ch01ta01.txt';
  input lotsize workhrs;
fvalue=finv(.9,2,23);
w2=sqrt(2*(fvalue));                              * Text eq 2.40a;
yhat=62.366+3.57*lotsize;                         * Text eq 228;
sdyhat=sqrt(2383.716*((1/25)+(((lotsize-70)**2)/19800)));   * Text eq 2.30;
bandhi=yhat+(w2*sdyhat);                          * Text eq 2.40;
bandlo=yhat-(w2*sdyhat);                          * Text eq 2.40;
proc sort;
 by lotsize;
 run;
proc gplot;
   symbol1 value=none color=black interpol=join;
   symbol2 value=none color=black interpol=join;
   plot bandhi*lotsize bandlo*lotsize  / overlay;
run;
```

SAS Output 2.8

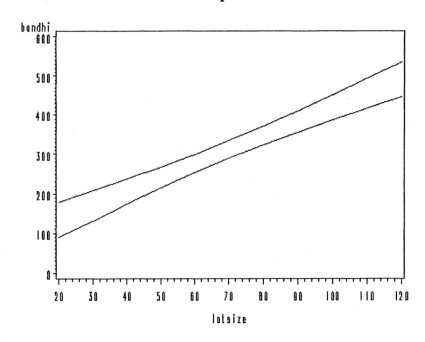

22

The coefficient of determination (R^2): From equation 2.72, text page 74, the coefficient of determination (R^2) is defined as Sum of Squares Regression / Sum of Squares Total; i.e.

$$R^2 = \frac{SSR}{SSTO} = 1 - \frac{SSE}{SSTO}$$

SAS prints the value of R^2 as part of the *proc reg* output (In SAS Output 2.2 above we see that *R-Square* = 0.8215). To compute R^2 directly, we refer to SAS Output 2.2 and find $SSR = 252,377.6$ and $SSTO = 307,203.0$. Thus $R^2 = 0.822$ and the correlation coefficient (r) = $+\sqrt{R^2}$ = 0.907. Note that here we have taken the positive square root of R^2 to find the correlation coefficient r because b_1 is positive indicating that the correlation coefficient is also positive.

Chapter 3 SAS®

Diagnostics and Remedial Measures

In Chapter 1 we learned that regression analysis involves using a sample of data values to find a linear model that will predict (the *dependent variables*) Y when the values of the (*independent variable*) X's are known. We also referred to the X's as *predictor variables*. We began with a sample of data where each sample unit provides an observation on Y (the dependent variable) and on one or more X's (the independent or predictor variables). Next, we formulated a model that we thought might specify the relationship between Y and the predictor variables. Once we decided on a model that we thought might be appropriate, we assumed the model was in fact an appropriate model for our purpose. We used the sample data to estimate parameters of the model and this estimation provided us with a fitted model. The adequacy of the estimates and of related statistical inferences depends on whether the specified model and the assumptions made in specifying the model are at least approximately true. In most applications, we do not know for certain in advance that the model we decide to use is appropriate for the specific analysis. In Chapter 3, the authors discuss methods for using the observed sample data to investigate the adequacy of a fitted regression model. A group of methods known as *regression diagnostics* are used for this purpose. When regression diagnostics identify inadequacies in the model, *remedial measures* can sometimes be used to correct or offset the inadequacies.

Many regression diagnostics are performed in terms of the residuals from the fitted model. Assuming the simple linear regression model is the specified model, the fitted model is denoted (equation 1.13, text page 21):

$$\hat{Y}_i = b_0 + b_1 X_i \qquad i = 1, \ldots, n$$

The estimates b_0 and b_1 are obtained from the sample data. Once they have been calculated and their numerical values obtained, we can substitute X_i into the fitted model and calculate the predicted value $\hat{Y}_i$ for the ith unit in the sample, $i = 1, 2, \ldots, n$. (See the example on text page 21). Ideally, the fitted model would predict the observed Y_i exactly for all $i = 1, 2, \ldots, n$ but this rarely happens in practice. Thus, there is usually a difference between the observed Y_i and its predicted value $\hat{Y}_i$. This difference is called the *residual*. The ith residual, denoted e_i, is defined as:

$$e_i = Y_i - \hat{Y}_i$$

In the discussion below, we use SAS to find the predicted values and the residuals for the sample data. We illustrate how to obtain graphs to perform graphical diagnostics discussed in the text. We then demonstrate how to produce regression diagnostics in terms of regression model residuals.

The Toluca Company Example Continued

Data File: ch01ta01.txt

We return to the Toluca Company example from Chapters 1 and 2 (text page 19, 46). We let the variable names *workhrs* denote work hours and *lotsize* denote lot size, and take these to be the dependent variable Y and the independent variable X, respectively, in the simple linear regression model. See Chapter 0, "Working with SAS," for instructions on opening the Toluca Company data file. After opening and printing ch01ta01.txt, the first 5 lines of the output should look like SAS Output 3.1.

SAS Output 3.1

Obs	lotsize	workhrs
1	80	399
2	30	121
3	50	221
4	90	376
5	70	361

Diagnostics for Predictor Variable: The authors begin by discussing graphical methods that can be used as diagnostic tools to see if there are any outlying X values in the sample. This is important because a few outlying X values could have too much influence on estimates of the model parameters and other aspects of the regression analysis. If the influence of a few outlying X values is dampened or eliminated altogether, we sometimes get a drastically different regression analysis.

We use SAS (Insert 3.1) to produce Figure 3.1 a-d shown on text page 101. A dot plot identical to text Figure 3.1a is not available in SAS but a bar chart (SAS Output 3.2) approximates that display of information. In the Toluca example there are 25 runs represented in the data file. Figure 3.1b, text page 101, shows a plot of *lotsize* against the run number. Run number is not available in the data file. However, the "*run+1;*" statement in Insert 3.1 creates a new variable named *run* which is a sequential number from 1 to 25 for each case. (Here, we assume that the original Toluca Company example is in order by run number.) We can now produce text Figure 3.1b as shown in SAS Output 3.2 below. The SAS procedure *proc univariate*, together with the related syntax shown in Insert 3.1, will produce a stem-and-leaf and box plots of *lotsize*. (SAS Output 3.2)

```
                            INSERT 3.1

data Toluca Company example;
   infile 'c:\alrm\chapter 1 data sets\ch01ta01.txt';
   input lotsize workhrs;
   run+1;
* text Figure 3.1a;
proc chart;
   vbar lotsize / levels=11;
* text Figure 3.1b;
proc plot;
   plot lotsize*run;
* text Figure 3.1c & d;
proc univariate plot;
   var lotsize;
run:
```

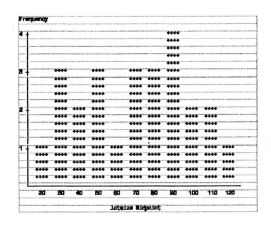

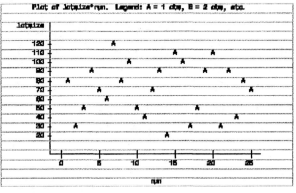

Diagnostics for Residuals: In Section 3.3, text page 103, the authors present informal diagnostic plots of *residuals*. Note that the authors make the distinction between *studentized* and *semistudentized* residuals. The authors point out that "both semistudentized residuals and studentized residuals can be very helpful in identifying outlying observations." SAS does not have an option to produce semistudentized residuals but it does produce studentized residuals. Studentized residuals are probably the more important of the two types of residuals. Therefore, we use studentized residuals in place of semistudentized residuals to present solutions to text chapter examples.

As before, we perform a regression analysis using the model:

$$workhrs = \beta_0 + \beta_1(lotsize) + \varepsilon$$

From Insert 3.2, we specify this regression model and request that unstandardized residuals be saved to a data set *step2* (*out = step2*) under a variable named *residual* (*r = residual*). To produce Figures 3.2 a, b, c, and d, text page 104, we use the syntax in Insert 3.2 (output not shown). The text presents numerous other diagnostic plots. For brevity, we do not present syntax for these other plots but relevant SAS statements are produced easily by a few modifications in the syntax shown in Inserts 3.1 and 3.2.

Correlation Test for Normality: On text page 111, the researchers for the Toluca Company calculate the coefficient of correlation between the ordered residuals and their expected value under normality. The expected value under normality of the kth smallest observation from a sample of size n is given by:

$$\sqrt{MSE}\left[z\left(\frac{k-0.375}{n+0.25}\right)\right]$$

The first step in obtaining the expected values of the ordered residuals under normality is to calculate $(k - 0.375)/(n + 0.25)$. (Insert 3.3) In the Toluca Company example, $n = 25$ and k is the rank of the ordered residuals. We rank the saved residuals (*residual*) using the *Proc Rank* procedure. The saved residuals are identified in the *proc rank* statement by the specification "*data = step2*". The "*out = step3*" option in the *proc rank* statement specifies the data file (*step3*) in which the ranks will be stored after the *proc ranks* statement has been executed. The *ranks ranke* statement creates another new variable, *ranke*, that is the variable name under which we chose to store the ranks of the variable *residual*, the unstandardized residual in the model with the predictor variable *lotsize* ($Y_i = \beta_0 + \beta_1 X_{i1} + \varepsilon_i$). In the next step of the program, we have written the statement "*data new*" followed by the statement "*set step3*". Thus, we have assigned the name "*new*" to this step and specified "*step3*" as the data file to be used. We use this approach so that we can carry out additional arithmetical operations involving the output (*ranke*) from the previous data step. Here, $(k - 0.375)/(n + 0.25)$ becomes $(ranke - 0.375)/(25 + 0.25)$. This quantity represents the cumulative probability ($0 \le$ cumulative probability ≤ 1.0) of the standard normal distribution. We now obtain a standard score (z) associated with the cumulative probability calculated above. In SAS, the *Probit* function is *quantile* function that returns a quantile from the standard normal distribution. It takes the form *Probit*(p), where $p =$ cumulative probability. For instance, if cumulative probability $= 0.50$ (*Probit*(0.50)) then $z = 0$ or if cumulative probability $= 0.8413$ (*Probit*(0.8413)) then $z = +1$. Thus, to obtain z in the Toluca Company example, we use

$$Probit\ ((ranke\ -0.375)/(54\ +0.25))$$

Finally, we multiply this quantity by the square root of $MSE = 2383.716$ of the first order model (Chapter 2 – SAS Output 2.2). From the above, we compute the *expected value* as:

$$ev = sqrt(2383.716) * Probit\ ((ranke\ -0.375)/(25\ +0.25)).$$

We can now obtain the coefficient of correlation between the residual and the expected value under normality (*ev*). The result of the correlation between *residual* and *ev* is 0.991 as reported on text page 115.

INSERT 3.3

```
data Toluca Company example;
  infile 'c:\alrm\chapter  1 data sets\ch01ta01.txt';
  input lotsize workhrs;
* Save residuals (/ r) in residual;
proc reg;
  model workhrs = lotsize / r;
  output out=step2 r=residual;
run;
* Rank the residuals and place the rank in "ranke";
proc rank data=step2 out=step3;
 var residual;
 ranks ranke;
run;
data new;
 set step3;
 ev = sqrt(2383.716) * Probit ((ranke - .375)/(25  + .25));
proc corr;
 var ev residual;
 run;
```

Test for Constancy of Error Variance - Brown-Forsythe: The authors present two formal tests for constancy of error variance. We begin with the *Brown-Forsythe test* using the Toluca Company example, text page 117. For simplicity, we divided the process into steps:

- Step 1: (Insert 3.4) Divide the 25 data-lines into two subsets (*group*): those with *lotsize* between 20 and 70 (if *lotsize*<71 then *group*=1) and those with *lotsize* > 70(*else group*=2).
- Step 2: Use *proc means* to obtain the *median residual values* $\tilde{e}_1$ (group 1) = –19.87 and $\tilde{e}_2$ (group 2) = –2.68.
- Step 3: Find *absdevmd*, the absolute value (*abs*) of the residual minus the median residual value for the respective group, d_{i1} and d_{i2}.
- Step 4: Find the means of *absdevmd* for the two groups, $\overline{d}_1 = 44.81$ and $\overline{d}_2 = 28.4$.

28

- Step 5: Find *sqrdev,* the square of *absdevmd* minus the respective group mean of *absdevmd.*
- Step 6: Find the sum of *sqrdev* for the two groups, $\sum \left(d_{11} - \bar{d}_1 \right)^2 = 12{,}567.86$ and $\sum \left(d_{12} - \bar{d}_2 \right)^2 = 9{,}621.54.$

We now have the required values to calculate the *Brown-Forsythe statistic*:

$$s^2 = \frac{12{,}567.86 + 9{,}621.54}{25 - 2} = 964.75$$

$$s = \sqrt{964.75} = 31.06$$

$$t_{BF}^* = \frac{44.81 - 28.4}{31.06\sqrt{\dfrac{1}{13} + \dfrac{1}{12}}} = 1.319$$

INSERT 3.4	INSERT 3.4 continued
| ```
data Toluca Company example;
  infile 'c:\alrm\chapter  1 data sets\ch01ta01.txt';
  input lotsize workhrs;
* Step 1;
if lotsize<71 then group=1;
 else group=2;
proc reg;
  model workhrs = lotsize / r;
  output out=step2 r=residual;
run;
* Step 2;
proc sort data=step2;
 by group;
proc means data=step2 median;
 by group;
  var residual;
run;
``` | ```
* Step 3;
data step2;
 set step2;
if group=1 then absdevmd=abs(residual -(-19.88));
if group=2 then absdevmd=abs(residual -(-2.68));
* Step 4;
proc means;
  by group;
   var absdevmd;
  run;
* Step 5;
data new;
  set step2;
if group=1 then sqrdev=(absdevmd-44.81)**2;
if group=2 then sqrdev=(absdevmd-29.45)**2;
run;
* Step 6;
proc means sum;
  by group;
   var sqrdev;
  run;
``` |

Test for Constancy of Error Variance - Breusch-Pagan Test: The second Test for *Constancy of Error Variance* considered by the authors is the *Breusch-Pagan Test* (text page 118). To compute the Breusch-Pagan test, we first save the residuals (*r=residual*) of the following model as shown in Insert 3.5:

$$workhrs = \beta_0 + \beta_1(lotsize) + \varepsilon$$

We now square *residual* and place the result in a variable named *ressqr* (e^2). The last step is to regress *ressqr* against *lotsize* in the usual manner. The last regression procedure in Insert 3.5 results in SAS Output 3.4 and contains $SSR^*=7,896,128$. SSE (54,825.459) is shown in SAS Output 3.5 and results from the first regression run in Insert 3.5. We now can find:

$$X_{BP}^2 = \frac{SSR^*}{2} \div \left(\frac{SSE}{n}\right)^2$$

where SSR^* for regressing e^2 (*ressqr*) on X

and

SSE for regressing Y on X

$$X_{BP}^2 = \frac{7,896,128}{2} \div \left(\frac{54,825}{25}\right)^2 = 0.821$$

INSERT 3.5

```
data Toluca Company example;
  infile 'c:\alrm\chapter  1 data sets\ch01ta01.txt';
  input lotsize workhrs;
proc reg;
  model workhrs = lotsize / r;
  output out=step2 r=residual;
data step2;
 set step2;
ressqr=residual**2;
proc reg;
  model ressqr = lotsize;
run;
```

SAS Output 3.4

Analysis of Variance					
Source	**DF**	**Sum of Squares**	**Mean Square**	**F Value**	**Pr > F**
Model	1	7896142	7896142	1.09	0.3070
Error	23	166395896	7234604		
Corrected Total	24	174292038			

Analysis of Variance					
Source	DF	Sum of Squares	Mean Square	F Value	Pr > F
Model	1	252378	252378	105.88	<.0001
Error	23	54825	2383.71562		
Corrected Total	24	307203			

Bank Example

Data File: ch03ta04.txt

F Test for Lack of Fit: On text pages 119-120, the authors discuss a formal test for determining if the regression function adequately fits the data. In an experiment involving 12 branch offices of a commercial bank, gifts were offered for customers who opened money market accounts. The value of the gift was proportional to the minimum deposit and there were 6 levels of minimum deposit used in the experiment. The researchers studied the relationship between the minimum deposit and value of the associated gift measured in terms of the size of minimum deposit (X) and the number of new accounts (Y) opened at each of the 12 branches. After opening and printing the Bank Example data (text Table 3.4), the first five lines of the output should look like SAS Output 3.6.

SAS Output 3.6

Obs	x	y
1	125	160
2	100	112
3	200	124
4	75	28
5	150	152

To compute the *Lack of Fit F** presented on text page 124, we switch to the *General Linear Model (GLM)* procedure. (Insert 3.6) We begin by fitting the full model to the data. This requires creating a new variable, *J*. As programmed in Insert 3.6, $j = 1$ if $X_1 = 75$, $j = 2$ if $X_2 = 100$ and so on. We now regress Y on X and J. Using the *class j* statement in *Proc GLM*, we specify that *J* is a classification variable. Thus, *J* identifies 6 discrete classifications based on the values of *X*. Insert 3.6 produces SAS Output 3.6 which shows that $SSE(F) = 1148.0$. From equation 3.16, text page 122, $SSE(F) = SSPE$. Therefore, $SSPE = 1148.0$. From equation 3.24, text page 124, we see that $SSLF = SSE(R) - SSPE$. If we were to regress Y on X, (reduced model) we would find that $SSE(R) = 14,741.6$ (not shown). Thus, $SSLF = 14,741.6 - 1148 = 13,593.6$. $SSLF = 13,593.6$ is given directly in SAS Output 3.6 under *Type III Sum of Squares* for *J*. Thus, regressing Y on X and J in *proc glm* gives us the quantities to find (text page 124):

$$F^* = \frac{SSLF}{c-2} \div \frac{SSPE}{n-c}$$

$$F^* = \frac{13{,}593.6}{4} \div \frac{1{,}148}{5} = 14.8$$

INSERT 3.6

```
* Full Model;
data Bank Example;
 infile 'c:\neter\ch03ta04.dat';
 input x y;
 if x=75 then j=1;
 if x=100 then j=2;
 if x=125 then j=3;
 if x=150 then j=4;
 if x=175 then j=5;
 if x=200 then j=6;
proc glm;
 class j;
 model y = x j;
run;
```

SAS Output 3.6

Source	DF	Sum of Squares	Mean Square	F Value	Pr > F
Model	5	18734.90909	3746.98182	16.32	0.0041
Error	5	1148.00000	229.60000		
Corrected Total	10	19882.90909			

R-Square	Coeff Var	Root MSE	y Mean
0.942262	12.94085	15.15256	117.0909

Source	DF	Type I SS	Mean Square	F Value	Pr > F
x	1	5141.33841	5141.33841	22.39	0.0052
j	4	13593.57068	3398.39267	14.80	0.0056

Source	DF	Type III SS	Mean Square	F Value	Pr > F
x	0	0.00000	.	.	.
j	4	13593.57068	3398.39267	14.80	0.0056

Box-Cox Transformations: SAS does not have a readily available option to generate *Box-Cox transformations* (text page 134). Insert 3.7 contains the necessary syntax in matrix language to produce Table 3.9 and Figure 3.17, text page 136. (SAS matrix language is covered in Chapter 5.) The text uses data on age (*X*) and plasma (*Y*) level of 25 children as presented in Table 3.8, text page 133, to illustrate the transformation procedure. The first 2 *"read all var"* statements set vectors **x** and **y** equal to *age* and *plasma*, respectively. This procedure uses a range of λ values starting at λ (*start*) = −2.3 and incrementing (*increment*) by a value of 0.1 for 34 iterations (*noiter*). A scatter plot of *SSE* as a function of λ is shown in SAS Output 3.7. The start, increment, and iteration values can be varied to accomplish different goals.

INSERT 3.7	INSERT 3.7 Continued
| ```
options ls=72 ps=30 nonumber nodate;
data plasma;
   infile 'c:\alrm\chapter 3 data sets\ch03ta08.txt';
   input age plasma;
run;
proc iml;
  use plasma;
  read all var {'plasma'} into y;
  read all var {'age'} into x;
  n = nrow(y);
  noiter = 34;
  start = -2.3;
  increment = 0.1;
  x = j(n,1,1)||x;
  hat =x*inv(x`*x)*x`;
  lambda = j(noiter,1);
  lambda[1,1] = start;
  do i = 2 to noiter;
   lambda[i,1]=lambda[i-1,1] + increment;
  end;
``` | ```
sse = j(noiter,1,0);
 do i = 1 to noiter;
  lamb = lambda[i];
       k2 = exp((1/n)*(j(1,n,1)*log(y)));
       k1 = 1/(lamb*k2**(lamb-1));
       if lamb > -1E-08 & lamb < 1E-08 then w =
k2*log(y);
       else w = k1*(y##lamb -j(n,1,1));
       e = (i(n) - hat)*w;
   sse[i] = e`*e;
end;
 print lambda [format = 5.1] sse [format = 10.4];
 xy = lambda||sse;
 call pgraf(xy,'*','Lambda','SSE','Plot for Box-Cox
Transformation');
quit;
``` |

SAS Output 3.7

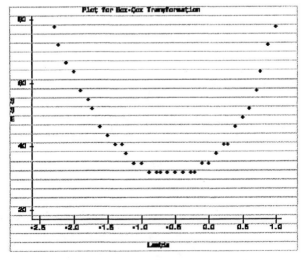

Exploration of Shape of Regression Function – Lowess Method: The authors state that "the name *lowess* stands for locally weighted regression scatter plot smoothing." The lowess procedure divides the independent variables into groups or *local neighborhoods* and then fits a regression model to each local neighborhood of data. The procedure uses weighted least squares to fit regression models for each neighborhood with the weights being a function of the distance of the X's for a given subject in a neighborhood from the center of the neighborhood. The final fitted model is a blending of these fitted models in each neighborhood. A *smoothing parameter* must be specified in executing the lowess method. The smoothing parameter specifies the proportion of the total number of observations to be included in each neighborhood. Thus, if the total sample size is 50 and the smoothing parameter is 0.10, then 5 observations will be included in each local neighborhood used in executing the lowess method. We can adjust the smoothing parameter and examine a plot of the fitted loess curves together with the residual sum of squares to choose a reasonable fitting smooth curve.

In Insert 3.8, we present SAS syntax to produce Figure 3.19a, text page 140. Using the Toluca Company example, we use *proc loess* to specify that the lowess method be used to fit a smoothed curve to the data. The *model* option *smooth* = 0.5 specifies that the model be fit with smoothing parameter 0.5. The total sample size in the Toluca Company example is $n = 25$, so $(0.5)(25) \approx$ 12 observations will be included in local neighborhood in the analysis. (SAS Output 3.9) We print out the first 5 lines of the *results* data set (SAS Output 3.10) and produce a scatter plot of Y versus X with the fitted loess curve. (SAS Output 3.11)

INSERT 3.8

```
data toluca;
  infile 'c:\alrm\chapter  1 data sets\ch01ta01.txt';
  input x y;
proc loess;
  model y = x / smooth = 0.5 residual;
  ods output OutputStatistics = Results;
run;
proc print data=results (obs=5);
proc sort;
 by pred;
 run;
goptions reset=all;
proc gplot;
   symbol1 color=black value=none interpol=join;
   symbol2 color=black value=circle;
   title 'Scatter Plot and Lowess Curve';
   plot Pred*x DepVar*x / overlay haxis=0 to 150 by 50 vaxis=100 to 600 by 100 ;
run;
```

SAS Output 3.9

Output from *proc loess* Fit Summary	
Fit Method	kd Tree
Blending	Linear
Number of Observations	25
Number of Fitting Points	11
kd Tree Bucket Size	2
Degree of Local Polynomials	1
Smoothing Parameter	0.50000
Points in Local Neighborhood	12
Residual Sum of Squares	46341

SAS Output 3.10

Obs	SmoothingParameter	Obs	x	DepVar	Pred	Residual
1	0.5	1	80.00000	399.00000	360.04835	38.95165
2	0.5	2	30.00000	121.00000	180.61465	-59.61465
3	0.5	3	50.00000	221.00000	214.46246	6.53754
4	0.5	4	90.00000	376.00000	390.43038	-14.43038
5	0.5	5	70.00000	361.00000	304.84581	56.15419

SAS Output 3.11

Scatter Plot and Lowess Curve
SmoothingParameter=0.5

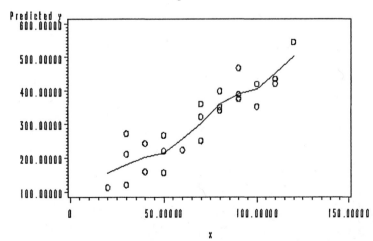

35

Chapter 4 SAS®

Simultaneous Inferences
and Other Topics in
Regression Analysis

The level of significance of a statistical test is the probability the test rejects the null hypothesis given that the null hypothesis is true. This probability is usually denoted α and is referred to as the probability of a Type I error. These concepts are in the context of a single application of the test. If the test is applied to multiple hypotheses we often want to make a simultaneous or joint statement about the probability of a Type I error given that all the hypotheses being tested are true. We call a set of tests under consideration a family of tests. If each test in a family of g tests is tested at level of significance α, the probability of making one or more Type I errors if all g hypotheses are true is α^* where:

$$\alpha^* \le 1 - (1-\alpha)^g$$

This inequality is a form of the Bonferroni Equality. In this statement, equality holds if the tests are mutually independent. It is apparent that as g increases α^* approaches 1 and, unless we take corrective action, the probability is close to one that we falsely reject at least one hypothesis even if they are all true. The right side of the statement is approximately equal to $g\alpha$ if α is small; i.e.,

$$1 - (1-\alpha)^g \approx g\alpha$$

Therefore, if we perform each of g tests at the α/g level we get:

$$1 - \left(1 - \frac{\alpha}{g}\right)^g \approx \alpha$$

This approach to multiple testing is referred to as the Bonferroni method or procedure. Using the Bonferroni procedure, we perform each of g tests at the α/g level and are assured that the probability is $\le \alpha$ of making one or more Type I errors if all g hypotheses are true. For example, if we perform 4 tests and use the $0.05/4 = 0.0125$ level of significance for each test, the probability we falsely reject one or more of the 4 corresponding hypotheses is less than or equal to 0.05.

If we are conducting a single two-sided test, we find critical values from the t-distribution by finding $t(1- \alpha/2;\ df)$ where df is the degrees of freedom of the test. Similarly, if we are conducting g two-sided multiple tests, we find critical values from the t-distribution by finding $t(1- \alpha/(2g);\ df)$ where df is the degrees of freedom of the test.

Analogous to multiple testing, we often want to set g confidence intervals with the knowledge that the confidence level is $\geq 100(1 - \alpha)$ percent that all g intervals bracket their respective objective. In the simple linear regression model, simultaneous (Bonferroni) confidence limits for estimating β_0 and β_1 are (equation 4.3, text page 156):

$$b_0 \pm Bs\{b_0\}$$

$$b_1 \pm Bs\{b_1\}$$

where $B = t(1 - \alpha/4; n - 2)$. Here $g = 2$ so that $\alpha/(2g) = \alpha/4$.

Section 4.2, text page 157, discusses simultaneous confidence intervals for the response means at different values of X. In addition to the Bonferroni procedure, the authors discuss the *Working-Hotelling* procedure. Section 4.3 discusses simultaneous prediction intervals for new observations. Section 4.4 presents the simple model for regression through the origin. Section 4.5 discusses the effects of measurement errors, while Section 4.6 presents inverse predictions.

More on The Toluca Company Example

Data File: ch01ta01.txt

We return to the Toluca Company example from Chapters 1, 2 and 3 (text page 19, 46, 156). See Chapter 0, "Working with SAS," for instructions on opening the Toluca Company data file. After opening and printing ch01ta01.dat, the first 5 lines of the output should look like SAS Output 4.1.

SAS Output 4.1

Obs	lotsize	workhrs
1	80	399
2	30	121
3	50	221
4	90	376
5	70	361

Bonferroni Joint Confidence Intervals for β_0 and β_1: On text page 156, the Toluca Company wants to estimate the 90 percent family confidence intervals for β_0 and β_1. To estimate the 90 percent confidence intervals for β_0 and β_1, the text reports $B = t(1 - 0.10/4; 23) = t(0.975; 23) = 2.069$. The t-value is actually based on $(1 - \alpha/(2g))$ where g is the number of joint intervals to be estimated. In the present example, we are estimating joint confidence intervals for β_0 and β_1 ($g=2$), so $B = t(1 - 0.10/(2*2); 23) = t(1 - 0.10/4; 23) = t(0.975; 23)$. In Insert 4.1, we use a *quantile function* to find $B = TINV(0.975, 23) = 2.068$. From SAS Output 1.3, we know b_0, $s\{b_0\}$, b_1, and $s\{b_1\}$. We then use equation 4.3, text page 156, and the Bonferroni procedure for developing joint confidence intervals for b_0 (*b0lo* and *b0hi*) and b_1 (*b1lo* and *b1hi*). The values for *b0lo*, *b0hi*, *b1lo*, and *b1hi* are written to each line in the Toluca Company date set. However, because the values are identical for each line, we need only to print the values for one observation. The (*obs=1*) option on the *proc print* statement causes only the first observation of the data set to be printed; it must be preceded by the *data=toluca* specification.

```
                        INSERT 4.1
data Toluca Company;
   infile 'c:\alrm\chapter  1 data sets\ch01ta01.txt';
   input lotsize workhrs;
b=tinv(.975,23);
b0lo=62.37-(b*26.18);
b0hi=62.37+(b*26.18);
b1lo=3.5702-(b*.3470);
b1hi=3.5702+(b*.3470);
run;
proc print data=toluca (obs=1);
 var b0lo b0hi b1lo b1hi;
run;
```

Simultaneous Estimation of Mean Responses: On text page 158, the Toluca Company wants to estimate the mean number of work hours required to produce lots of 30, 65, and 100 units where the family confidence coefficient is 0.90. In Chapter 2, we presented the steps to find the confidence interval for $E\{Y_h\}$ when lot size (X_h) = 65 units and the confidence band for the entire regression line. We repeat these steps here with fewer notations and more emphasis on SAS computational statements.

a. Working-Hotelling Procedure: We first find the 90 percent confidence interval for work hours when lot size (X_h) = 65 units. We use the Working-Hotelling $1-\alpha$ band (equations 2.40 and 2.40a, text page 61, and equation 2.30, text page 53) to establish boundary values for (X_h) = 65 units:

$$\hat{Y}_h \pm Ws\{\hat{Y}_h\}$$

where

$$W^2 = 2\,F(1-\alpha;\,2,\,n-2)$$

and

$$s^2\{\hat{Y}_h\} = MSE\left[\frac{1}{n} + \frac{(X_h - \bar{X})^2}{\sum(X_i - \bar{X})^2}\right]$$

In Insert 4.2 – Method 1, we find the *F*-value (*fvalue*) associated with α = 0.90 and 2, 23 degrees of freedom. We then find $W = \sqrt{2 * fvalue}$. Next, we find $\hat{Y}_h$ (*yhat*) when X_h = 65 units using previously found parameter estimates (SAS Output 1.3), b_0 = 62.366 and b_1 = 3.570. We find $s\{\hat{Y}_h\}$ (*sdyhat*), the square root of text equation 2.30. Finally, we compute $\hat{Y}_h \pm$

$Ws\{\hat{Y}_h\}$ placing the lower boundary of the confidence interval in a variable named *bandlo* and the upper boundary of the confidence interval in a variable named *bandhi*. Execution of Insert 4.2 will print the 90 percent confidence interval for work hours when $X_h = 65$ units. After the syntax is working properly, as in Insert 4.2 – Method 1, we can rewrite the syntax in a more compact form if desired. Insert 4.2 Method 1 can be reduced to Method 2.

To find the confidence intervals for $X_h = 30$ and $X_h = 100$, simply change the computation of *yhat* and *sdyhat* in Insert 4.2 – Method 1 to reflect 30 or 100 units instead of 65 units. Using this method, old values for *bandlo* and *bandhi* will be overwritten each time you change the value of X_h. If this is undesirable, the syntax in Insert 4.3 will find the 90 percent confidence interval for each value of X_h. These intervals will be placed in unique variables, i.e., *bandlo30*, *bandhi30*, *bandlo65*, *bandhi65*, *bandlo00*, and *bandhi00*.

INSERT 4.2

```
* Method 1;
data Toluca;
 infile 'c:\alrm\chapter  1 data sets\ch01ta01.txt';
  input lotsize workhrs;

fvalue=finv(.9,2,23);
w=sqrt(2*fvalue);
yhat=62.366+(3.570*65);
sdyhat=sqrt(2383.716*((1/25)+((65-70)**2/19800)));
bandlo=yhat-(w*sdyhat);
bandhi=yhat+(w*sdyhat);
run;
proc print data=toluca (obs=1);
 var bandlo bandhi;
run;

* Method 2;
data Toluca;
  infile 'c:\alrm\chapter  1 data sets\ch01ta01.txt';
  input lotsize workhrs;
bandlo=(62.366+(3.570*65))-(sqrt(2*finv(.9,2,23))*sqrt(2383.716*((1/25)+((65-70)**2/19800))));
bandhi=(62.366+(3.570*65))+(sqrt(2*finv(.9,2,23))*sqrt(2383.716*((1/25)+((65-70)**2/19800))));
run;
proc print data=toluca (obs=1);
 var bandlo bandhi;
run;
```

```
                        INSERT 4.3
data Toluca Company;
  infile 'c:\alrm\chapter  1 data sets\ch01ta01.txt';
  input lotsize workhrs;
fvalue=finv(.9,2,23);
w=sqrt(2*fvalue);
yhat30=62.366+(3.570*30);
sdyhat30=sqrt(2383.716*((1/25)+((30-70)**2/19800)));
bandlo30=yhat30-(w*sdyhat30);
bandhi30=yhat30+(w*sdyhat30);
yhat65=62.366+(3.570*65);
sdyhat65=sqrt(2383.716*((1/25)+((65-70)**2/19800)));
bandlo65=yhat65-(w*sdyhat65);
bandhi65=yhat65+(w*sdyhat65);
yhat100=62.366+(3.570*100);
sdyhat00=sqrt(2383.716*((1/25)+((100-70)**2/19800)));
bandlo00=yhat100-(w*sdyhat00);
bandhi00=yhat100+(w*sdyhat00);
proc print data=toluca (obs=1);
 var bandlo30 bandhi30 bandlo65 bandhi65 bandlo00 bandhi00;
run;
```

b. Bonferroni Procedure: On text page 159, the authors use data from the Toluca Company example and the Bonferroni Procedure to estimate the mean number of work hours required to produce lots of 30, 65, and 100 units where the family confidence coefficient is 0.90. The Bonferroni confidence limits are:

$$\hat{Y}_h \pm Bs\{\hat{Y}_h\}$$

where:

$$B = t(1 - \alpha/2g; n-2)$$

and g is the number of joint confidence intervals to be estimated. In this example, there are three intervals in the family. Thus, $B = t(1 - 0.10/(2*3); 23) = t(1 - 0.10/6; 23) = t(1 - 0.01666; 23) = t(0.9833, 23)$. We can use the *TINV* function as before to find $B = TINV(0.98333, 23) = 2.263$. Computation of $s\{\hat{Y}_h\}$ is the same as for the Working-Hotelling procedure above including Insert 4.2. Thus, to use the Bonferroni Procedure to find the 90 percent confidence interval for work hours when $X_h = 65$ units, we use text equations 4.7 and 4.7a as in Insert 4.4 to find the upper (*bandhi*) and lower (*bandlo*) confidence limits. Confidence limits for $X_h = 30$ and $X_h = 100$ can be generalized from Inserts 4.2 and 4.3.

Prediction Intervals for New Observations: The Toluca Company now wants to predict the required work hours for lots of 80 and 100 units. The researchers first obtained S and B (text page 160) to determine which prediction interval was smaller. In Insert 4.5, we use the *TINV* and *FINV quantile* functions to find $S = 2.616$ (equation 4.8a, text page 160) and $B = 2.398$ (text equation 4.9a). Because the Bonferroni procedure produces the tighter limits, we use B to establish the 95% family coefficients.

We now calculate $\hat{Y}_h$ for 80 (*yhat80*) and 100 (*yhat100*) units. We then use equation 2.38a, text page 59, to find $s\{pred\}$ for 80 (*sdpre80*) and 100 (*sdpre00*) units. Finally, we use text equation 4.9, text page 160, to find the lower and upper 95% confidence limits for 80 (*bandlo80* and *bandhi80*, respectively) and 100 (*bandlo00* and *bandhi00*, respectively) units.

Inverse Predictions: The authors present an example of inverse prediction on text page 168. We do not have access to the source data file, however, and will use the Toluca Company for illustration. Suppose the auditing department of the Toluca Company's corporate home office requested that lot size $\hat{X}_{h(new)}$ be estimated with a 95% confidence interval when the work hours for a run is known to be $Y_{h(new)} = 300$. Although the accountants at the Toluca Company failed to appreciate the need for such a calculation, they complied with the request.

In Insert 4.6, we first find $\hat{X}_{h(new)} = Y_{h(new)} - 62.366 / 3.570 = 66.56$ using equation 4.31, text page 168, and parameter estimates from SAS Output 1.3. For equation 4.32, text page 169, we find t using $TINV(1-0.05/2, 23) = 2.07$ and place the result in *tvalue*. Using text equation 4.32a, we find $s^2\{predX\}$ (*sdprex*) = 13.95. Finally, we use text equation 4.32 to find:

$$\hat{X}_{h(new)} \pm t(1-\alpha/2;\ n-2)s\{predX\}$$

$$66.56 \pm 2.07 * 13.95$$

$$37.70 < \hat{X}_{h(new)} < 95.42$$

The auditors at the corporate home office conclude with 95% confidence that the number of hours required to produce the lot size of 300 units is between 37.70 and 95.42 hours. The auditors calculate the error to be 66.56/((95.42-37.70)/2) or approximately $\pm 43\%$.

INSERT 4.6

```
data Toluca Company;
   infile 'c:\alrm\chapter  1 data sets\ch01ta01.txt';
   input lotsize workhrs;
xhnew=(300-62.366)/3.570;
tvalue=tinv(1-.05/2,23);
sdprex=sqrt((2383.716/(3.570**2)*(1+(1/25)+((xhnew-70)**2/19800))));
prexlo=xhnew-(tvalue*sdprex);
prexhi=xhnew+(tvalue*sdprex);
proc print data=toluca (obs=1);
 var xhnew prexlo prexhi;
run;
```

The Charles Plumbing Supplies Example

Data File: ch04ta02.txt

Regression through the Origin: From text page 162, the Charles Plumbing Supplies Company studied the relationship between the number of work units performed (*X*) and the total variable labor costs (*Y*). After opening and printing ch04ta02.dat, the first 5 lines should appear as SAS Output 4.2. To perform regression through the origin, we use the *noint* option on the model statement to suppress the intercept term. Confidence limits are also requested (*clb*) on the model statement in Insert 4.7. This results in SAS Output 4.3 and is consistent with text pages 163-164.

SAS Output 4.2

Obs	x	y
1	20	114
2	196	921
3	115	560
4	50	245
5	122	575

INSERT 4.7

```
data Charles Plumbing;
  infile 'c:\alrm\chapter  4 data sets\ch04ta02.txt';
  input x y;
proc reg;
 model y = x / noint clb;
 run;
```

SAS Output 4.3

Analysis of Variance					
Source	DF	Sum of Squares	Mean Square	F Value	Pr > F
Model	1	4191980	4191980	18762.5	<.0001
Error	11	2457.65933	223.42358		
Uncorrected Total	12	4194438			

Parameter Estimates							
Variable	DF	Parameter Estimate	Standard Error	t Value	Pr > |t|	95% Confidence Limits	
x	1	4.68527	0.03421	136.98	<.0001	4.60999	4.76056

Parameter Estimates							
Variable	DF	Parameter Estimate	Standard Error	t Value	Pr > |t|	95% Confidence Limits	
x	1	4.68527	0.03421	136.98	<.0001	4.60999	4.76056

Chapter 5 SAS®

Matrix Approach to Simple
Linear Regression Analysis

The text begins Chapter 5 by presenting an overview of the algebra for vectors and matrices. In Chapter 5 of this manual, we introduce SAS/IML software through *proc iml*, a procedure that allows the user to write SAS statements for performing vector/matrix algebra. We follow examples in the text and show how to write SAS programs to perform operations such as matrix addition, subtraction and multiplication. Following the overview of matrix algebra, the text expresses the simple linear regression model as a matrix equation and then repeats the steps followed in Chapter 1 only now these steps are described using matrix algebra.

Again, we are given Y_i and X_i for a sample of n subjects and we want to find a "best fitting" model for predicting Y given X. In Chapter 1, the simple linear regression model for the ith subject, $i = 1, \dots, n$ was written as :

$$Y_i = \beta_0 + \beta_1 X_i + \varepsilon_i$$

To summarize the key results of the matrix formulation, let the capital bold-faced letter $\mathbf{Y}$ represent the nx1 column vector containing the data for the dependent variable and the bold-faced letter $\mathbf{X}$ represent the nx2 matrix with the first column being a column vector of ones and the second column being the column vector of data for the independent variable; i.e.,

$$\mathbf{Y} = \begin{bmatrix} Y_1 \\ Y_2 \\ \vdots \\ Y_n \end{bmatrix} \quad \text{and} \quad \mathbf{X} = \begin{bmatrix} 1 & X_1 \\ 1 & X_2 \\ \vdots & \vdots \\ 1 & X_n \end{bmatrix}$$

Further, let:

$$\boldsymbol{\beta} = \begin{bmatrix} \beta_0 \\ \beta_1 \end{bmatrix} \quad \text{and} \quad \boldsymbol{\varepsilon} = \begin{bmatrix} \varepsilon_1 \\ \varepsilon_2 \\ \vdots \\ \varepsilon_n \end{bmatrix}$$

Using this notation, the simple linear regression model can be written:

$$\mathbf{Y} = \mathbf{X}\boldsymbol{\beta} + \boldsymbol{\varepsilon}$$

A few key results of the matrix formulation of the regression model are:

(1.) Matrix solution for the regression coefficients: $\mathbf{b} = \begin{bmatrix} b_0 \\ b_1 \end{bmatrix} = (\mathbf{X'X})^{-1}\mathbf{X'Y}$

(2.) Matrix solution for the predicted values: $\hat{\mathbf{Y}} = \mathbf{Xb} = \mathbf{X}(\mathbf{X'X})^{-1}\mathbf{X'Y} = \mathbf{HY}$

where $\mathbf{H} = \mathbf{X}(\mathbf{X'X})^{-1}\mathbf{X'}$ is called the *hat matrix.*

(3.) Matrix solution for the fitted model residuals: $\mathbf{e} = \mathbf{Y} - \hat{\mathbf{Y}} = \mathbf{Y} - \mathbf{Xb} = (\mathbf{I} - \mathbf{H})\mathbf{Y}$

From these equations, we begin to see that the computations required for regression analysis can be written as matrix expressions. As stated earlier, IML is a SAS procedure that performs matrix algebra. Most of our needs are met by SAS syntax that is more user friendly than that specific to IML, but occasionally we need to perform computations that are not directly available, and IML provides a convenient way to write programs according to our specifications. Moreover, it is instructional to write matrix language programs that perform regression analysis. We illustrate the use of IML by presenting SAS statements to compute (1.) – (3.) and other key matrix expressions for the regression analysis of data given in the Toluca Company example. We begin by introducing some basic mathematical operators used in the SAS/IML software.

Using *proc iml* for Matrix Operations: We initiate the IML procedure by writing *proc iml.* Next, we use IML syntax to write statements for performing one or more matrix operations. We end the procedure by writing *quit.* As in previous chapters, we terminate all SAS statements with a semi-colon. In the following, we illustrate some of the IML statements that are useful for programming a regression analysis.

A SAS program for finding the transpose of a matrix **A** and a vector **C** is given in Insert 5.1. In this program, the numerical values of the matrix **A** and the vector **C** are specified directly in the *iml* procedure. To accomplish this, we choose a name for the matrix and set that name equal to a quantity in braces. The SAS program does not differentiate between upper and lower case names and does not accept bold-face notation. Therefore, in SAS code we can write A = { } or a = { } to specify the matrix **A**. The matrix is specified by rows with row elements separated by one or more spaces and complete rows separated by a comma. It is not necessary to begin each row of a matrix on a new line of the program but it is helpful for proof-reading to do so and to line up the columns as we have done in specifying A and C in Insert 5.1. Thus, the three lines of code that immediately follow the *proc iml* statement specify the 3x2 matrix **A** and the next three lines specify the 3x1 vector **C**.

Insert 5.1 finds the transpose of **A** and **C**. The transpose operator is denoted by the back-quote character. Thus, the statement *AT = A`* exchanges the rows and columns of matrix **A** and stores the results in a matrix named **AT**. Similarly, the statement *CT = C`* exchanges the rows and columns of matrix **C** and stores the results in a matrix named **CT**. Alternatively, one can use AT=T(A) or CT=T(C) to find the transpose of **A** and **C**, respectively. The print statement in the next to last line of Insert 5.1 instructs SAS to print matrices **A, C, AT**, and **CT** (SAS Output 5.1). These results are given on text page 178.

SAS Output 5.1

A	
2	5
7	10
3	4

AT		
2	7	3
5	10	4

C
4
7
10

CT		
4	7	10

Insert 5.2 shows SAS/IML syntax for text examples of matrix addition, subtraction, and multiplication, and finding the inverse of a matrix. The output is not presented for brevity. The symbol + denotes the addition operator and the symbol − denotes the subtraction operator. The asterisk symbol * denotes the multiplication operator. The IML function *inv(A)* finds the inverse of the matrix **A**. We have named the result *Ainv* but we could have chosen another convenient name. For example, we could have called the result *d* by writing *d = inv(A)*.

47

Simple Linear Regression in Matrix Terms

The Toluca Company Example

Data File: 'ch01ta01.txt'

We use matrix methods to perform a simple linear regression analysis of the Toluca Company example (text page 19). The required steps are presented in text Chapter 5 and are not detailed here except in Insert 5.3. We begin the SAS program with the statement *data TolucaCo*. We assigned the arbitrary name *TolucaCo* to the data. Next, we identified the data file with the statement *infile* 'c:\alrm\chapter 1 data sets\ch01ta01.txt'. The statement *input lotsize workhrs* assigns names to the columns of data in the data file and stores the numerical data under these names. Excerpts from the **Y** vector (*workhrs*) and **X** matrix (a column of ones and a column of numerical values of *lotsize*) are given in equation 5.61 page 200 of the text. The *run* statement activates the *data* step and all the data are entered.

The *proc iml* statement initiates the IML procedure. The statement *read all var {'workhrs'} into y* stores the column of data named *workhrs* in the matrix (vector) *y*. Similarly, the next statement stores the column of data labeled *lotsize* in the matrix (vector) *x*. In this program we used lower case letters to denote matrices.

The statement $n = nrow(x)$ assigns the value of the number of rows in the matrix *x* to variable label *n*. The statement $p = ncol(x)$ assigns the value of the number of columns in the matrix *x* to variable label *p*.

The function $j(a,b,c)$ creates an *axb* matrix whose elements all have the value *c*. Thus, $j(n,1,1)$ creates an *n*x1 matrix whose elements all have the value '1'.

The expression $c = a\|b$ concatenates the columns of the matrices *a* and *b* and stores the result in the matrix *c*. The result is a matrix whose first set of columns are the columns of *a* and whose next set of columns are the columns of *b*. This is called horizontal concatenation.

In Insert 5.3, the statement $x = j(n,1,1)\|x$ creates an *n*x1 matrix of ones, concatenates this created matrix with *x*, and stores the result in a new matrix named *x*. This results in a matrix *x* whose first column is a vector of ones and second column is a vector containing the data for *lotsize*. Here we have used *x* on both sides of the equal sign. This means that the "old" variable *x* is modified and the variable label *x* is used as the name for the "new" variable. The "old" variable is no longer associated with the name *x*.

The remaining statements are easy to follow. The statement $xpx = x`*x$ computes **X'X** and stores the result as a matrix named *xpx*; the statement $xpy = x`*y$ computes **X'Y** and stores the result as a matrix named *xpy*, and so on.

The print statement specifies printing **Y** (eq 5.61, text page 200), **X** (eq 5.61, text page 200), **Ŷ** (eq 5.71, text page 202), and **E** (**e**, eq 5.76, text page 203). Matrices **XPXINV** ($(X'X)^{-1}$, eq 5.64, text page 201), **B** (eq 5.60, text page 200), **YPY** ($Y'Y$, text page 205), **CORRECT** (corrected

SSTO), **SSTO** (eq 5.83 and 5.89a, text page 204), **SSR** (**b'X'Y**, text page 205), **SSE** (**Y'Y−b'X'Y**, text page 205), **MSE, VARB** ($s^2\{b\}$, eq 5.94, text page 207), **XH, VARYHATH** ($s^2\{\hat{Y}\}$, eq 5.95 and 598, text page 208) are also presented. SAS Output 5.2 displays excerpts of output to illustrate the results.

<table>
<tr><td>

INSERT 5.3

```
data TolucaCo;
  infile 'a:\ch01ta01.dat';
  input lotsize workhrs;
run;
proc iml;
  use ch01p0021;
  read all var {'workhrs'} into y;
  read all var {'lotsize'} into x;
  n = nrow(x);
  p = ncol(x);
  x = j(n,1,1)||x;
  xpx = x`*x;
  xpy = x`*y;
  xpxinv = inv(x`*x);
  b = inv(x`*x)*x`*y;
```

</td><td>

INSERT 5.3 continued

```
yhat = x*b;
  e = y - yhat;
  ypy = y`*y;
  correct = y`*j(n,n,1)*y/n;
  sstot = y`*y - y`*j(n,n,1)*y/n;
  ssr = b`*x`*y;
  sse = y`*y - b`*x`*y;
  mse = sse/(n-p);
  varb = mse*inv(x`*x);
  xh ={1, 65};
  varyhath = xh`*varb*xh;
print y x yhat e, xpx xpy, xpxinv,
  b, ypy correct sstot ssr sse mse, varb, xh,
varyhath;
quit;
```

</td></tr>
</table>

SAS Output 5.2

Y	X		YHAT	E
399	1	80	347.98202	51.01798
121	1	30	169.47192	-48.47192
221	1	50	240.87596	-19.87596
376	1	90	383.68404	-7.68404
361	1	70	312.28	48.72

XPXINV	
0.2874747	-0.003535
-0.003535	0.0000505

B
62.365859
3.570202

XH
1
65

VARB	
685.25805	-8.427277
-8.427277	0.1203897

VARYHATH	YPY	CORRECT	SSTOT	SSR	SSE	MSE
98.358367	2745173	2437970	307203.04	2690347.5	54825.459	2383.7156

Chapter 6 SAS®

Multiple Regression I

In many applications using regression analysis the model may involve more than one predictor variable. We call such a model the *multiple regression model* or the *general linear regression model*. We can write the general linear regression model as:

$$Y_i = \beta_0 + \beta_1 X_{i1} + \beta_2 X_{i2} + \dots + \beta_{p-1} X_{i,p-1} + \varepsilon_i$$

where:

$\beta_0, \beta_1, \dots, \beta_{p-1}$ are unknown model parameters with fixed values

$X_{i1}, \dots, X_{i,p-1}$ are known constants

ε_i are independent $N(0, \sigma^2)$ random variables

$i = 1, \dots, n$

In matrix notation, the general linear regression model is:

$$\mathbf{Y} = \mathbf{X}\boldsymbol{\beta} + \boldsymbol{\varepsilon}$$

where:

$\mathbf{Y}$ is the *nx1* vector of response observations on the dependent variable

$\boldsymbol{\beta}$ is the *px1* vector of unknown model parameters

$\mathbf{X}$ is the *nxp* matrix of known constants

$\boldsymbol{\varepsilon}$ is a vector of independent $N(0, \sigma^2)$ random variables

Chapter 6 focuses on the special case where $p = 3$ and therefore only two predictor variables are in the general linear regression model. In the following, we give illustrative SAS syntax for performing a statistical analysis of the general linear regression model with $p = 3$.

Dwaine Studios, Inc. Example

Data File: ch06fi05.txt

Multiple Regression with Two Predictor Variables: Dwaine Studios, Inc. has locations in 21 medium size cities and is considering expansion into additional cities (text page 236). The company wants to predict sales (Y) in a community using the number of persons greater than 15 years in the community ($X1$) and per capita disposable income in the community ($X2$). See Chapter 0, "Working with SAS," for instructions on opening the Dwaine Studios data file and naming the variables $X1$, $X2$, and Y. After printing ch06fi05.txt, the first 5 lines of output should look like SAS Output 6.1.

SAS Output 6.1

Obs	x1	x2	y
1	68.5	16.7	174.4
2	45.2	16.8	164.4
3	91.3	18.2	244.2
4	47.8	16.3	154.6
5	46.9	17.3	181.6

The researchers began with the first order regression model. (Insert 6.1) Results are shown in SAS Output 6.2 and are consistent with Figure 6.5, text page 237. In Insert 6.1, we also present the matrix language syntax for estimation of the parameters. Notice that the **X** matrix on text page 237 contains a column of 1's and this column is not in the Dwaine Studios data file. Thus, in Insert 6.1 we first compute a variable, *X0* that is equal to 1 for each of the 21 cities in the data file. We then enter *proc iml* and set column vectors **CON, A, B**, and **Y** equal to *X0, X1, X2*, and *Y*, respectively. Next, we concatenate the **CON, A**, and **B** column vectors into the **X** matrix such that the **X** matrix is now as shown in equation 6.70, text page 237. (Although the method of constructing the **X** matrix in Insert 6.1 is different from the method presented in Chapter 5, both methods achieve the same outcome.) Next, we find **X'X** (*xtx*), **X'Y** (*xty*), $(X'X)^{-1}$ (*xtxinv*), and **b** $= (X'X)^{-1}X'Y$. The *"print b"* statement will print the parameter estimates. Although we do not display the output from the *"print b"* statement, the estimates obtained through the matrix language program are equal to the estimates shown in SAS Output 6.2. We did not print **X'X**, **X'Y**, and $(X'X)^{-1}$, but we could do so and compare the numerical output to the text results shown on page 238-239 and the SAS regression analysis shown in SAS Output 6.2.

SAS Output 6.2

Analysis of Variance					
Source	DF	Sum of Squares	Mean Square	F Value	Pr > F
Model	2	24015	12008	99.10	<.0001
Error	18	2180.92741	121.16263		
Corrected Total	20	26196			

Root MSE	11.00739	R-Square	0.9167
Dependent Mean	181.90476	Adj R-Sq	0.9075
Coeff Var	6.05118		

51

Parameter Estimates					
Variable	DF	Parameter Estimate	Standard Error	t Value	Pr > |t|
Intercept	1	-68.85707	60.01695	-1.15	0.2663
x1	1	1.45456	0.21178	6.87	<.0001
x2	1	9.36550	4.06396	2.30	0.0333

INSERT 6.1

```
* Regresson analysis;
data dwaine;
  infile 'c:\alrm\chapter  6 data sets\ch06fi05.txt';
  input x1 x2 y;
proc reg;
 model y=x1 x2;
run;

* Matrix format;
data dwaine;
  infile 'c:\alrm\chapter  6 data sets\ch06fi05.txt';
  input x1 x2 y;
x0=1;
proc iml;
  use dwaine;
  read all var {'y'} into y;
  read all var {'x1'} into a;
  read all var {'x2'} into b;
  read all var {'x0'} into con;
x = (con||a||b);
xtx=x`*x;
xty=x`*y;
xtxinv=inv(x`*x);
b=inv(x`*x)*x`*y;
print b;
quit;
```

To produce the three-dimensional scatter plot shown in Figure 6.6(a)), text page 238, we use *proc g3d* from the SAS/Graph package (Insert 6.2). We first produce the three-dimensional scatter plot (SAS Output 6.3), and then rotate the plot by 45° in the second *proc g3d* (SAS Output 6.4). Optionally, we can rotate and tilt the plot in the same *proc g3d* statement, e.g., *scatter x2*x1=y / rotate=45 tilt=45;*.

INSERT 6.2

```
* Produce text Figure 6.6 3-D graphs;
data dwaine;
  infile 'c:\alrm\chapter 6 data sets\ch06fi05.txt';
  input x1 x2 y;
proc g3d data=dwaine;
  scatter x2*x1=y / shape='pyramid' caxis=black noneedle;
title1 'Basic scatter plot';
run;
proc g3d data=dwaine;
  scatter x2*x1=y / shape='pyramid' caxis=black noneedle rotate=45;
title1 'Basic scatter plot';
run;

* Produce text Figures 6.8 and 6.9;
data dwaine;
  infile 'c:\alrm\chapter 6 data sets\ch06fi05.txt';
  input x1 x2 y;
proc reg;
 model y=x1 x2/ r p;
 output out=results r=residual p=yhat;
run;
data results;
 set results;
absresid=abs(residual);
x1x2=x1*x2;
proc plot;
* Produce text Figures 6.8 a-d;
 plot residual*yhat;
 plot residual*x1;
 plot residual*x2;
 plot residual*x1x2;
* Produce text Figures 6.9 a;
 plot absresid*yhat="*";
 run;
* Produce text Figures 6.9 b;
proc univariate normal plot;
 var residual;
run;
```

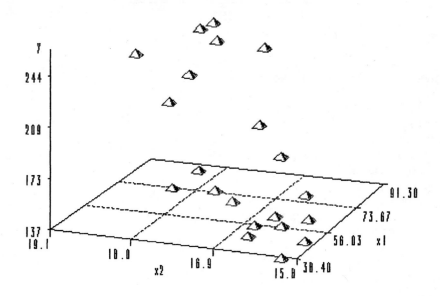

Basic scatter plot

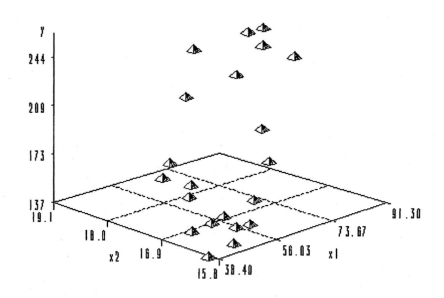

Basic scatter plot

Analysis of Appropriateness of Model: Inspection of Figures 6.8 and 6.9, text pages 242-243, reveals that we need to produce the residuals and predicted values of Y based on the first order regression model. In Insert 6.2 above, we save the unstandardized residuals (*resesidual*) and predicted values (*yhat*) in the output file named *results* (*output out=results*). We also compute *absresid*, the absolute value of the residuals, and *x1x2*, the interaction term between *X1* and *X1*. *Proc plot* can now produce text Figures 6.8 a-d and 6.9a. *Proc univariate* is used to produce the normal probability plot shown in text Figure 6.9b.

On text page 243, the authors calculate the coefficient of correlation between the ordered residuals and their expected value under normality for the Dwaine Studios example. The expected value under normality of the kth smallest observation in a sample of size n is given by:

$$\sqrt{MSE}\left[z\left(\frac{k-0.375}{n+0.25}\right)\right]$$

The first step in obtaining the expected value of the ordered residuals under normality is to calculate $(k - 0.375)/(n + 0.25)$. (Insert 6.3) In the Dwaine Studios example, $n = 21$. Since k is the rank of the ordered residuals, we obtain the first order regression model and save the residuals (*residual*). We then use *proc rank* to create new variable (*ranke*) that is based on the value of *residual*. Thus, $(k - 0.375)/(n + 0.25)$ becomes $(ranke - 0.375)/(21 + 0.25)$. This quantity represents the cumulative probability ($0 \leq$ cumulative probability ≤ 1.0) of the standard normal distribution. We now obtain a standard score (z) associated with the cumulative probability calculated above. In SAS, the *Probit* function is a *quantile function* and takes the form *Probit(p)*, where $p =$ cumulative probability. For instance, if cumulative probability $= 0.50$ (*Probit(0.50)*) then $z = 0$. If cumulative probability $= 0.8413$ (*Probit(0.8413)*) then $z = +1$. Thus, to obtain z we use *Probit ((ranke − 0.375)/(21 + 0.25))*. Finally, we multiply this quantity by the square root of *MSE* = 121.163 of the first order model (SAS Output 6.2). From the above, we compute the expected value (*ev*) as:

$$ev = sqrt(121.163) * Probit ((ranke − 0.375)/(21 + 0.25)).$$

We can now obtain the coefficient of correlation between the residual and the expected value under normality (*ev*). The result of the correlation between *ev* and *residual* is 0.979 and is within rounding error of the correlation (0.980) reported on text page 243. The interpolated critical value from text Table B.6 for $n = 21$ and $\alpha = 0.05$ is 0.9525. The obtained correlation coefficient (0.980) is greater than the table value (0.9525), and we conclude that the distribution of the error terms appears to be reasonably close to a normal distribution.

On text page 244, the researchers tested the null hypothesis H_0: $\beta_1 = 0$ and $\beta_2 = 0$. The test result is found in the output of the first order regression model above (SAS Output 6.2). Because the p-value (<0.000) associated with $F = 99.103$ is less than the desired significance level ($\alpha = 0.05$), we conclude that sales are related to target population (*X1*) and disposable income (*X2*).

We can use the ANOVA table in SAS Output 6.2 to find the *Coefficient of Multiple Determination* using equation 6.40, text page 226:

$$R^2 = \frac{SSR}{SSTO} = \frac{24,015.28}{26,196.21} = 0.917$$

$R^2 = 0.917$ is also given in the Model Summary table of SAS Output 6.2. Additionally, $R = 0.957$ and *Adjusted R Square* = 0.907 are presented in the Model Summary table.

Estimation of Regression Parameters: On text page 245, the researchers wish to jointly estimate β_1 and β_2 with a family confidence coefficient of 0.90. The confidence limits for β_k are given by equations 6.52 and 6.52a, text page 228:

$$b_k \pm Bs\{b_k\}$$

$$b_k \pm t(1-\alpha/2g; n-p)s\{b_k\}$$

where g = the number of intervals to be estimated and p = the number of parameters in the regression model. In the present discussion, we wish to estimate $g = 2$ intervals and there are 3 parameters in the regression model. We use the *quantile function TINV*(probability, df) where "probability" = $(1-0.90/2g)$ and df = $n - p$ or TINV(0.975, 18). Thus, in Insert 6.4, we find t and

56

place its value in *tvalue*. We use the parameter estimates shown in SAS Output 6.2 (*b1* = 1.455 and *b2* = 9.366) and the standard deviation of each estimate (shown as "Standard Error" in SAS Output 6.2, $s\{b_1\}$ = 0.212 and $s\{b_2\}$ = 4.06) to find the lower (*b1lo* and *b2lo*) and upper (*b1hi* and *b2hi*) confidence limit for each estimated parameter. We print the limits using *proc print data=dwaine (obs=1);*.

INSERT 6.4

```
data intervals;
tvalue = tinv(.975,18);
b1lo=1.45455-(tvalue*.211178);
b1hi=1.45455+(tvalue*.211178);
b2lo=9.3655-(tvalue*4.06395);
b2hi=9.3655+(tvalue*4.06395);
proc print;
 var b1lo b1hi b2lo b2hi;
run;
```

Estimation of Mean Response: On text page 245 the authors wish to estimate the expected mean sales when the target population is 65.4 thousand persons and the per capita disposable income is 17.6 thousand dollars. They also estimate a 95 percent confidence interval for $E\{Y_h\}$. The $1-\alpha$ confidence limits for $E\{Y_h\}$ are given by text equations 6.59, text page 229:

$$\hat{Y}_h \pm t(1-\alpha/2;n-p)s\{\hat{Y}_h\}$$

where:

$$s^2\{\hat{Y}_h\}= MSE\left(X_h'\left(X'X\right)^{-1}X_h\right)$$

In chapter 2 we found the confidence intervals for $E\{Y_h\}$ where there was one predictor variable in the model. However, as the number of predictor variables increases, the task of finding the variance of $\hat{Y}_h$ becomes increasingly laborious. Although equation 2.30, text page 53, can be generalized to the case of multiple regression (*p* > 2), the authors present the formulation of the variance of $\hat{Y}_h$ in matrix notation in text equation 6.58 shown above. We use this matrix formulation and *proc iml* to solve the current problem.

The matrix program shown in Insert 6.5 is an extension of the matrix program shown in Insert 6.1 above. We find *t* (*tvalue*) prior to entering the matrix program and after entering set the matrix **tvalue** to the variable *tvalue*. We then find $s^2\{\hat{Y}_h\}$ using the above notation. *MSE* = 121.163 is given in SAS Output 6.2 above. X_h' is defined in Insert 6.5 as shown on text page 246. We find $\hat{Y}_h$ (*yhath*) using text equation 6.55, text page 229. Finally, we compute and print the lower (*yhathlo*=185.29) and upper (*yhathlo*=196.91) confidence limits.

57

```
data dwaine;
  infile 'c:\alrm\chapter 6 data sets\ch06fi05.txt';
  input x1 x2 y;
* Find t(1-alpha/2,n-p) from text equation 6.59;
tvalue=tinv(.975,18);
* Write a column of 1's into the Dwaine Studios data file;
x0=1;
proc iml;
  use dwaine;
  read all var {'y'} into y;
  read all var {'x1'} into a;
  read all var {'x2'} into b;
  read all var {'x0'} into con;
  read all var {'tvalue'} into tvalue;
x = (con||a||b);
xtx=x`*x;
xty=x`*y;
* Fine the b matrix using text equation 6.25;
b=inv(x`*x)*x`*y;
xtxinv=inv(x`*x);
* Define xh matrix per text page 246;
xh={1,65.4,17.6};
* Find standard deviation of yhat(new) per text equation 6.58;
sxh=sqrt(121.163*((xh`)*(xtxinv)*xh));
* Find yhat(new)using text equation 6.56;
yhath=xh`*b;
* Find confidence interval using text equation 6.59;
yhathlo=yhath-(tvalue*sxh);
yhathhi=yhath+(tvalue*sxh);
low = yhathlo[1,1];
high = yhathhi[1,1];
print low high;
quit;
```

Prediction Limits for New Observations: Dwaine Studios now wants to predict sales for two new cities (text page 247). Here, X_{h1} and X_{h2} equal 65.4 and 17.6, respectively, for City A and 53.1 and 17.7, respectively, for City B. To determine which prediction intervals (Scheffe versus Bonferroni) are preferable here, we find S and B using equations 6.65a and 6.66a, text page 231, where g = 2 and $1-\alpha = 0.90$. In Insert 6.6, we compute S (s) and B (b) and print the two values. We find $S = 2.29$ and $B = 2.10$ so that the Bonferroni limits are tighter here and therefore more efficient.

```
data dwaine;
  infile 'c:\alrm\chapter 6 data sets\ch06fi05.txt';
  input x1 x2 y;
s=sqrt(2*finv(.9,2,18));
b=tinv(1-(.1/(2*2)),18);
proc print data=dwaine (obs=1);
 var s b;
run;
```

To find the simultaneous Bonferroni prediction limits for city A, we use (equation 6.66, text page 231):

$$\hat{Y}_h \pm Bs\{pred\}$$

where (text equations 6.66a, 6.63a, and 6.58) :

$$B = t(1 - \alpha/2g; n - p)$$

$$s^2\{pred\} = MSE + s^2\{\hat{Y}_h\}$$

$$s^2\{\hat{Y}_h\} = MSE\left(\mathbf{X'_h}\left(\mathbf{X'X}\right)^{-1}\mathbf{X_h}\right)$$

From Insert 6.7, we find the lower (*yhathlo*=167.71) and upper (*yhathhi*=214.49) 90% confidence limits for city A. We simply replace "*xh* = {1, 65.4, 17.6}." with "*xh* = {1, 53.1, 17.7}." in Insert 6.7and find the 90 percent confidence interval for city B: *yhathlo* = 149.08 and *yhathi* = 199.21.

INSERT 6.7

```
data dwaine;
  infile 'c:\alrm\chapter  6 data sets\ch06fi05.txt';
  input x1 x2 y;
* Find b(1-alpha/2g,n-p) from text equation 6.59;
b=tinv(1-(.1/(2*2)),18);
* Write a column of 1's into the Dwaine Studios data file;
x0=1;
proc iml;
  use dwaine;
  read all var {'y'} into y;
  read all var {'x1'} into a;
  read all var {'x2'} into b;
  read all var {'x0'} into con;
  read all var {'b'} into bon;
* Define the X matrix;
x = (con||a||b);
xtx=x`*x;
xpx=t(x)*x;
xty=x`*y;
* Fine the b matrix using text equation 6.25;
b=inv(t(x)*x)*t(x)*y;
xtxinv=inv(t(x)*x);
* Define xh matrix as on text page 247;
xh={1, 65.4, 17.6};
* Find standard deviation of new observation per text equation 6.58;
spre=sqrt(121.163+(121.163*(t(xh)*(xtxinv)*xh)));
* Find yhat(new);
yhath=t(xh)*b;
* Find confidence interval using text equation 6.66;
yhathlo=yhath-(bon*spre);
yhathhi=yhath+(bon*spre);
low = yhathlo[1,1];
high = yhathhi[1,1];
print low high;
quit;
```

Chapter 7 SAS®

Multiple Regression II

We use *sum of squared deviations* or *sum of squares* of various types as a quantitative measurement of variability that is attributable to different sources associated with parameters specified in the model. In previous chapters we defined the sum of squared deviations of the observations on the dependent variable Y_i from the overall sample mean $\overline{Y}$ as the (corrected) *total sum of squares*, denoted *SSTO*. We defined the sum of squared deviations of the fitted (predicted) values $\hat{Y}_i$ from $\overline{Y}$ as the *regression sum of squares* denoted *SSR*. We defined the sum of squared deviations of the observations Y_i from the fitted values $\hat{Y}_i$ as the *error sum of squares* denoted *SSE*. We found that:

$$SSTO = SSR + SSE$$

Suppose we fit a general linear model with $p - 1$ predictor variables and obtain *SST*, *SSR* and *SSE*. Now, suppose we add one or more predictor variables to the model and fit this "new" model. The observations Y_i will remain unchanged as will the mean $\overline{Y}$, but the $\hat{Y}_i$ with this new model may be different from the $\hat{Y}_i$ based on the first model. Thus, the numerical values of *SSR* and *SSE* will likely change. In fact, *SSR* will increase and, correspondingly, *SSE* will decrease by the same magnitude as the increase in *SSR*. We call this magnitude of change the *extra sum of squares* due to the predictor variable(s) that were added to the model. To avoid confusion we need additional notation to identify the variables that are in the model when *SSR* and *SSE* are computed.

It is easier to explain the extra sum of squares notation in terms of a specific example. Suppose predictor variables X_1, X_2 and X_3 are available for fitting a general linear regression model. If we fit a model with only X_1 in the model, we denote the sums of squares for regression and error as $SSR(X_1)$ and $SSE(X_1)$, respectively. If we fit a model that includes the predictor variables X_1 and X_2, we write $SSR(X_1, X_2)$ and $SSE(X_1, X_2)$ to denote the regression and error sums of squares, respectively. If we fit a model that includes X_1 and then add X_2 to the model and refit, the additional or extra regression sum of squares due to adding the additional predictor variable is denoted $SSR(X_2|X_1)$ and the reduction or extra error sum of squares is denoted $SSE(X_2|X_1)$. Similarly, if we fit a model that includes all three predictor variables, we write $SSR(X_1, X_2, X_3)$ and $SSE(X_1, X_2, X_3)$ to denote the regression and error sums of squares, respectively. If we fit a model that includes X_1 and then add X_2 and X_3 to the model and refit, the additional or extra regression sum of squares due to adding both additional predictor variables is denoted $SSR(X_2, X_3 |X_1)$ and the reduction or extra error sum of squares is denoted $SSE(X_2, X_3 |X_1)$. The extra sum of squares is also known as the *partial sum of squares*. It is obvious that we need to specify the predictor variables as being already in the model or as being added to the model when we discuss extra (partial) sums of squares.

In Chapter 7, the authors discuss methods used to make statistical inferences when adding and deleting predictor variables to the general linear regression model. The problem of deciding the merits of adding or deleting predictor variables in choosing an appropriate regression model is of particular interest. The extra sum of squares concept plays a fundamental role in the decision making process.

The Body Fat Example

Data File: ch07ta01.txt

Extra Sum of Squares: The text presents an excellent discussion of extra sum of squares. "Extra sum of squares measure the marginal reduction in error sum of squares when one or several predictor variables are added to the regression model, given that other predictor variables are already in the model." Many statistical packages use slightly different terminology to denote two major types of sum of squares: Type I and Type III. Type I sum of squares examines the sequential incremental improvement in the fit of the model as each independent variable is added to the model. Thus, the order of entry of variables into the model will affect the Type I sum of squares. Type I sum of squares is also called the hierarchical decomposition of sum of squares method. In contrast, Type III sum of squares for a particular predictor variable is the amount of increase in the regression sum of squares (decrease in error sum of squares) due to that predictor variable after all other terms in the model have been included. Thus, the Type III sums of squares for a given predictor is the reduction in error sum of squares achieved when that predictor variable is the last predictor variable added to the model. We will see how Type I and III sum of squares are used in the following examples.

On text page 256, researchers in the Body Fat example wished to develop a regression model to predict body fat from three predictor variables: triceps skin fold thickness, thigh circumference, and midarm circumference. They considered four regression models: (1.) body fat (Y) as a linear function of triceps skin fold thickness alone, (2.) body fat as a linear function of thigh circumference alone, (3.) body fat as a linear function of the two predictor variables triceps skin fold and thigh circumference, and (4.) body fat as a linear function of all three predictor variables.

See Chapter 0, "Working with SAS," for instructions on opening the Body Fat example (text Table 7.1) and renaming the variables $X1$, $X2$, $X3$, and Y. After opening ch07ta01.txt, the first 5 lines of output should look like SAS Output 7.1.

SAS Output 7.1

Obs	x1	x2	x3	y
1	19.5	43.1	29.1	11.9
2	24.7	49.8	28.2	22.8
3	30.7	51.9	37.0	18.7
4	29.8	54.3	31.1	20.1
5	19.1	42.2	30.9	12.9

In Insert 7.1, we regressed Y on $X1$ in the first analysis (SAS Output 7.2). Under the output labeled "*Analysis of Variance*", we see a column labeled "*Source*", under which we see "*Model*", "*Error*", and "*Corrected Total*". "*Source*" refers to source of variation. For our purposes here, "*Model*" means "*Regression Model*", or simply "*Regression.*" Under the column labeled "*DF*" we see the degrees of freedom associated with the sum of squares for each "*Source*": 1 degree(s) of freedom for regression, 18 for error, and 19 for the corrected total. Under the column labeled "*Sum of Squares*" we see that, $SSR(X1) = 352.270$, $SSE(X1) = 143.120$, and $SSTO = 495.390$. In the second analysis, we regressed Y on $X1$ and $X2$ (SAS Output 7.3). In this model, there are 2 degrees of freedom for regression and 17 for error. Also, $SSR(X1, X2) = 385.439$ and $SSE(X1, X2) = 109.951$. The difference

$$SSE(X2|X1) = SSE(X1) - SSE(X1, X2) = 143.120 - 109.951 = 33.169$$

is referred to as the *extra reduction in error sum of squares* achieved by adding $X2$ to a model that has $X1$ as the only predictor variable. Correspondingly,

$$SSR(X2|X1) = SSR(X1, X2) - SSR(X1) = 385.439 - 352.270 = 33.169$$

is the *extra increase in the regression sum of squares* achieved by adding $X2$ to a model that has $X1$ as the only predictor variable.

Note that $SSR(X2|X1)$ is not given in SAS Output 7.2 and 7.3. This is because, by default, SAS *proc reg* output displays only the regression and error sum of squares for a model specified in a SAS "*model*" statement and does not give the incremental change due to adding predictor variables to the model. To directly produce the extra sum of squares, we can use the *ss1* option in the model statement to request the Type I sum of squares. The syntax in Insert 7.2 requests Type I sum of squares based on a model with $X1$ and $X2$ as predictor variables (SAS Output 7.4). The Type I sum of squares $SSR(X2|X1) = 33.169$ here is the same as found earlier.

INSERT 7.1

```
data bodyfat;
  infile 'c:\alrm\chapter 7 data sets\ch07ta01.txt';
  input x1 x2 x3 y;
proc reg;
 model y=x1;
 model y=x1 x2;
run;
```

SAS Output 7.2

Y on $X1$

Analysis of Variance					
Source	DF	Sum of Squares	Mean Square	F Value	Pr > F
Model	1	352.26980	352.26980	44.30	<.0001
Error	18	143.11970	7.95109		
Corrected Total	19	495.38950			

SAS Output 7.3

Y on $X1$ and $X2$

Analysis of Variance					
Source	DF	Sum of Squares	Mean Square	F Value	Pr > F
Model	2	385.43871	192.71935	29.80	<.0001
Error	17	109.95079	6.46769		
Corrected Total	19	495.38950			

INSERT 7.2

```
data bodyfat;
  infile 'c:\alrm\chapter  7 data sets\ch07ta01.txt';
  input x1 x2 x3 y;
proc reg;
 model y=x1 x2 / ss1;
run;
```

SAS Output 7.4

Parameter Estimates						
Variable	DF	Parameter Estimate	Standard Error	t Value	Pr > |t|	Type I SS
Intercept	1	-19.17425	8.36064	-2.29	0.0348	8156.76050
x1	1	0.22235	0.30344	0.73	0.4737	352.26980
x2	1	0.65942	0.29119	2.26	0.0369	33.16891

Test whether a single $\beta_k = 0$: On page 264, the researchers wished to test whether $X3$ should be dropped from the model that contained $X1$, $X2$, and $X3$. That is, they wanted to test H_0: $\beta_3 = 0$. From Chapter 6, we know that we can use a t-test to test this hypothesis. The result, shown in SAS Output 7.5 (Parameter Estimates table), indicates $t = -1.37$ for $X3$ with p-value for a two-sided test equal to 0.1896. Based on this finding, we do not reject H_0 and drop $X3$ from the model. We get an equivalent F-test as follows. We regress Y on $X1$, $X2$, and $X3$ (Insert 7.3). Results are shown in SAS Output 7.5. Under the column labeled "*Sum of Squares*" we see that, $SSR(X1, X2, X3) = 396.985$, $SSE(X1, X2, X3) = 98.405$, and $SSTO = 495.390$.

Recalling from SAS Output 7.3 above that $SSR(X1, X2) = 385.439$, we find:

$$SSR(X3|X1, X2) = SSR(X1, X2, X3) - SSR(X1, X2) = 396.985 - 385.439 = 11.546.$$

On adding $X3$ to a model that already contains $X1$ and $X2$, we must estimate one additional parameter, so the partial sum of squares $SSR(X3|X1, X2)$ has one degree of freedom. Recalling from SAS Output 7.5 that $SSE(X1, X2, X3) = 98.405$ with 16 degrees of freedom, we find on substituting in equation 7.15, text page 264:

$$F^* = \frac{11.546}{1} \div \frac{98.405}{16} = 1.88.$$

Earlier, we found $t = -1.37$ for $X3$. We know that $t^2 = F$, and this is easily verified in the present example because $(-1.37)^2 = 1.88$.

Alternatively, we can obtain both Type I and Type III sums of squares using *proc glm* with $X1$, $X2$, and $X3$ as predictors. (Insert 7.3 & SAS Output 7.6). Using *proc glm*, we get by default both Type I and Type III sums of squares as part of the output. Note that the Type I and Type III sums of squares will always be the same for the last predictor variable that is added to the model. Thus, because $X3$ was the last variable listed in specifying the predictor variables in the model statement, tests of the effect of $X3$ based on Type I and Type III sums of squares are identical, SAS Output 7.6 Type I and III sum of squares = 11.454. However, when the statistical test is not based on the last variable in the model, test statistics based on Type I and Type III sum of squares test different hypotheses. Assume that we wish to determine if $X2$ should be dropped from the above model. From the regression procedure using Type III sum of squares (Insert 7.3 - SAS Output 7.5), we see that $t = -1.11$ for $X2$. This is a test of the effect of $X2$ controlling for all other variables in the model, i.e., $X1$ and $X3$. From the *GLM* procedure Type I sum of squares (Insert 7.3 - SAS Output 7.6) we see that $F = 5.39$ for $X2$. This is a test of the effect of $X2$ controlling for all variables that preceded it in the list of predictor variables specified in the model statement, i.e., controlling for $X1$. ($t^2 = F$, but $(-1.106)^2 \neq 5.393$). So, in the $X2$ line of the analysis of variance, using Type III sum of squares, the F-test ($F = 1.22 = (-1.106)^2 = t^2$) is a test of the statistical significance of $SSR(X2|X1, X3)$, or, equivalently, $SSE(X2|X1, X3)$. On the other hand, the F-test ($F = 5.39$) in the $X2$ line of the analysis of variance using Type I sum of squares tests the significance of $SSR(X2|X1)$. Note that SAS *GLM* prints both Type I and Type III sums of squares as part of the output by default.

Finally, we call attention to the t values for the first regression analysis from Insert 7.3 (SAS Output 7.5) in which we regress Y on $X1$, $X2$, and $X3$. In the second regression analysis of Insert 7.3, we use the same regression model as in the first regression analysis only now we specify that Type I sum of squares (*ss1*) be included in the output. Results of this maneuver are shown in SAS Output 7.7. Now we compare the t-values of SAS Output 7.5 to the t-values of SAS Output 7.7 where we requested Type I sum of squares and verify that the values are identical. Although the test for $X2$ (SAS Output 7.7) displays Type I sum of squares on the same line, the statistical test is not testing $SSR(X2|X1)$, but rather it is testing $SSR(X2|X1, X3)$, a Type III sum of squares or, equivalently, a Type I sum of squares for a model with $X2$ being the last predictor variable to enter the model. This illustrates a very important point: to properly interpret the output of a compute software program, we must know the type of sum of squares used in the analysis.

INSERT 7.3

```
data bodyfat;
  infile 'a:\ch07ta01.dat';
  input x1 x2 x3 y;

proc reg;
 model y=x1 x2 x3;
run;

proc reg;
 model y=x1 x2 x3 / ss1;
run;

proc glm;
 model y=x1 x2 x3;
run;
```

SAS Output 7.5

SAS Regression

Analysis of Variance					
Source	DF	Sum of Squares	Mean Square	F Value	Pr > F
Model	3	396.98461	132.32820	21.52	<.0001
Error	16	98.40489	6.15031		
Corrected Total	19	495.38950			

65

Parameter Estimates					
Variable	DF	Parameter Estimate	Standard Error	t Value	Pr > |t|
Intercept	1	117.08469	99.78240	1.17	0.2578
x1	1	4.33409	3.01551	1.44	0.1699
x2	1	-2.85685	2.58202	-1.11	0.2849
x3	1	-2.18606	1.59550	-1.37	0.1896

SAS Output 7.6

SAS GLM

Source	DF	Sum of Squares	Mean Square	F Value	Pr > F
Model	3	396.9846118	132.3282039	21.52	<.0001
Error	16	98.4048882	6.1503055		
Corrected Total	19	495.3895000			

Source	DF	Type I SS	Mean Square	F Value	Pr > F
x1	1	352.2697968	352.2697968	57.28	<.0001
x2	1	33.1689128	33.1689128	5.39	0.0337
x3	1	11.5459022	11.5459022	1.88	0.1896

Source	DF	Type III SS	Mean Square	F Value	Pr > F
x1	1	12.70489278	12.70489278	2.07	0.1699
x2	1	7.52927788	7.52927788	1.22	0.2849
x3	1	11.54590217	11.54590217	1.88	0.1896

Parameter	Estimate	Standard Error	t Value	Pr > |t|
Intercept	117.0846948	99.78240295	1.17	0.2578
x1	4.3340920	3.01551136	1.44	0.1699
x2	-2.8568479	2.58201527	-1.11	0.2849
x3	-2.1860603	1.59549900	-1.37	0.1896

SAS Output 7.7

SAS Regression and display of Type I SS

Parameter Estimates						
Variable	DF	Parameter Estimate	Standard Error	t Value	Pr > |t|	Type I SS
Intercept	1	117.08469	99.78240	1.17	0.2578	8156.76050
x1	1	4.33409	3.01551	1.44	0.1699	352.26980
x2	1	-2.85685	2.58202	-1.11	0.2849	33.16891
x3	1	-2.18606	1.59550	-1.37	0.1896	11.54590

Test whether several $\beta_k = 0$: Continuing with the body fat example (text page 265), the researchers wished to determine if, in a single step, $X2$ and $X3$ should be dropped from the model that contained $X1$, $X2$, and $X3$. This is equivalent to testing the hypothesis $H_0: \beta_2 = \beta_3 = 0$. From SAS Output 7.6 above, we have the necessary information to find equation 7.18, text page 265:

$$F^* = \frac{SSR(X2 \mid X1) + SSR(X3 \mid X1, X2)}{2} \div MSE(X1, X2, X3)$$

$$F^* = \frac{33.169 + 11.546}{2} \div 6.15 = 3.63$$

In the above expression, we used the fact that

$$SSR(X2|X1) + SSR(X3|X1, X2) = SSR(X2, X3|X1)$$

Alternatively, there is an easier way to test $H_0: \beta_2 = \beta_3 = 0$. We use the *test* option that is available within *proc reg* (Insert 7.4). The *test* statement tests user-defined hypotheses about the parameters estimated in the preceding *model* statement. In Insert 7.4 we test the hypothesis that the coefficients for both $X2$ and $X3$ equal 0. We specify in the *test* statement a label ($x2x3$) for SAS to use in the output to identify the user-defined test. The test statistic $F = 3.64$ (SAS Output 7.8) represents the incremental effect of adding $X2$ and $X3$ to a model that contains $X1$, i.e., $SSR(X2, X3|X1)$.

```
                              INSERT 7.4
data bodyfat;
 infile 'c:\alrm\chapter 7 data sets\ch07ta01.txt';
  input x1 x2 x3 y;
proc reg;
 model y=x1 x2 x3;
 x2x3: test x2=x3=0;
run;
```

Test x2x3 Results for Dependent Variable y				
Source	DF	Mean Square	F Value	Pr > F
Numerator	2	22.35741	3.64	0.0500
Denominator	16	6.15031		

Coefficients of Partial Determination and Correlation: In the body fat example (text page 270) we obtain first the *Coefficients of Partial Correlation* for *X2* given *X1* is in the model, $r_{Y2|1}$ (Insert 7.5). This results in $r_{Y2|1} = 0.4814$ (output not shown). We now can find the *Coefficient of Partial Determination* by squaring the *Coefficient of Partial Correlation*, $0.4814^2 = 0.2317$.

INSERT 7.5

```
data bodyfat;
  infile 'c:\alrm\chapter 7 data sets\ch07ta01.txt';
  input x1 x2 x3 y;
proc corr;
var y x2;
 partial x1;
 run;
```

Dwaine Studios, Inc. Example

Data File: ch06fi05.txt

Standardized Multiple Regression Model: To produce the regression model with standardized coefficients, we specified the regression procedure with the model to regress *Y* on *X1* and *X2* using the Dwaine Studios, Inc. data (text page 276). We also requested descriptive statistics using the *simple* option in the *proc reg* statement. The *simple* option displays the sum, mean, variance, standard deviation, and uncorrected sum of squares for each variable used in *proc reg*. Partial output for this procedure is shown in SAS Output 7.9. Notice that we also used the *stb* option to request *standardized beta coefficients*. With the descriptive statistics, we can use equation 7.53, text page 276, to shift the *standardized regression coefficients* back to the *unstandardized regression coefficients*.

$$b_1 = \left(\frac{s_y}{s_1}\right)b_1^* = \frac{36.1913}{18.620}(.748) = 1.455$$

SAS Output 7.9

			Descriptive Statistics		
Variable	**Sum**	**Mean**	**Uncorrected SS**	**Variance**	**Standard Deviation**
Intercept	21.00000	1.00000	21.00000	0	0
x1	1302.40000	62.01905	87708	346.71662	18.62033
x2	360.00000	17.14286	6190.26000	0.94157	0.97035
y	3820.00000	181.90476	721072	1309.81048	36.19130

			Parameter Estimates			
Variable	**DF**	**Parameter Estimate**	**Standard Error**	**t Value**	**Pr > |t|**	**Standardized Estimate**
Intercept	1	-68.85707	60.01695	-1.15	0.2663	0
x1	1	1.45456	0.21178	6.87	<.0001	0.74837
x2	1	9.36550	4.06396	2.30	0.0333	0.25110

Multicollinearity and Its Effects: When multiple predictor variables in a general linear regression model are correlated among themselves, we say there is *intercorrelation* or *multicollinearity* among these variables (text page 279). When there is multicollinearity among a set of predictor variables being considered for a general linear regression model, the extra regression (error) sum of squares for a given predictor or set of predictor variables will be different depending on which other predictor variables are in the model. On the other hand, if there is no multicollinearity among the predictor variables, the extra regression sum of squares will be the same no matter which other variables are included.

Work Crew Productivity Example

On text page 279, the authors present the results of an experiment on the effect of work crew size (*X1*) and level of pay (*X2*) on productivity (*Y*). The text presents results from a correlation matrix of the 3 variables and Table 7.7, the output of 3 separate regression models. We present the SAS syntax (Insert 7.7) for this example as a further illustration. The rationale closely follows that in the body fat example.

```
                          INSERT 7.7
data work;
 infile 'c:\alrm\chapter 7 data sets\ch07ta06.txt';
 input x1 x2 y;

proc corr;
 var x1 x2 y;
proc reg;
 model y=x1 x2;
 model y=x1;
 model y=x2;
 run;
```

More on the Body Fat Example

Data File: ch07ta01.txt

On text page 283, the authors revisit the Body Fat example (text page 256) to illustrate the effect of *multicollinearity*. In Insert 7.8, we explore the relation between the predictor variables by producing *scatter plots* and a *correlation matrix* (SAS Output 7.10) of $X1$, $X2$, and $X3$. The coefficient of correlation between $X1$ and $X2$ is $r_{12} = 0.9238$ and although $X3$ is not strongly correlated with $X1$ and $X2$ individually, the multiple R^2 when $X3$ is regressed on $X1$ and $X2$ is 0.990 (Insert 7.11). Thus, there is a strong correlation between $X3$ and a linear function of $X1$ and $X2$. To see the change in parameter estimates as variables are added to the model, we regressed Y on $X1$ and $X2$ (SAS Output 7.12) and Y on $X1$, $X2$, and $X3$ (SAS Output 7.13). As noted in the text, the regression coefficient for $X2$ changes in the two analyses from 0.659 to -2.857 due to the intercorrelations of the predictor variables. In Chapter 10, the authors explore formal methods that we can use to detect multicollinearity.

```
                          INSERT 7.8
data bodyfat;
 infile 'c:\alrm\chapter 7 data sets\ch07ta01.txt';
 input x1 x2 x3 y;

 proc plot;
 plot x1*x2;
 plot x1*y;
 plot x2*y;
run;

 proc corr;
 var x1 x2 y;
 run;

 proc reg;
 model x3=x2 x1;
 model y=x1 x2;
 model y=x1 x2 x3;
 run;
```

SAS Output 7.10

Pearson Correlation Coefficients, N = 20 Prob > |r| under H0: Rho=0			
	x1	**x2**	**y**
x1	1.00000	0.92384 <.0001	0.84327 <.0001
x2	0.92384 <.0001	1.00000	0.87809 <.0001
y	0.84327 <.0001	0.87809 <.0001	1.00000

SAS Output 7.11

X3 regressed on X1 and X2

Root MSE	0.37699	**R-Square**	0.9904
Dependent Mean	27.62000	**Adj R-Sq**	0.9893
Coeff Var	1.36491		

SAS Output 7.12

Y regressed on X1 and X2

Parameter Estimates					
Variable	**DF**	**Parameter Estimate**	**Standard Error**	**t Value**	**Pr > |t|**
Intercept	1	-19.17425	8.36064	-2.29	0.0348
x1	1	0.22235	0.30344	0.73	0.4737
x2	1	0.65942	0.29119	2.26	0.0369

SAS Output 7.13

Y regressed on X1, X2, and X3

Parameter Estimates					
Variable	**DF**	**Parameter Estimate**	**Standard Error**	**t Value**	**Pr > |t|**
Intercept	1	117.08469	99.78240	1.17	0.2578
x1	1	4.33409	3.01551	1.44	0.1699
x2	1	-2.85685	2.58202	-1.11	0.2849
x3	1	-2.18606	1.59550	-1.37	0.1896

Chapter 8 SAS®

Regression Models for Quantitative and Qualitative Predictors

In this chapter, the authors discuss some special types of general linear regression models. They discuss *polynomial regression models, interaction regression models, regression models with qualitative predictor variables,* and *more complex regression models.* Recall the simple linear regression model is of the form:

$$Y_i = \beta_0 + \beta_1 X_i + \varepsilon_i$$

This model is said to be a *first-order model* because the predictor variable appears in the model expressed to the first power; i.e., $X_i^1 = X_i$. Similarly, the model:

$$Y_i = \beta_0 + \beta_1 X_{i1} + \beta_2 X_{i2} + \varepsilon_i$$

is a first-order model because the two predictor variables X_{i1} and X_{i2} each appear in the model expressed to the first power.

 The model:

$$Y_i = \beta_0 + \beta_1 x_i + \beta_2 x_i^2 + \varepsilon_i$$

where $x_i = X_i - \overline{X}$ is the *centered predictor variable*, is called a *polynomial regression model.* It is referred to as a *second-order model with one predictor* because the single predictor variable appears in the model expressed to the second power; i.e., x_i^2. Similarly, the polynomial regression model:

$$Y_i = \beta_0 + \beta_1 x_i + \beta_2 x_i^2 + \beta_3 x_i^3 + \varepsilon_i$$

where $x_i = X_i - \overline{X}$, is a *third-order model with one predictor variable.* The regression model:

$$Y_i = \beta_0 + \beta_1 x_{i1} + \beta_2 x_{i2} + \beta_{11} x_{i1}^2 + \beta_{22} x_{i2}^2 + \beta_{12} x_{i1} x_{i2} + \varepsilon_i$$

where $x_{i1} = X_{i1} - \overline{X}_1$ and $x_{i2} = X_{i2} - \overline{X}_2$, is a *second-order model with two predictor variables.* The cross-product term represents the *interaction effect* between x_1 and x_2. This nomenclature extends to models with any number of predictor variables.

The authors use indicator variables that take on values 0 and 1 in a regression model to identify classes or categories of a qualitative variable. They use several examples to illustrate how to fit and interpret a variety of higher-order models including models where some of the predictor variables are quantitative and others are qualitative.

The Power Cell Example

Data File: ch08ta01.txt

Using data from Table 8.1, text page 300, researchers studied the relationship between charge rate and temperature on the life of a power cell. There were three levels (0.6, 1.0, and 1.4 amperes) of charge rate ($X1$) and three levels (10, 20, and 30° C) of ambient temperature ($X2$). The dependent variable (Y) was defined as the number of discharge-charge cycles that a power cell underwent before failure. See Chapter 0, "Working with SAS," for instructions on opening the Power Cell data file and naming the variables Y, $X1$, and $X2$.

The researchers were uncertain of the expected response function, but decided to fit a second-order polynomial regression model. They first centered $X1$ and $X2$ around their respective means and scaled them using the absolute difference between adjacent levels of the variable. From Insert 8.1, we compute the coded values for $X1$ and $X2$ (*x1code* and *x2code*, respectively), the square of the coded values (*x1code2* and *x2code2*, respectively), and the cross-product of $X1$ and $X2$ (*x12inter*). After executing the syntax in Insert 8.1, the printed output should look like SAS Output 8.1 and is consistent with Table 8.1, text page 300.

```
                        INSERT 8.1
data power;
  infile 'c:\alrm\chapter 8 data sets\ch08ta01.txt';
  input y x1 x2;
x1code=(x1-1)/.4;
x2code=(x2-20)/10;
x1code2=x1code**2;
x2code2=x2code**2;
x12inter=x1code*x2code;
run;
proc print;
run;
```

SAS Output 8.1

Obs	y	x1	x2	x1code	x2code	x1code2	x2code2	x12inter
1	150	0.6	10	-1	-1	1	1	1
2	86	1.0	10	0	-1	0	1	0
3	49	1.4	10	1	-1	1	1	-1
4	288	0.6	20	-1	0	1	0	0
5	157	1.0	20	0	0	0	0	0

Fitting the Model: We begin by obtaining the regression results of the second-order polynomial regression model and saving the unstandardized predicted values (*yhat*) and residuals (*residual*) in a file named *"results"* for future use (Insert 8.2). Partial results are shown in SAS Output 8.2. The Unstandardized Coefficients displayed under the column labeled "Parameter Estimate" in SAS Output 8.2 are the estimated regression coefficients shown in equation 8.16, text page 301.

```
                                INSERT 8.2
data power;
  infile 'c:\alrm\chapter 8 data sets\ch08ta01.txt';
  input y x1 x2;
x1code=(x1-1)/.4;
x2code=(x2-20)/10;
x1code2=x1code**2;
x2code2=x2code**2;
x12inter=x1code*x2code;
run;
proc reg;
 model y=x1code x1code2 x2code x2code2 x12inter / r p;
 output out=results r=residual p=yhat;
run;
```

SAS Output 8.2

Analysis of Variance

Source	DF	Sum of Squares	Mean Square	F Value	Pr > F
Model	5	55366	11073	10.57	0.0109
Error	5	5240.43860	1048.08772		
Corrected Total	10	60606			

Root MSE	32.37418	R-Square	0.9135
Dependent Mean	172.00000	Adj R-Sq	0.8271
Coeff Var	18.82220		

Parameter Estimates

Variable	DF	Parameter Estimate	Standard Error	t Value	Pr > |t|
Intercept	1	162.84211	16.60761	9.81	0.0002
x1code	1	-55.83333	13.21670	-4.22	0.0083
x1code2	1	27.39474	20.34008	1.35	0.2359
x2code	1	75.50000	13.21670	5.71	0.0023
x2code2	1	-10.60526	20.34008	-0.52	0.6244
x12inter	1	11.50000	16.18709	0.71	0.5092

Residual Plots: In this section, we discuss SAS code for obtaining some simple graphical plots that are useful in assessing the adequacy of a fitted model. For brevity, we do not present the output plots. To produce the plots in Figure 8.5, text page 303, we continue with the SAS code shown in Insert 8.3 and request the residuals and predicted values of the current model using the *r* and *p* options in the *model* statement. The residual and predicted values are now available to be used within *proc reg* by specifying "*r.*" and "*p.*". That is, to produce text Figure 8.5 (a), we request a plot of *r.* (residual values) by *p.* (predicted values) using the code "*plot r.*p.*". To replicate the axes of text Figure 8.5 (a), we set the values of the horizontal axis (*haxis*) to range from 0 to 300 while incrementing by 100 units and we set the values of the vertical axis (*vaxis*) to range from 40 to 60 while incrementing by 20 units. Next we plot the residual values (*r.*) against *x1code* and *x2code*, and set the axes to correspond to those shown in Figure 8.5. Finally, we request a normal probability plot (*npp.*) for the residual values (*r.*).

INSERT 8.3

```
data power;
  infile 'c:\alrm\chapter  8 data sets\ch08ta01.txt';
  input  y x1 x2;
x1code=(x1-1)/.4;
x2code=(x2-20)/10;
x1code2=x1code**2;
x2code2=x2code**2;
x12inter=x1code*x2code;
run;
proc reg;
  model y=x1code x1code2 x2code x2code2 x12inter / r p;
  plot r.*p. / haxis=0 to 300 by 100 vaxis=-40 to 60 by 20;
  plot r.*x1code / haxis=-2 to 2 by 1 vaxis=-40 to 60 by 20;
  plot r.*x2code / haxis=-2 to 2 by 1 vaxis=-40 to 60 by 20;
  plot npp.*r.;
run;
```

Further Aspects of the Power Cell Example

Data File: ch08ta05.txt

Correlation Test for Normality: On text page 301, the researchers in the Power Cell example calculated the coefficient of correlation between the ordered residuals and their expected value under normality. The expected value under normality of the kth smallest observation from a sample of size n is given by:

$$\sqrt{MSE}\left[z\left(\frac{k - 0.375}{n + 0.25} \right) \right]$$

We find $MSE = 1048.088$ in SAS Output 8.2 above and here $n = 11$. In Chapter 3 we give a detailed description of the steps necessary to obtain z and k. In Insert 8.4, we regress Y on $X1$, $X2$, and $X1X2$ and save the residual values in a variable named *residual*. We then rank the residuals and place the rank in a variable named *rank*. We then compute *ev*, the expected value under normality and correlate this variable with *residual*. The resulting correlation is 0.984, as reported on the text page 301.

```
                              INSERT 8.4

data soap;
  infile 'c:\alrm\chapter  8 data sets\ch08ta05.txt';
  input  y x1 x2;
x1x2=x1*x2;
proc reg;
 model y=x1 x2 x1x2/ r;
 output out=results r=residual;
* Rank the residuals and place the rank in "rank";
proc rank data=results out=results;
 var residual;
 ranks rank;
run;
data new;
 set results;
 ev = sqrt(430.611) * Probit ((rank - .375)/(27  + .25));
proc corr;
 var ev residual;
 run;
```

Test of Fit: To perform a formal test of goodness of fit of the regression model, we use *proc glm* as detailed in Insert 8.5. We first compute a new variable (*J*) to represent each of the 9 combinations of *X1* and *X2*. We then regress *Y* on *J*, *x1code*, *x1code2*, *x2code*, *x2code2*, and *x12inter*. Additionally, in the *proc glm* statement we define *J* as a *class* variable. Thus, *J* identifies 9 discrete classifications based on the values of *X1* and *X2*. This produces the full model including $SSE(F)$= 1404.6 as shown in SAS Output 8.3 a. From equation 3.16, text page 122, $SSE(F) = SSPE$. Thus, $SSPE = 1404.6$. To produce the reduced model, we drop *J* from the previous model and refit the model in the second *proc glm* statement in Insert 8.5. This results in SAS Output 8.3 b. where $SSE(R) = 5240.4$. From equation 3.24, text page 123, $SSLF = SSE - SSPE$, or $SSLF = 5240.4 - 1404.6 = 3835.8$. In the present example, $n = 11$, $c = 9$ discrete classifications, and $p = 6$ parameters in the reduced model. We can now find:

$$F^* = \frac{SSLF}{c-p} \div \frac{SSPE}{n-c} = \frac{3835.8}{3} \div \frac{1404.6}{2} = 1.82$$

The *F*-value = 1.82 is identical to the Lack of Fit *F*-value reported on text page 302. The associated probability level of *F* is >0.05, and we conclude that the second-order polynomial regression is a reasonably good fitting model.

```
                            INSERT 8.5

data power;
  infile 'c:\alrm\chapter  8 data sets\ch08ta01.txt';
  input  y x1 x2;
x1code=(x1-1)/.4;
x2code=(x2-20)/10;
x1code2=x1code**2;
x2code2=x2code**2;
x12inter=x1code*x2code;
  if x2=10 & x1=0.6 then j=1;
  if x2=10 & x1=1.0 then j=2;
  if x2=10 & x1=1.4 then j=3;
  if x2=20 & x1=0.6 then j=4;
  if x2=20 & x1=1.0 then j=5;
  if x2=20 & x1=1.4 then j=6;
  if x2=30 & x1=0.6 then j=7;
  if x2=30 & x1=1.0 then j=8;
  if x2=30 & x1=1.4 then j=9;
proc glm;
 class j;
 model y=j x1code x1code2 x2code x2code2 x12inter;
run;
proc glm;
 model y=x1code x1code2 x2code x2code2 x12inter;
run;
```

SAS Output 8.3

a. Full Model

Source	DF	Sum of Squares	Mean Square	F Value	Pr > F
Model	8	59201.33333	7400.16667	10.54	0.0895
Error	2	1404.66667	702.33333		
Corrected Total	10	60606.00000			

b. Reduced Model

Source	DF	Sum of Squares	Mean Square	F Value	Pr > F
Model	5	55365.56140	11073.11228	10.57	0.0109
Error	5	5240.43860	1048.08772		
Corrected Total	10	60606.00000			

77

Coefficient of Multiple Determination: The coefficient of multiple correlation ($R = 0.956$), the coefficient of multiple determination ($R^2 = 0.914$) and the adjusted coefficient of multiple correlation ($Adj\ R^2 = 0.827$) are given in SAS Output 8.2 above.

Partial F-Test: The researchers now want to test whether a first-order model is appropriate. That is, they wish to decide if *x1code2*, *x2code2*, and *x12inter* can be dropped from the model; they test H_0: $\beta_{11} = \beta_{22} = \beta_{12} = 0$. In Insert 8.6, we use the *test* option available with the *proc reg* statement. That is, we request a test that *x1code2=x2code2=x12inter=0*. The "*b11b22b12:*" portion of the *test* statement is a label to identify the test on output. Partial results are shown in SAS Output 8.4. Consistent with text page 304, the *F*-value associated with this user defined test is 0.78 and the associated probability level ("Pr>F") is 0.5527. Thus, we fail to reject H_0: $\beta_{11} = \beta_{22} = \beta_{12} = 0$ and conclude that the first-order model is adequate.

INSERT 8.6

```
data power;
  infile 'c:\alrm\chapter 8 data sets\ch08ta01.txt';
  input y x1 x2;
x1code=(x1-1)/.4;
x2code=(x2-20)/10;
x1code2=x1code**2;
x2code2=x2code**2;
x12inter=x1code*x2code;
proc reg;
 model y=x1code x2code x1code2 x2code2 x12inter;
  b11b22b12: test x1code2=x2code2=x12inter=0;
run;
```

SAS Output 8.4

Test b11b22b12 Results for Dependent Variable y				
Source	DF	Mean Square	F Value	Pr > F
Numerator	3	819.96491	0.78	0.5527
Denominator	5	1048.08772		

First-Order Model: Based on the above analysis, the researchers chose to fit the first-order model:

$$Y_i = \beta_0 + \beta_1 x_{i1} + \beta_2 x_{i2} + \varepsilon_i$$

From Insert 8.7 we use *"proc reg"* together with *"model y = x1code x2code"* to produce the first-order model with results as shown in SAS Output 8.5(a). The regression coefficients are identical to those reported in equation 8.18, text page 304, and to the estimates shown below in SAS Output 8.5; i.e. (standardized) $b_1 = -55.83333$ and $b_2 = 75.50000$ with standard errors $s\{b_1\} = s\{b_2\} = 12.66584$.

INSERT 8.7

```
data power;
  infile 'c:\alrm\chapter 8 data sets\ch08ta01.txt';
  input y x1 x2;
x1code=(x1-1)/.4;
x2code=(x2-20)/10;
x1code2=x1code**2;
x2code2=x2code**2;
x12inter=x1code*x2code;
proc reg;
 model y=x1code x2code ;
 model y = x1 x2 /clb alpha = 0.10;
run;
```

SAS Output 8.5

(a)

Parameter Estimates					
Variable	DF	Parameter Estimate	Standard Error	t Value	Pr > |t|
Intercept	1	172.00000	9.35435	18.39	<.0001
x1code	1	-55.83333	12.66584	-4.41	0.0023
x2code	1	75.50000	12.66584	5.96	0.0003

(b)

Parameter Estimates							
Variable	DF	Parameter Estimate	Standard Error	t Value	Pr > |t|	95% Confidence Limits	
Intercept	1	160.58333	41.61545	3.86	0.0048	64.61793	256.54874
x1	1	-139.58333	31.66461	-4.41	0.0023	-212.60206	-66.56461
x2	1	7.55000	1.26658	5.96	0.0003	4.62925	10.47075

Fitted First-Order Model in Terms of *X1* and *X2*: In Insert 8.7, we add a second *model* statement to *proc reg* to obtain a fitted first order regression model in terms of the original (untransformed) predictor variables *X1* and *X2*. The resulting estimated coefficients are shown in SAS Output 8.5(b) and in equation 8.19, text page 304; i.e., (unstandardized) $b_1' = -139.58333$ and $b_2' = 7.55000$ with respective standard errors $s\{b_1'\} = 31.66461$ and $s\{b_2'\} = 1.26658$.

Estimation of Confidence Limits for Regression Coefficients: The researcher now wished to estimate the Bonferroni 90% confidence limits in terms of the original predictor variables. We apply the confidence limits formula for β_1 and β_2 given in equation 6.52, text page 228. In Insert 8.8, we first find $B = t(1 - \alpha/2g; n - p)$. In this case, $\alpha = 0.10$ with two confidence intervals being estimated ($g = 2$), $n = 11$, and there are 3 parameters in the model ($p = 3$). Thus, in Insert 8.8 we find $B = tinv(1-(0.10/4), 8) = 2.306$. We obtain the numerical values $s\{b_1'\} = 31.66461$ and $s\{b_2'\} = 1.26658$ directly from SAS Output 8.5(b). Alternatively, we can find $s\{b_1'\}$ (*sdb1*) and $s\{b_2'\}$ (*sdb2*) from $s\{b_1\}$ and $s\{b_2\}$ as shown on text page 305. Finally, we use equation 6.52, text page 258 to find the two Bonferroni simultaneous 90% confidence intervals:

$$-212.63 \le \beta_1 \le -66.55$$

and

$$4.63 \le \beta_2 \le 10.47 .$$

These confidence limits are equal to those printed as part of the output [SAS Output 8.5(b)] for the second *model* statement of Insert 8.7. In this second *model* statement, we use the option *clb* with *alpha* = 0.05. The confidence intervals calculated through the *cbl* option are separate confidence intevals rather than simultaneous confidence intervals. However, we can adjust α through the *alpha* option to get simultaneous confidence intervals. We use *alpha* = 0.05 to get 90% confidence intervals because:

$$1 - \alpha/2g = 1 - 0.10/4 = 1 - 0.05/2$$

and, hence, the separate 95% confidence intervals are identical to the simultaneous 90% confidence intervals.

INSERT 8.8

```
data intervals;
x1code=(x1-1)/.4;
x2code=(x2-20)/10;
B=tinv(1-(.1/4),8);
sdb1=(1/.4)*12.67;
sdb2=(1/10)*12.67;
b1lo=-139.59-(b*sdb1);
b1hi=-139.59+(b*sdb1);
b2lo=7.55-(b*sdb2);
b2hi=7.55+(b*sdb2);
run;
proc print;
 var b1lo b1hi b2lo b2hi;
 run;
```

Insurance Innovation Example

Data File: ch08ta02.txt

An economist studied the relationship between the adoption of new insurance innovations and the type of insurance company and the size of the insurance firm (text page 313, 317). The response variable (time to adoption) is coded Y, and the predictor variables are coded $X1$ for size of the firm and $X2$ for type of firm where $X2 = 0$ for mutual companies and $X2 = 1$ for stock companies. See Chapter 0, "Working with SAS," for instructions on opening ch08ta02.txt and naming variables Y, $X1$, and $X2$. After opening and printing the Insurance Innovation example, the first 5 lines in the output should look like SAS Output 8.6.

SAS Output 8.6

Obs	y	x1	x2
1	17	151	0
2	26	92	0
3	21	175	0
4	30	31	0
5	22	104	0

In Insert 8.9, we produce SAS Output 8.7, which replicates Figure 8.12, text page 318. To draw separate regression lines for mutual and stock companies, we use the interpolate (*interplo*) option on *symbol* statements to specify that a regression line (*rl*) be drawn. The first character to be used to plot $X2$ values is the *circle* and the second value of $X2$ is plotted using the *dot* character. We also include an optional plot title and set the horizontal and vertical axes to replicate the axes shown in Figure 8.12, text page 318.

```
                        INSERT 8.9

data insurance;
  infile 'c:\alrm\chapter  8 data sets\ch08ta02.txt';
  input  y x1 x2;
goptions reset=all;
proc gplot;
  title1 'Text Figure 8.12';
   symbol1 value=circle color=black interpol=rl;
   symbol2 value=dot color=black interpol=rl;
   footnote h=1.5 j=c 'Mutual Firm [circle] vs Stock Firm [dot]';
    plot y*x1=x2 / haxis=0 to 350 by 50 vaxis=0 to 40 by 5 nolegend;
run;
```

Text Figure 8.12

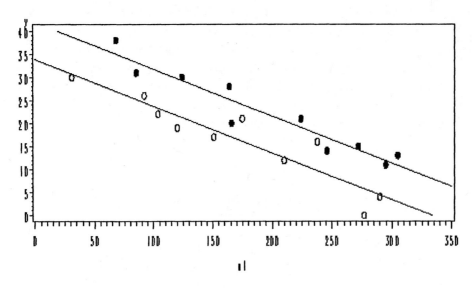

Mutual Firm [circle] vs Stock Firm [dot]

The economist was most interested in the effect of the type of insurance firm (*X2*) on the response variable. In Insert 8.10 we request the first-order model including confidence limits (*clb*) for the coefficients in the regression model. Consistent with text page 316, the 95% confidence interval for β_2 is displayed as $4.977 \leq \beta_2 \leq 11.134$. (SAS Output 8.8) This is not a simultaneous confidence interval like the intervals calculated in the previous example. Because the interval does not include zero, we conclude H_a: $\beta_2 \neq 0$, or that the type of insurance firm has a significant effect on the response variable. We can also reach this conclusion based on the probability level ($Pr>|t| = 0.000$) of the *t*-value ($t = 5.52$) associated with *X2*.

```
INSERT 8.10

data insurance;
  infile 'c:\alrm\chapter  8 data sets\ch08ta02.txt';
  input  y x1 x2;
proc reg;
  model y = x1 x2 / clb;
run;
```

SAS Output 8.8

Parameter Estimates							
Variable	DF	Parameter Estimate	Standard Error	t Value	Pr > |t|	95% Confidence Limits	
Intercept	1	33.87407	1.81386	18.68	<.0001	30.04716	37.70098
x1	1	-0.10174	0.00889	-11.44	<.0001	-0.12050	-0.08298
x2	1	8.05547	1.45911	5.52	<.0001	4.97703	11.13391

On text page 326, the economist wanted to test the hypothesis that there is no interaction between size (*X1*) and type (*X2*) of insurance company. Thus, the economist considered a regression model (equation 8.49, text page 324) that included the interaction term *X1X2*, the cross-product of *X1* and *X2*. In Insert 8.11, we compute *X1X2* and then fit the regression model with *X1*, *X2*, and *X1X2* as predictors. We test the hypothesis that there is no interaction in terms of β_3; i.e., we test H_0: $\beta_3 = 0$ against H_a: $\beta_3 \neq 0$. From SAS Output 8.9 and consistent with text page 326, we see the *t*-value associated with *X1X2* is $t = -0.02$, $p = 0.9821$. Thus, the economist concluded that the interaction is not statistically significant and, hence, the interaction term should be dropped from the model.

```
INSERT 8.11

data insurance;
  infile 'c:\alrm\chapter  8 data sets\ch08ta02.txt';
  input  y x1 x2;
  x1x2=x1*x2;
proc reg;
  model y = x1 x2 x1x2;
run;
```

SAS Output 8.9

Parameter Estimates					
Variable	DF	Parameter Estimate	Standard Error	t Value	Pr > |t|
Intercept	1	33.83837	2.44065	13.86	<.0001
x1	1	-0.10153	0.01305	-7.78	<.0001
x2	1	8.13125	3.65405	2.23	0.0408
x1x2	1	-0.00041714	0.01833	-0.02	0.9821

Soap Production Lines Example

Data File: ch08ta05.txt

A company studied the relationship between production line speed (*X1*) and the amount of scrap (*Y*) generated on their 2 production lines (*X2*). The data are presented in Table 8.5, text page 330. After opening and printing ch08ta05.txt, the first 5 lines of output should look like SAS Output 8.10.

SAS Output 8.10

Obs	y	x1	x2
1	218	100	1
2	248	125	1
3	360	220	1
4	351	205	1
5	470	300	1

Based on Figure 8.16, text page 331, the researchers decided to fit a model with an interaction term:

$$Y_i = \beta_0 + \beta_1 X_{i1} + \beta_2 X_{i2} + \beta_3 X_{i1} X_{i2} + \varepsilon_i$$

Tentative Model: To produce the tentative model, we first compute a new variable, *X1X2*, the cross-product of *X1* and *X2*. (Insert 8.12)   We then fit a model using *Y* as the dependent variable and *X1*, *X2*, and *X1X2* as independent variables and save the unstandardized residuals (*residual*) and predicted values (*yhat*). This produces SAS Output 8.11, which is consistent with Table 8.6(a), text page 332. In Insert 8.13, we produce text Figure 8.16, text page 331, using *proc plot* to plot *Y* against *X1* while displaying different symbols for the two values of *X2*.

```
INSERT 8.12

data soap;
  infile 'c:\alrm\chapter  8 data sets\ch08ta05.txt';
  input  y x1 x2;
x1x2=x1*x2;
proc reg;
 model y=x1 x2 x1x2/ r p;
 output out=results r=residual p=yhat;
run;
```

SAS Output 8.11

Analysis of Variance					
Source	DF	Sum of Squares	Mean Square	F Value	Pr > F
Model	3	169165	56388	130.95	<.0001
Error	23	9904.05692	430.61117		
Corrected Total	26	179069			

Parameter Estimates					
Variable	DF	Parameter Estimate	Standard Error	t Value	Pr > |t|
Intercept	1	7.57446	20.86970	0.36	0.7200
x1	1	1.32205	0.09262	14.27	<.0001
x2	1	90.39086	28.34573	3.19	0.0041
x1x2	1	-0.17666	0.12884	-1.37	0.1835

INSERT 8.13

```
data soap;
  infile 'c:\alrm\chapter  8 data sets\ch08ta05.txt';
  input  y x1 x2;
proc plot;
  plot y * x1 = x2;
run;
```

Diagnostics: The researchers in the Soap Production Lines Example next calculated the coefficient of correlation between the ordered residuals and their expected values under normality. The expected value under normality of the kth smallest observation from a sample of n is given by:

$$\sqrt{MSE}\left[z\left(\frac{k-0.375}{n+0.25}\right)\right]$$

We find $MSE = 430.611$ in SAS Output 8.11 above and $n = 27$ in the Soap Production Lines Example. In Chapter 3 we give a detailed description of the steps necessary to obtain z and k. In Insert 8.14 we rank the residuals and place the rank in a variable named *rank*. We then compute the expected value under normality (*ev*) and correlate this variable with *residual*. The resulting correlation is 0.990, as reported on text page 331. From text Table B.6, the critical value for $\alpha = .05$ and $n = 28$ is 0.962. The calculated value is greater than the critical value found in the table, so we conclude there is sufficient evidence to support the assumption of normality of error terms.

```
                          INSERT 8.14

data soap;
  infile 'c:\alrm\chapter  8 data sets\ch08ta05.txt';
  input  y x1 x2;
x1x2=x1*x2;
proc reg;
 model y=x1 x2 x1x2/ r;
 output out=results r=residual;
* Rank the residuals and place the rank in "rank";
proc rank data=results out=results;
 var residual;
 ranks rank;
run;
data new;
 set results;
 ev = sqrt(430.611) * Probit ((rank - .375)/(27  + .25));
proc corr;
 var ev residual;
 run;
```

Continuing on text page 331, the authors formally test for constancy of error variance using the *Brown-Forsythe test* as described in Chapter 3. For simplicity, we again divide the process into steps:

- Step 1: (Insert 8.15) Divide the data file into two subsets. In the Soap Production Lines Example this has been done based on the value of $X2$ where $X2 = 1$ if production line = 1 and $X2 = 0$ if production line = 2. We also regress Y on $X1$, $X2$, and $X1X2$ and save the residual values (*residual*).
- Step 2: Use *proc means* with the *median* option to obtain the median residual values $\tilde{e}_1$ (Production line = 1, therefore, $X2 = 1$) = 5.4959 and $\tilde{e}_2$ (Production line = 2, therefore, $X2 = 0$) = −6.2203.
- Step 3: Find *absdevmd*, the absolute value (*abs*) of the residual minus the median residual value for the respective group, d_{i1} and d_{i2}.
- Step 4: Find the means of *absdevmd* for the two groups, $\bar{d}_1$=16.132 and $\bar{d}_2$ = 12.648.
- Step 5: Find *sqrdev*, the square of *absdevmd* minus the respective group mean of *absdevmd*.
- Step 6: Find the sum of *sqrdev* for the two production lines, $\sum(d_{11} - \bar{d}_1)^2 =$ 2952.19 and $\sum(d_{12} - \bar{d}_2)^2 = 2045.82$.

We now have the values required to calculate the Brown-Forsythe statistic:

$$s^2 = \frac{2,952.20 + 2,04582}{27 - 2} = 199.921$$

$$s = \sqrt{199.921} = 14.139$$

$$t_{BF}^* = \frac{16.132 - 12.648}{14.139\sqrt{\dfrac{1}{15} + \dfrac{1}{12}}} = 0.636$$

Using $\alpha = 0.05$ and $df = 25$, the critical value of t can be obtained using a quantile function such as " $tvalue = tinv(0.975, 25)$ ", which will return 2.059. Since $t_{BF}^* = 0.636 < 2.059$, we conclude there is no significant lack of constancy of error variance.

INSERT 8.15	INSERT 8.15 continued
| ```
data soap;
  infile 'c:\alrm\chapter 8 data sets\ch08ta05.txt';
  input  y x1 x2;
x1x2=x1*x2;
* Step 1;
proc reg;
 model y=x1 x2 x1x2/ r;
 output out=results r=residual;
* Step 2;
proc sort;
 by x2;
proc means median data=results;
 by x2;
 var residual;
* Step 3;
data results;
 set results;
if x2=0 then absdevmd=abs(residual-(-6.2203));
if x2=1 then absdevmd=abs(residual-(5.4959));
* Step 4;
proc means;
 var absdevmd;
 by x2;
``` | ```
* Step 5;
data results;
 set results;
if x2=0 then sqrdev=(absdevmd-12.6482)**2;
if x2=1 then sqrdev=(absdevmd-16.1321)**2;
* Step 6;
proc means data=results sum;
 by x2;
  var sqrdev;
run;

* Find tvalue;
data results;
 set results;
tvalue=tinv(0.975, 25);
proc print data=results (obs=1);
 var tvalue;
 run;
``` |

Inferences about Two Regression Lines: The researchers now wish to test the identity of the regression lines for the two soap production lines. The null and alternative hypotheses, respectively, are:

$$H_0 : \beta_2 = \beta_3 = 0$$

and

$$H_a : not\ both\ \beta_2 = 0\ and\ \beta_3 = 0$$

To test the null hypothesis using equation 7.27, text page 267, we need to produce text Table 8.6(b), text page 332, which lists Type I sum of squares. We request Type I sum of squares with the *ss1* option available on the *proc reg* statement (Insert 8.16). SAS Output 8.12 is the result of this *proc reg* procedure where we regressed Y on $X1$, $X2$, and $X1X2$. We can now use equation 8.56a, text page 333, to find $F^* = 22.65$. Alternatively, we can request a test in the *proc reg* procedure that the regression coefficients for $X2$ and $X1X2$ are both equal 0 (*x2x1x2: test x2=x1x2=0;*). The "*x2x1x2:*" portion of the statement is a label to identify the test in the output. From SAS Output 8.12, we see that the F-value for $H_0 : \beta_2 = \beta_3 = 0$ is $F = 22.647$. This is a test statistic for evaluating the effect of adding $X2$ and $X1X3$ to a model that already contains $X1$; this is, we test the hypothesis that the coefficients for $X2$ and $X1X2|X1$ are simultaneously zero. Since the computed F-value (22.65) is greater than the table F-value (5.67) when $\alpha = 0.01$, we conclude that the regression functions are not the same for the two production lines.

INSERT 8.16

```
data soap;
 infile 'c:\alrm\chapter 8 data sets\ch08ta05.txt';
 input  y x1 x2;
x1x2=x1*x2;
proc reg;
 model y=x1 x2 x1x2 / ss1 clb;
x2x1x2: test x2=x1x2=0;
run;
```

SAS Output 8.12

Analysis of Variance					
Source	**DF**	**Sum of Squares**	**Mean Square**	**F Value**	**Pr > F**
Model	3	169165	56388	130.95	<.0001
Error	23	9904.05692	430.61117		
Corrected Total	26	179069			

Parameter Estimates						
Variable	**DF**	**Parameter Estimate**	**Standard Error**	**t Value**	**Pr > |t|**	**Type I SS**
Intercept	1	7.57446	20.86970	0.36	0.7200	2687271
x1	1	1.32205	0.09262	14.27	<.0001	149661
x2	1	90.39086	28.34573	3.19	0.0041	18694
x1x2	1	-0.17666	0.12884	-1.37	0.1835	809.62258

Test x2x1x2 Results for Dependent Variable y				
Source	DF	Mean Square	F Value	Pr > F
Numerator	2	9751.85064	22.65	<.0001
Denominator	23	430.61117		

The researcher also wanted to test whether the slopes of the regression lines for the two production lines were identical. The null and alternative hypotheses are:

$$H_0 : \beta_3 = 0$$

and

$$H_a : \beta_3 \neq 0$$

We can test the null hypothesis by equation 7.25, text page 267, or the partial F-test, text equation 7.24. Table 8.6(b), text page 332, and SAS Output 8.12 above provide the necessary information to find $F^* = 1.88$ (text equation 8.57a). Equivalently, we can use SAS Output 8.12 (Parameter Estimates Table) to obtain $t = -1.371$ for $X1X2$. Remember that $t^2 = F$ and $(-1.371)^2 = 1.88$. Thus, the t-statistic in SAS Output 8.12 using Type III sum of squares is a partial test of $X1X2|X1, X2$. Since the probability level 0.1835 is greater than the required $\alpha = 0.01$ we do not reject H_0: the slopes of the two regression functions are identical. Finally, to produce the 95% confidence limits for β_2, we use the *clb* option in Insert 8.16.

Chapter **9** SAS®

Building the Regression Model I: Model Selection and Validation

This chapter presents numerous strategies that are often used to build a regression model. These tactics include making use of theoretical relationships and knowledge gained from previous studies as well as information contained in a set of data available for analysis. We frequently have a large set of potential predictor variables and we wish to choose a subset of these variables to build a regression model that is efficient in its use of predictor variables. This process involves checking for outlying observations as well as investigating the contributions of potential predictor variables. The problem is complicated by the fact that there may be multicollinearity among the predictor variables. We may visually inspect relationships that are sometimes revealed in well-planned graphs. The authors discuss the criteria commonly used to select the predictor variables for inclusion or exclusion from the model. They then discuss three basic ways to validate a regression model once the selection process is in its final stages (see text page 369). These techniques are illustrated with well-chosen numerical examples.

Surgical Unit Example

Data File: ch09ta01.txt

A hospital studied the survival time (Y) of patients who had undergone a liver operation. The predictor or explanatory variables for the predictive regression model were blood clotting score ($X1$), prognostic index ($X2$), enzyme function score ($X3$), and liver function score ($X4$). The researchers also extracted from the preoperative evaluation the patients age ($X5$), sex ($X6$), and history of alcohol use, coded as two indicator variables, $X7$ (moderate use) and $X8$ (severe use). A representation of the Surgical Unit data set is shown in Table 9.1, text page 350. See Chapter 0, "Working with SAS," for instructions on opening the Surgical Unit data file and naming the variables $X1$, $X2$, $X3$, $X4$, $X5$, $X6$, $X7$, $X8$, Y, and *ylog*. After opening and printing ch09ta01.txt, the first 5 lines of printed output should look like SAS Output 9.1.

SAS Output 9.1

Obs	x1	x2	x3	x4	x5	x6	x7	x8	y	ylog
1	6.7	62	81	2.59	50	0	1	0	695	6.54
2	5.1	59	66	1.70	39	0	0	0	403	6.00
3	7.4	57	83	2.16	55	0	0	0	710	6.57
4	6.5	73	41	2.01	48	0	0	0	349	5.85
5	7.8	65	115	4.30	45	0	0	1	2343	7.76

The researchers selected a random sample of 108 patients. Because the researchers intended to validate the final model, the sample was split into a model-building set (first 54 patients) and a validation set (second 54 patients). To simplify the illustration, the authors used only the first four available predictors variables. We begin by regressing Y on $X1$, $X2$, $X3$, and $X4$, and saving the unstandardized residuals and predicted values. In Insert 9.1 we produce the regression analysis including plots shown in Figures 9.2 (a and b), text page 351 (SAS Output 9.2). In Insert 9.1, the "*plot r.*p.;*" statement instructs SAS to plot the residuals against the predicted values. The "*plot npp.*r.;*" statement requests a normal probability plot of the residuals (*r.*). Plot (a) of SAS Output 9.2 suggests both nonconstant error variances and curvature. That is, the variability of the residuals appears to increase as the magnitude of the predicted values increases. Plot (b) suggests the distribution of the residuals may not be a member of the family of normal distributions.

INSERT 9.1

```
data surgical;
  infile 'c:\alrm\chapter  9 data sets\ch09ta01.txt';
  input x1 x2 x3 x4 x5 x6 x7 x8 y ylog;
proc reg;
 model y = x1 x2 x3 x4;
  output out=results r=residual p=yhat;
   plot r.*p.;
   plot npp.*r. ;
 run;
```

SAS Output 9.2

Plot (a)

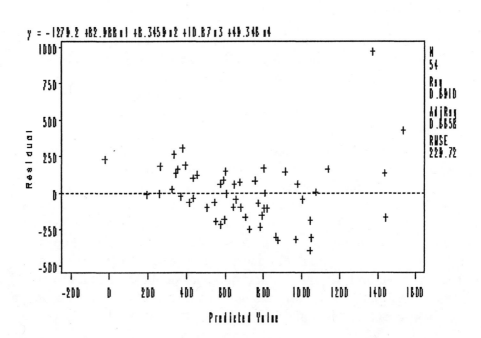

Predicted Value

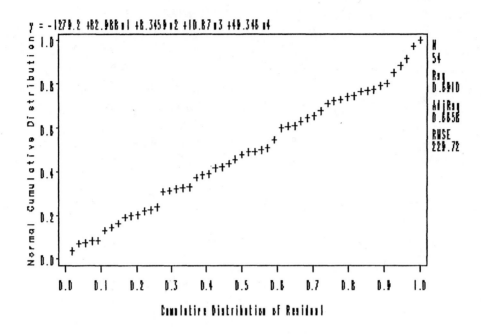

Because of the above findings, the investigator performed a logarithmic transformation of *Y*. We assigned the variable name *ylog* to the transformed data. Based on various diagnostics, the investigator then decided to use *ylog* as the response variable; to use *X1*, *X2*, *X3*, and *X4* as predictors; and to not include any interaction terms in the linear regression model. Using the first *proc reg* statement in Insert 9.2, we produce the regression analysis including plots shown in Figures 9.2 (c and d), text page 351 (SAS Output 9.2 – Plots c & d). Plot (c) of SAS Output 9.2 suggests both constant error variances and a first order model may be appropriate. Plot (d) suggests the distribution of the residuals more nearly normal compared to the residuals of the model for the untransformed *Y*.

The investigator also produced a correlation matrix and a scatter plot matrix to investigate pairwise correlations among the variables used in the model. We produce a correlation matrix for the variables in the *model* statement of the regression procedure by including the *corr* option in the regression statement; i.e., by writing *proc reg corr* as in the second *proc reg* statement in Insert 9.2 (Output not shown). An alternative way to produce a correlation matrix is to use the *proc corr* statement. We use the *proc corr* statement in Insert 9.2 to produce the correlation matrix shown SAS Output 9.3 and in Figure 9.3, text page 352. The *var* statement following the *proc corr* statement specifies the variables used in the correlation matrix. The *proc corr* statement produces the simple statistics (*n*, mean, standard deviation, sum, minimum, and maximum) as well as the correlation matrix. The correlation matrix output is shown in SAS Output 9.3. The correlation coefficients shown in the correlation matrix are Pearson's correlation coefficients. From the correlation matrix, we see that Pearson's correlation coefficient is 0.2462 for *ylog* and *X1*, 0.4699 for *ylog* and *X2*, and so on. There is redundancy in the correlation information shown above and below the main diagonal of the correlation matrix because the correlation between *ylog* and *X1* is the same as the correlation between *X1* and *ylog*, and so on. A *p*-value is given below each correlation coefficient. The *p*-value corresponds to a test of the null hypothesis that the true correlation coefficient corresponding to the relevant sample estimate is zero; i.e., $\rho = 0$ for the indicated pair of variables. Thus, the correlation coefficient = 0.50242, with *p*-value = 0.0001, between *X1* and *X4* is indicative that the true correlation coefficient for *X1* and *X4* is significantly greater than 0.

We use the *proc plot* statement shown in Insert 9.2 to produce the scatter plots between pairs of variables specified in the *plot* statement. The SAS output for the scatter plots are not organized in the concise format shown in the text and are not shown here for brevity.

```
                          INSERT 9.2
data surgical;
  infile 'c:\alrm\chapter  9 data sets\ch09ta01.txt';
  input x1 x2 x3 x4 x5 x6 x7 x8 y ylog;

proc reg;
 model ylog = x1 x2 x3 x4;
  output out=results r=residual p=yhat;
   plot r.*p.;
   plot npp.*r. ;

proc reg corr;
  model ylog = x1 x2 x3 x4;

proc corr;
  var ylog x1 x2 x3 x4;

proc plot;
  plot ylog*(x1 x2 x3 x4)  x1*(x2 x3 x4)  x2*(x3 x4)  x3*x4;
run;
```

SAS Output 9.2 continued

Plot (c)

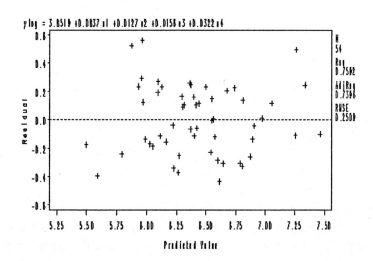

<h1>Plot (d)</h1>

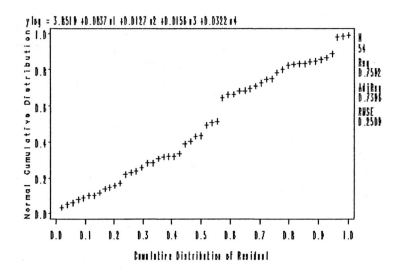

<h2>SAS Output 9.3</h2>

Correlation					
Variable	**x1**	**x2**	**x3**	**x4**	**ylog**
x1	1.0000	0.0901	-0.1496	0.5024	0.2462
x2	0.0901	1.0000	-0.0236	0.3690	0.4699
x3	-0.1496	-0.0236	1.0000	0.4164	0.6539
x4	0.5024	0.3690	0.4164	1.0000	0.6493
ylog	0.2462	0.4699	0.6539	0.6493	1.0000

<h1>Criteria for Model Selection</h1>

The next phase of the authors' model-building procedure calls for examining "whether all of the potential predictor variables are needed or whether a subset of them is adequate." We can choose from a rather long list of criteria for selecting or eliminating predicator variables.

In *proc reg*, the *selection=option* specifies the method used to select the regression model, where the option can be *forward* (or *f*), *backward* (or *b*), *stepwise*, *maxr*, *minr*, *rsquare*, *adjrsq*, *cp* or *none* (include all predictor variables). The default method is *none*.

All-Possible-Regressions. The *All-Possible-Regressions Procedure* compares the results obtained by fitting all possible subsets of potential predictor variables that could be used to build regression models. Based on various criteria, a small number of regression models are selected for further evaluation. The number of candidate models, however, increases rapidly as the number of predictor variables increases. A study with only 3 predicator variables has $2^3 = 8$ possible first-order regression models, whereas one with 7 predictor variables has $2^7 = 128$. Thus, the all-possible-regressions procedure can be used when the number of potential predictor variables is relatively small, but becomes impractical for use when the number of predictor variables is very large.

In *proc reg*, the *selection=rsquare* option produces R^2, the proportion of variability attributable to the regression model, for all possible subsets of predictor variables that are specified in the *model* statement. For example, the SAS syntax given in Insert 9.3 produces SAS Output 9.4. The SAS output is arranged from largest to smallest values of R^2 for all one (predictor) variable models, then from largest to smallest values of R^2 for all two variable models, and so on. As noted in the text, the largest R^2 will always be realized for the model with all the predictor variables included. If we can, we want to choose a model with a smaller number of predictor variables but one with an R^2 reasonably close to the maximum.

SAS Output 9.4

Number in Model	R-Square	Variables in Model
1	0.4276	x3
1	0.4215	x4
1	0.2208	x2
1	0.0606	x1
2	0.6633	x2 x3
2	0.5995	x3 x4
2	0.5486	x1 x3
2	0.4830	x2 x4
2	0.4301	x1 x4
2	0.2627	x1 x2
3	0.7573	x1 x2 x3
3	0.7178	x2 x3 x4
3	0.6121	x1 x3 x4
3	0.4870	x1 x2 x4
4	0.7592	x1 x2 x3 x4

```
                        INSERT 9.3

data surgical;
  infile 'c:\alrm\chapter  9 data sets\ch09ta01.txt';
  input x1 x2 x3 x4 x5 x6 x7 x8 y ylog;
proc reg;
 model ylog = x1 x2 x3 x4 / selection = rsquare;
run;
```

The SSE_p, R_p^2, $R_{a,p}^2$, C_p, AIC_p, SBC_p, and $PRESS_p$ values for all-possible-regression models are presented in Table 9.2, text page 353, for the Surgical Unit example when regressing *ylog* on *X1*, *X2*, *X3*, *X4*. In Insert 9.4, we request that the selection of possible models based on the adjusted R-square criterion (*selection=adjrsq*) and that R-square (*rsquare*), mean square error (*mse*), sum of squares error (*sse*), Mallows' C_p (*cp*), and Akaike's information criterion (*aic*) be printed for each model. A summary of the selection criteria shown in Table 9.2, text page 353, and those available as an option in the *model* statement using *proc reg* is displayed here in the table below Insert 9.4. The Schwarz Bayesian Criterion (SBC_p) shown in Table 9.2, text page 353, is not available in SAS.

Output for the first *proc reg* procedure in Insert 9.4 is presented in SAS Output 9.5. When we request residuals as an option in the *model* statement, SAS prints the $PRESS_p$ statistic after listing the residuals. The second *proc reg* statement of Insert 9.4 illustrates syntax that produces a listing of the residuals and the $PRESS_p$ statistic for an individual model (Output not shown).The *best=*option is used with the *rsquare*, *adjrsq*, and *cp* model selection options. If *selection=cp* or *selection=adjrsq* is specified, the *best=*option specifies the maximum number of subset models to be displayed. For *selection=rsquare*, the *best=*option sets the maximum number of subset models for each number of predictor variables. That is, if *best=*2 and there are 4 predictors variables in the *model* statement, the two best models based on *rsquare* will be displayed for all possible models with 1 predictor variable (only one model is available to be displayed), all possible models with 2 predictor variable, all possible models with 3 predictor variables, and all possible models with 4 predictor variables. If the *best=*option is omitted and there are less than 11 predictors variables, all possible subsets will be evaluated. If the *best=*option is omitted and the number of predictor variables is greater than 10, the maximum number of subsets will be equal to the number of predictor variables. The third *proc reg* statement and the corresponding *model* statement illustrates the use of the *best=*3 option (Output not shown).

```
                        INSERT  9.4
data surgical;
  infile 'c:\alrm\chapter  9 data sets\ch09ta01.txt';
  input x1 x2 x3 x4 x5 x6 x7 x8 y ylog;
proc reg;
  model ylog = x1 x2 x3 x4 / selection = adjrsq rsquare sse mse cp aic;

proc reg;
  model ylog = x1 x2 x3 x4 / r;

proc reg;
 model ylog = x1 x2 x3 x4 / selection =  adjrsq best=3;
run;
```

Table 9.1

Text Table 9.2 Heading	SAS Output Heading (See SAS Output 9.5)	Comment
SSE_p	SSE	R_p^2 varies inversely with SSE_p
Not shown	MSE	
R_p^2	R-Square	
$R_{a,p}^2$	Adjusted R Square	
C_p	C(p)	
AIC_p	AIC	
SBC_p	Not available in SAS	
$PRESS_p$	Not computed for all possible regressions but is computed for one model at a time	Specify p or r as an option in the model statement and Press statistic is printed at the end of the list of predicted values and residuals.

SAS Output 9.5

Number in Model	Adjusted R-Square	R-Square	C(p)	AIC	MSE	SSE	Variables in Model
3	0.7427	0.7573	3.3905	-146.1609	0.06217	3.10854	x1 x2 x3
4	0.7396	0.7592	5.0000	-144.5895	0.06294	3.08396	x1 x2 x3 x4
3	0.7009	0.7178	11.4237	-138.0232	0.07228	3.61413	x2 x3 x4
2	0.6501	0.6633	20.5197	-130.4833	0.08456	4.31249	x2 x3
3	0.5889	0.6121	32.9320	-120.8442	0.09936	4.96782	x1 x3 x4
2	0.5838	0.5995	33.5041	-121.1126	0.10058	5.12970	x3 x4
2	0.5309	0.5486	43.8517	-114.6583	0.11335	5.78096	x1 x3
2	0.4627	0.4830	57.2149	-107.3236	0.12984	6.62201	x2 x4
3	0.4562	0.4870	58.3917	-105.7477	0.13140	6.57020	x1 x2 x4
1	0.4166	0.4276	66.4889	-103.8269	0.14099	7.33157	x3

Number in Model	Adjusted R-Square	R-Square	C(p)	AIC	MSE	SSE	Variables in Model
1	0.4104	0.4215	67.7148	-103.2615	0.14248	7.40873	x4
2	0.4078	0.4301	67.9721	-102.0669	0.14312	7.29905	x1 x4
2	0.2338	0.2627	102.0313	-88.1622	0.18515	9.44267	x1 x2
1	0.2059	0.2208	108.5558	-87.1781	0.19191	9.97918	x2
1	0.0425	0.0606	141.1639	-77.0788	0.23137	12.03147	x1

Automatic Search Procedures

SAS offers nine methods of variable selection: *forward* (or *f*), *backward* (or *b*), *stepwise*, *maxr*, *minr*, *rsquare*, *adjrsq*, *cp* or *none* (include all predictor variables). We illustrate below stepwise selection, forward selection, and backward selection.

Stepwise Selection: The *Stepwise* selection method first selects the best predictor variable (the predictor variable with the largest partial correlation with the response variable). In each step thereafter, a variable is either added or one of the variables already entered is dropped from the model. Additional variables are added to the model based on their incremental improvement to the reduction in error sum of squares. Variables are dropped from the model if their test for entry is not significant given the other variables in the model. The default and modifiable α limits for adding or dropping a variable are 0.50 and 0.15, respectively. The α limits for selecting variables for entry in the model (*sle*) and for selecting variables to stay in the model (*sls*) can be specified directly using the *sle* and *sls* options as shown in Insert 9.5. Consistent with text page 366, we have set *sle*=.1 and *sls*=.15.

The Stepwise method is used in Insert 9.5 to regress *ylog* on all eight X variables from the Surgical Unit Example with results shown in SAS Output 9.8. From SAS Output 9.8, we see that $X3$ was entered at step 1 because $X3$ had the highest partial F-value of all variables under consideration. At step 2, $X2$ had the highest partial F-value of the variables not in the model and was added to the model at step 2. Of the variables not in the model after step 2, $X8$ has the highest partial F-value and was added to the model at step 3. Again, of the variables not in the model after step 3, $X1$ has the highest partial F-value and was added to the model at step 4. Beginning with step 2, the search algorithm compares the probability level of the partial correlation of the variables in the equation to the α level needed to remove a variable from the model. A variable is removed from the model if the p-value associated with the partial F-value is greater than the predetermined α level to remove. In the present example, none of the variables was removed from the model. Notice that the p-value associated with each predictor variable not in the model after step 4 (Stepwise Selection: Step 5 – SAS Output 9.8) is greater than the α level needed to add that variable to the model. Thus, the stepping process terminates with $X3$, $X2$, $X8$, and $X1$ in the final model.

INSERT 9.5	INSERT 9.5 continued

data surgical;
  infile 'c:\alrm\chapter 9 data sets\ch09ta01.txt';
  input x1 x2 x3 x4 x5 x6 x7 x8 y ylog;
	proc reg;
	  model ylog = x1 x2 x3 x4 x5 x6 x7 x8/
	   selection = stepwise
	    sle=.1 sls=.15 details=all ;
	  run;

SAS Output 9.8

Stepwise Selection: Step 1

Statistics for Entry DF = 1, 52				
Variable	**Tolerance**	**Model R-Square**	**F Value**	**Pr > F**
x1	1.000000	0.0606	3.35	0.0727
x2	1.000000	0.2208	14.74	0.0003
x3	1.000000	0.4276	38.84	<.0001
x4	1.000000	0.4215	37.89	<.0001
x5	1.000000	0.0210	1.12	0.2956
x6	1.000000	0.0538	2.96	0.0913
x7	1.000000	0.0160	0.85	0.3617
x8	1.000000	0.1390	8.39	0.0055

Stepwise Selection: Step 2

Statistics for Entry DF = 1, 51				
Variable	**Tolerance**	**Model R-Square**	**F Value**	**Pr > F**
x1	0.977610	0.5486	13.68	0.0005
x2	0.999443	0.6633	35.70	<.0001
x4	0.826591	0.5995	21.89	<.0001
x5	0.999834	0.4462	1.72	0.1960
x6	0.980475	0.4478	1.86	0.1781
x7	0.992766	0.4326	0.46	0.5025
x8	0.986198	0.5164	9.37	0.0035

Stepwise Selection: Step 3

	Statistics for Entry DF = 1, 50			
Variable	**Tolerance**	**Model R-Square**	**F Value**	**Pr > F**
x1	0.970108	0.7573	19.37	<.0001
x4	0.682979	0.7178	9.66	0.0031
x5	0.997531	0.6761	1.98	0.1651
x6	0.964220	0.6697	0.97	0.3283
x7	0.977269	0.6810	2.77	0.1023
x8	0.979642	0.7780	25.85	<.0001

Stepwise Selection: Step 4

	Statistics for Entry DF = 1, 49			
Variable	**Tolerance**	**Model R-Square**	**F Value**	**Pr > F**
x1	0.906955	0.8299	14.93	0.0003
x4	0.671275	0.8144	9.61	0.0032
x5	0.983952	0.7836	1.25	0.2687
x6	0.959734	0.7888	2.49	0.1211
x7	0.728783	0.7800	0.43	0.5158

Stepwise Selection: Step 5

	Statistics for Entry DF = 1, 48			
Variable	**Tolerance**	**Model R-Square**	**F Value**	**Pr > F**
x4	0.390402	0.8331	0.94	0.3380
x5	0.983823	0.8358	1.73	0.1944
x6	0.954457	0.8374	2.23	0.1418
x7	0.728750	0.8317	0.51	0.4782

Forward Elimination: The *Forward* method begins with no predictor variables in the model. In each step, the predictor variable with the largest partial *F*-value is entered in the model. This method continues to add variables with significant ($p<\alpha$ to add) partial *F*-values until none of the partial *F*-values is significant. To request the *forward* method in the present example, we simply replace *selection=stepwise* with *selection=forward* in the *model* statement in Insert 9.5. In the present example, the progression of predictor variable entry is the same as the *Stepwise* method shown in SAS Output 9.8. However, in contrast to the stepwise method, the *Forward* method makes no attempt to remove variables from the model. Since the *Stepwise* procedure above did not remove any predictor variables from the model and the *Forward* method makes no attempt to remove variables, the result of the *Stepwise* method (SAS Output 9.8) and *Forward* method (Output not shown) are the same in the present example, although this will not always be true.

Backward Elimination: The *Backward* method begins with all predictor variables in the model. Beginning with the one with the smallest partial *F*-value, the *backward* method then removes each predictor variable that does not have a significant ($p>\alpha$ to remove) partial *F*-value. To request the *backward* method in the present example, we simply replace *selection=stepwise* with *selection=backward* in the *model* statement in Insert 9.5. Consistent with text page 366, we have set *sle*=.1 and *sls*=.15. From SAS Output 9.9 we see that at step 1 all eight *X* variables are entered in the model. Because the probability level ("Pr>F") of *X4* is greater than the α needed to remove a variable and is greater than other probability levels of variables in the model, *X4* is removed from the model at step 2. *X7* and *X5* are then removed at steps 2 and 3, respectively, based on their significance levels. At Backward Elimination: Step 4 (SAS Output 9.9), the significance level for *X6* is 0.1418, which is less than the α to remove (*sls*=.15). Thus, *X6* is retained in the model. The final model contains *X1*, *X2*, *X3*, *X6*, and *X8*.

SAS Output 9.9

Backward Elimination: Step 1

Variable	Statistics for Removal DF = 1, 45			
	Partial R-Square	Model R-Square	F Value	Pr > F
x1	0.0248	0.8213	7.26	0.0099
x2	0.1632	0.6829	47.73	<.0001
x3	0.2335	0.6126	68.29	<.0001
x4	0.0001	0.8460	0.03	0.8645
x5	0.0057	0.8404	1.68	0.2016
x6	0.0066	0.8396	1.92	0.1725
x7	0.0025	0.8436	0.74	0.3957
x8	0.0660	0.7801	19.31	<.0001

Backward Elimination: Step 2

	Statistics for Removal DF = 1, 46			
Variable	**Partial R-Square**	**Model R-Square**	**F Value**	**Pr > F**
x1	0.0489	0.7971	14.62	0.0004
x2	0.2087	0.6373	62.35	<.0001
x3	0.3981	0.4479	118.93	<.0001
x5	0.0068	0.8392	2.04	0.1597
x6	0.0076	0.8384	2.26	0.1392
x7	0.0026	0.8434	0.77	0.3835
x8	0.0659	0.7801	19.70	<.0001

Backward Elimination: Step 3

	Statistics for Removal DF = 1, 47			
Variable	**Partial R-Square**	**Model R-Square**	**F Value**	**Pr > F**
x1	0.0491	0.7944	14.72	0.0004
x2	0.2157	0.6277	64.75	<.0001
x3	0.3966	0.4468	119.07	<.0001
x5	0.0060	0.8374	1.80	0.1862
x6	0.0076	0.8358	2.29	0.1369
x8	0.0703	0.7731	21.10	<.0001

Backward Elimination: Step 4

	Statistics for Removal DF = 1, 48			
Variable	**Partial R-Square**	**Model R-Square**	**F Value**	**Pr > F**
x1	0.0487	0.7888	14.37	0.0004
x2	0.2207	0.6167	65.17	<.0001
x3	0.3966	0.4409	117.10	<.0001
x6	0.0076	0.8299	2.23	0.1418
x8	0.0761	0.7614	22.47	<.0001

Model Validation: The Surgical Unit Example Continued

Data File: ch09ta05.txt

Continuing with the Surgical Unit example, the researchers identified three candidate models using the model-selection criteria presented above: SSE_p, R_p^2, $R_{a,p}^2$, C_p, AIC_p, SBC_p, and $PRESS_p$. The researchers now wish to estimate the predictive capability of the three models using the *mean squared prediction error* ($MSPR$) described on text page 370. Model 1, given on text page 373, uses *ylog* as the response variable and *X1*, *X2*, *X3*, and *X8* as predictor variables.

In Insert 9.6, we first fit the regression model based on Model 1 using the Surgical Unit training data set presented in Table 9.1, text page 350. The estimated regression coefficients are given in SAS Output 9.10. We now open the Surgical Unit validation data set presented in Table 9.5, text page 374. After printing ch09ta05.txt, the first 5 lines of the data file should look like SAS Output 9.12.

INSERT 9.6

```
data surgical;
 infile 'c:\alrm\chapter 9 data sets\ch09ta01.txt';
 input x1 x2 x3 x4 x5 x6 x7 x8 y ylog;

proc reg;
 model ylog=x1 x2 x3 x8;
 run;
```

SAS Output 9.10

		Parameter Estimates			
Variable	**DF**	**Parameter Estimate**	**Standard Error**	**t Value**	**Pr > |t|**
Intercept	1	3.85242	0.19270	19.99	<.0001
x1	1	0.07332	0.01897	3.86	0.0003
x2	1	0.01419	0.00173	8.20	<.0001
x3	1	0.01545	0.00140	11.07	<.0001
x8	1	0.35297	0.07719	4.57	<.0001

SAS Output 9.11

Obs	x1	x2	x3	x4	x5	x6	x7	x8	y	ylog
1	7.1	23	78	1.93	45	0	1	0	302	5.710
2	4.9	66	91	3.05	34	1	0	0	767	6.642
3	6.4	90	35	1.06	39	1	0	1	487	6.188
4	5.7	35	70	2.13	68	1	0	0	242	5.489
5	6.1	42	69	2.25	70	0	0	1	705	6.558

103

From fitted results corresponding to Model 1 in equation 9.21, text page 373, we use the syntax in Insert 9.7 to compute $\hat{Y}_i$ ($m1$ is used to denote the predicted value for model 1), the predicted value of the ith observation in the validation data set using the regression coefficients (SAS Output 9.10) estimated from the training data set (Insert 9.6). We then find $devsq=\left(Y_i-\hat{Y}_i\right)^2$, where Y_i is the value of the response variable of the ith observation in the validation data. Finally, we use *proc means sum* to find the $\sum_{i=1}^{n^*}\left(Y_i-\hat{Y}_i\right)^2$ (SAS Output 9.13). We find by hand calculator $\sum_{i=1}^{n^*}\left(Y_i-\hat{Y}_i\right)^2 / n^* = 4.1773269 / 54 = 0.07735$.

We repeat the above process by estimating regression functions for Models 2 and 3 on text page 373 based on the Surgical Unit training data set and finding $\sum_{i=1}^{n^*}\left(Y_i-\hat{Y}_i\right)^2 / n^*$ when using the validation data set. This results in $MSPR = 0.0764$ and 0.0794, respectively. From the relative closeness of the three $MSPR$ values, the researchers concluded that the three candidate models performed equally in terms of predictive accuracy. Model 3 was discarded due to reversal of the sign of a regression coefficient from the training data set to the validation data set. Models 1 and 2 preformed equally well in the validation study. The final selection of Model 1 was based on Occam's razor; Model 1 required one less parameter than Model 2.

INSERT 9.7

```
data surgical;
 infile 'c:\alrm\chapter 9 data sets\ch09ta05.txt';
  input x1 x2 x3 x4 x5 x6 x7 x8 y ylog;
   m1=3.85241+(0.07332*x1)+(0.01418*x2)+(0.01545*x3)+(0.35296*x8);
 devsq=(ylog-m1)**2;
 proc means sum;
  var devsq;
 run;
```

SAS Output 9.13

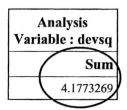

Analysis Variable : devsq
Sum
4.1773269

Building the Regression Model II: Diagnostics

This chapter is concerned with diagnostic techniques for identifying inadequacies of a fitted regression model. Graphical methods were used in previous chapters as an aid in deciding whether to use higher-order terms for a predictor variable in the model or whether to add predictor variables that have not been included in the model. Chapter 10 discusses graphs that can provide useful model-building information about a predictor variable, given that other predictor variables are already in the model. When we fit a regression model to a set of data, some observations may be outlying or extreme relative to most of the data. As shown in Figure 10.5, text page 391, an observation may contain a Y value, one or more X values, or both X and Y values that are outliers. Outlying observations may have too much influence on the estimates of the parameters of the model and, therefore, may require remedial steps. The authors discuss methods for detecting outliers and excessively influential observations. They also discuss remedial steps for dealing with these observations. In Section 10.5, text page 406, the authors continue their discussion of problems created when there is multicollinearity among the predictor variables of a regression model. They present informal methods of detecting multicollinearity problems on text page 407. They then define the variance inflation factor and show how it is used in more formal ways to check for serious multicollinearity among the predictor variables.

The Life Insurance Example

Data File: ch10ta01.txt

Table 10.1, text page 387, presents data from a study of the relationship between amount of insurance carried (Y) and annual income ($X1$) and a measure of risk aversion ($X2$). After opening the Life Insurance data file and naming the variables $X1$, $X2$, and Y, the first 5 lines of output should look like SAS Output 10.1.

SAS Output 10.1

Obs	x1	x2	y
1	45.010	6	91
2	57.204	4	162
3	26.852	5	11
4	66.290	7	240
5	40.964	5	73

We begin by regressing Y on $X1$ and $X2$ and generating the unstandardized residuals by using the r option on the *proc reg model* statement. (Insert 10.1) From within the *proc reg* procedure we add a title and then plot the residual values on the y-axis and $X2$ on the x-axis (*plot r.*x1;*) to produce text Figure 10.3 (a), text page 387, (SAS Output 10.2). To produce Figure 10.3 (b), we regress Y on $X2$ and save the unstandardized residual values in a variable named *res1* in data set *results*. We then regress $X1$ on $X2$ and save the unstandardized residuals in a variable named *res2* in the *results* data set. We use *proc gplot* to request a scatter plot with *res1* [$e(Y \mid X_2)$] on the y-axis and *res2* [$e(X_1 \mid X_2)$] on the x-axis. We also request that a dot (*value=dot*) be used to plot the data points, that a regression line (*interpol=rl*) be drawn through the data points, and that a vertical reference line be drawn from position 0 of the vertical axis (*vref=0*). We also define the values of the x and y axes. This produces SAS Output 10.3.

INSERT 10.1	INSERT 10.1 continued
``data life;
infile 'c:\alrm\chapter 10 data sets\ch10ta01.txt';
input x1 x2 y;
goptions reset=all;
proc reg;
model y=x1 x2 / r;
title1 'Text Figure 10.3 a';
plot r.*x1;
run;
proc reg;
model y=x2 / r;
output out=results r=res1;```	```proc reg;
model x1=x2 / r;
output out=results r=res2;
run;
proc gplot;
title1 'Text Figure 10.3 b';
symbol1 value=dot interpol=rl;
plot res1*res2 / haxis=-25 to 50 by 25 vaxis=-150
to 300 by 50 vref=0;
run;```

INSERT 10.1
``data life;
infile 'c:\alrm\chapter 10 data sets\ch10ta01.txt';
input x1 x2 y;
goptions reset=all;
proc reg;
model y=x1 x2 / r;
title1 'Text Figure 10.3 a';
plot r.*x1;
run;
proc reg;
model y=x2 / r;
output out=results r=res1;
proc reg;
model x1=x2 / r;
output out=results r=res2;
run;
proc gplot;
title1 'Text Figure 10.3 b';
symbol1 value=dot interpol=rl;
plot res1*res2 / haxis=-25 to 50 by 25 vaxis=-150 to 300 by 50 vref=0;
run;```

SAS Output 10.2

Text Figure 10.3 a

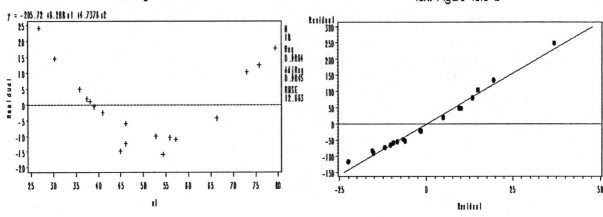

SAS Output 10.3

Text Figure 10.3 b

The Body Fat Example

Data File: ch07ta01.txt

We use the Body Fat example (Table 7.1, text page 257) to illustrate all regression diagnostics. See Chapter 0, "Working with SAS," for instructions on opening the Body Fat example and naming the variables $X1$, $X2$, $X3$, and Y. After opening ch07ta01.txt, the first 5 lines of output should look like SAS Output 10.4.

SAS Output 10.4

Obs	x1	x2	x3	y
1	19.5	43.1	29.1	11.9
2	24.7	49.8	28.2	22.8
3	30.7	51.9	37.0	18.7
4	29.8	54.3	31.1	20.1
5	19.1	42.2	30.9	12.9

Added-Variable Plots: We begin by regressing body fat (Y) on triceps skinfold thickness ($X1$) and thigh circumference ($X2$) (Insert 10.2). This results in the regression analysis shown in SAS Output 10.5. The fitted regression function is:

$$\hat{Y} = -19.174 + 0.222X_1 + 0.659X_2$$

INSERT 10.2

```
data bodyfat;
  infile 'c:\alrm\chapter 7 data sets\ch07ta01.txt';
  input x1 x2 x3 y;
proc reg;
 model y=x1 x2 ;
run;
```

Analysis of Variance					
Source	DF	Sum of Squares	Mean Square	F Value	Pr > F
Model	2	385.43871	192.71935	29.80	<.0001
Error	17	109.95079	6.46769		
Corrected Total	19	495.38950			

Parameter Estimates					
Variable	DF	Parameter Estimate	Standard Error	t Value	Pr > |t|
Intercept	1	-19.17425	8.36064	-2.29	0.0348
x1	1	0.22235	0.30344	0.73	0.4737
x2	1	0.65942	0.29119	2.26	0.0369

In Insert 10.3, we regress Y on $X1$ and $X2$ and use the r option on the *model* statement to request the residuals for this model. We save the residuals in a file named *results* using the statement *output out=results r=residual*. We specify the title of the plot and request that the residuals be plotted against $X1$ (*plot r.*x1*). We also request a second plot within the *proc reg* procedure. Because graphs are not created until a *run* statement is executed, we issue a *run* statement following the first plot statement. If we omitted the *run* statement that precedes the title for the second plot, the title for the second plot would overwrite the title for the first plot and be used for both plots. In the *plot r.*x2* statement, we request that the residual values be plotted against $X2$. The above maneuvers produce Figures 10.4 (a) and (c), text page 389 (SAS Output 10.6).

```
                          INSERT 10.3

data bodyfat;
  infile 'c:\alrm\chapter  7 data sets\ch07ta01.txt';
  input x1 x2 x3 y;
goptions reset=all;
proc reg;
 model y=x1 x2 / r ;
 output out=results r=residual;
 title 'Text Figure 10.4 a';
 plot r.*x1 ;
run;
 title 'Text Figure 10.4 c';
 plot r.*x2 ;
run;
```

SAS Output 10.6

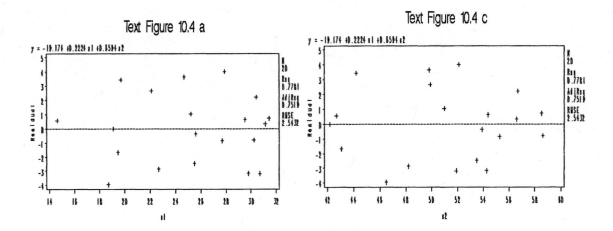

To produce Figures 10.4 b and d, text page 389 (SAS Output 10.7), we regress *Y* on *X1*, *Y* on *X2*, *X1* on *X2*, and *X2* on *X1*, and for each model save the residual values as *ryx1*, *ryx2*, *rx1x2*, and *rx2x1*, respectively (Insert 10.4). Since the intent of each regression analysis was only to generate residuals, we used the *noprint* option to suppress printing of standard output. We then use *proc gplot* to produce text Figures 10.4 b and d. Notice we requested that a *dot* be used to display data points on the graph and that a regression line be drawn through the data points (*symbol value=dot interpol=rl;*).

```
                              INSERT 10.4

proc reg;
 model y=x1 / r noprint;
 output out=results r=ryx1;
proc reg;
 model y=x2 / r noprint;
 output out=results r=ryx2;
proc reg;
 model x1=x2 / r noprint;
 output out=results r=rx1x2;
proc reg;
 model x2=x1 / r noprint;
 output out=results r=rx2x1;
 run;
proc gplot;
 title 'Text Figure 10.4 b';
 symbol value=dot interpol=rl;
 plot ryx2*rx1x2;
run;
proc gplot;
 title 'Text Figure 10.4 d';
 symbol value=dot interpol=rl;
 plot ryx1*rx2x1;
run;
```

109

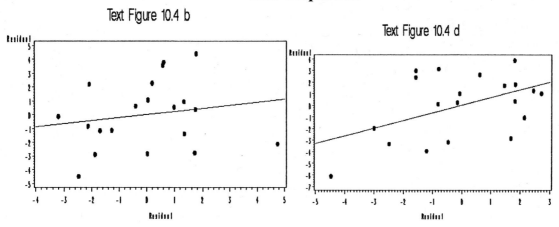

Text Figure 10.4 b

Text Figure 10.4 d

Illustration of Hat Matrix

Data File: ch10ta02.txt

On text page 393, the authors present a small data set to illustrate the relationship between the response variable and the predictor variables. After opening ch10ta02.txt and naming the variables *Y*, *X1*, and *X2*, the 4 lines of output should look like SAS Output 10.8. In Insert 10.5, we create a new variable named *X0* that is equal 1 for each line of data in the input file (there are 4 lines of data in this example). The data stored under the variable *X0* will be used to create a column of ones in the **X** matrix. We convert the data stored under the variable y to matrix (vector) form and call the resulting matrix y by issuing the statement *"read all var {'y'} into y"*. We create the matrix **X** whose columns are made up of the data stored under the variables *X0*, *X1*, and *X2*, respectively, by issuing the statement *"read all var {'x0' 'x1' 'x2'} into x"*. We then use the matrix syntax of *proc iml* and program various text equations to find the **B** (parameter estimates), **H** (hat matrix), **YHAT** (predicted values), and **VARCOVAR** (variance-covariance) matrices. Although the syntax only prints **VARCOVAR**, a *print* statement for any matrix can be added at any location in the matrix program.

SAS Output 10.8

Obs	y	x1	x2
1	301	14	25
2	327	19	32
3	246	12	22
4	187	11	15

INSERT 10.5	INSERT 10.5 continued

data hat;
  infile 'c:\alrm\chapter 10 data sets\ch10ta02.txt';
  input y x1 x2;
x0=1;
proc iml;
  use hat;
* Define Y matrix;
  read all var {'y'} into y;
* Create X matrix;
  read all var {'x0' 'x1' 'x2'} into x;
	* Find the B matrix using text equation 6.25;
	  b=inv(t(x)*x)*t(x)*y;
	* Find hat matrix (hat) by text equation 10.10;
	  hat=x*inv(t(x)*x)*t(x);
	* Find fitted values of y (yhat) by text equation 10.11;
	  yhat=hat*y;
	* Create a 4x4 idenity (I) matrix;
	  i=i(4);
	* Find variance-covariance matrix using text equation 10.13;
	  varcovar=574.9*(i-hat);
	  print varcovar;
	quit;

Formal Diagnostics: The remainder of the diagnostics on text pages 390-410 deals with *residuals, deleted residuals, leverage values, DFFITS, DFBETAS, Cook's distance measure*, and *variance inflation factors*. To produce these statistics we request the appropriate regression model as before. The options listed in SAS Output 10.9 are available on the *proc reg model* statement and will display the statistics shown under the Output heading. Using the options listed in SAS Output 10.10, you can save the selected statistics to a SAS data set.

SAS Output 10.9

SAS *proc reg **model*** statement option	Output
influence	Unstandardized residuals, studentized deleted residuals, leverage values, covariance ratio, DFFITS, DFBETAS
r	Unstandardized residual values
vif	Variance inflation factors

SAS Output 10.10

SAS *proc reg **output*** statement option	Description
Cookd=varname	Saves Cook's D influence statistics into *varname*
dffits=varname	Saves *dffits* value into *varname*
h=varname	Saves leverage values into *varname*
r=varname	Saves residual values into *varname*
rstudent=varname	Saves studentized deleted residual values into *varname*
student=varname	Saves studentized residual values into *varname*

111

Identifying Outlying *Y* Observations – Studentized Deleted Residuals: On text page 396 the researchers wished to examine whether there were outlying *Y* observations when using a regression model with the two predictor variables *X1* and *X2*. In Insert 10.6 – Step 1, we obtain a first order regression model with predictors *X1* and *X2* and request *influence* statistics. As noted in SAS Output 10.11, this results in the output of residuals (*residual*), studentized deleted residuals (*Rstudent*), leverage values (*Hat Diag H*), DFFITS, and DFBETAS for the current model. The first, third, and second columns of SAS Output 10.11 are consistent with the first, second and third columns, respectively, of Table 10.3, text page 397. Next, we save the studentized deleted residuals in a variable named *sdr*. (Insert 10.6 – Step 1)

The researchers now wish to determine if case 13 is an outlier based on the *studentized deleted residual* (text page 397). Using SAS it is just as easy to test for outliers for each of the 20 cases. Here, $\alpha = 0.10$, $n = 20$, and p = number of independent variables in the model (*X0*, *X1*, and *X2*) = 3 so that the appropriate Bonferroni critical value is $t(1-\alpha/2n;\ n-p-1) = t(1-0.10/2*20;\ 20-3-1) = t(0.9975,16)$. In Insert 10.6 – Step 2, we use a *quantile* function to find *tvalue* = $t(0.9975;16)$ = 3.2519. We then compare the absolute value of the *studentized deleted residual* (*abs(sdr)*) to *tvalue*. If the absolute value of the *studentized deleted residual* is greater than *tvalue*, we set *outlier* = 1, otherwise *outlier* = 0. Consistent with text page 398, all data lines with *outlier* = 1 should be considered outliers and, conversely, all data lines with *outlier* = 0 should be considered not to be outliers. If we were dealing with a large data file, more than 100 data lines, for example, we could search the entire data file by using the *where* option on the *proc print* statement to find all data lines with *outlier* = 1. We use *case+1* in the fourth line of Insert 10.6 to create a case number for each of the 20 cases. Thus we can print *case* for all cases *where outlier=1;* that is, if *outlier=1* we print the case number.

INSERT 10.6

```
* Step 1;
data bodyfat;
  infile 'c:\alrm\chapter  7 data sets\ch07ta01.txt';
  input x1 x2 x3 y;
  case+1;
proc reg;
 model y = x1 x2 / influence ;
 output out=results r=residual rstudent=sdr ;
 run;

* Step 2;
data results;
 set results;
tvalue=tinv(.9975,16);
if (abs(sd)) > tvalue then outlier=1;
if (abs(sdr)) <= tvalue then outlier=0;
proc print data=results;
 where outlier=1;
 var case;
 run;
```

112

SAS Output 10.11

			Output Statistics					
						DFBETAS		
Obs	Residual	RStudent	Hat Diag H	Cov Ratio	DFFITS	Intercept	x1	x2
1	-1.6827	-0.7300	0.2010	1.3607	-0.3661	-0.3052	-0.1315	0.2320
2	3.6429	1.5343	0.0589	0.8443	0.3838	0.1726	0.1150	-0.1426
3	-3.1760	-1.6543	0.3719	1.1892	-1.2731	-0.8471	-1.1825	1.0669
4	-3.1585	-1.3485	0.1109	0.9768	-0.4763	-0.1016	-0.2935	0.1961
5	-0.000289	-0.000127	0.2480	1.5951	-0.0001	-0.0001	-0.0000	0.0001

Identifying Outlying *X* Observations - Hat Matrix Leverage Values: To identify cases with potentially outlying *X* values, the researchers in the Body Fat example plotted triceps skinfold thickness (*X1*) against thigh circumference (*X2*) and identified each line by a case number (text page 399). SAS does not have a readily available way to include the case number on a scatter plot. In Insert 10.7 we produce that scatter plot of *X1* and *X2* using *proc gplot*. From SAS Output 10.12 we observe that two cases appear to be outlying with respect to the *X* values. A visual inspection indicates that the circled dot in the lower left corner of the plot is the only dot that has an *X1* value less than 16. The other circled dot appears to have an *X1* value greater than 30 and an *X2* value less than 54. To find the case numbers corresponding to these dots, we use the *where* option on the *proc print* statement to print the case number for each selected case. Thus, a case number is printed if the condition "*where* (x1< 16) or (x1 > 30 *and* x2 < 54" is satisfied.

INSERT 10.7	INSERT 10.7 continued

data bodyfat;
  infile 'c:\alrm\chapter 7 data sets\ch07ta01.txt';
  input x1 x2 x3 y;
  case+1;
goptions reset=all;
proc gplot;
 symbol value=dot;
	plot x2*x1;
	 run;
	proc print ;
	 where (x1< 16) or (x1 > 30 and x2 < 54);
	 var case;
	 run;

SAS Output 10.12

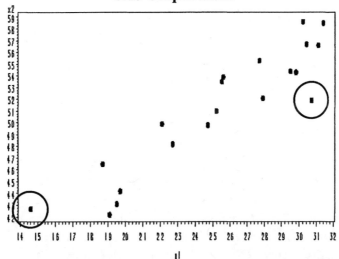

113

The ith diagonal element (h_{ii}) of the hat matrix is a measure of the distance between the point represented by the geometric coordinates of the X values for the ith case and the point represented by the coordinates of the means (the center of the distribution) of the X values for all n cases. The larger the value of h_{ii}, the farther the ith case is from the center of the distribution and the larger the effect of the X values for that case on the corresponding fitted value $(\hat{y}_i)$. Thus, h_{ii} is referred to as the *leverage* of the ith case. In Insert 10.8, we regress Y on $X1$ and $X2$ and request that leverage values be saved in a variable named *lev*. The values of *lev* are identical to the leverage values (h_{ii}) reported in column 2 of text Table 10.3.

The text defines a "large" leverage value as one that is more than twice the magnitude of the mean leverage value, defined as:

$$\bar{h} = \frac{p}{n}$$

where p is the number of independent variables (including the intercept term) and n is the number of cases. So, the leverage value (*lev*) is considered large in this example if *lev* > 2*(3/20). If so, we set *hilev* = 1, otherwise *hilev* = 0. We use the *where* option in the *proc print* procedure to list cases that the have *hilev*=1. After executing Insert 10.8, we see that case numbers 3 and 15 have high leverage values as indicated by *hilev* = 1 (output not shown).

INSERT 10.8

```
data bodyfat;
  infile 'c:\alrm\chapter  7 data sets\ch07ta01.txt';
  input x1 x2 x3 y;
  case+1;
proc reg;
 model y = x1 x2 / vif;
 output out=results h=lev;
 run;
data results;
 set results;
if lev > 2*(3/20)then hilev = 1;
if lev <= 2*(3/20) then hilev = 0;
proc print;
 where hilev=1;
 var case;
run;
```

Identifying Influential Cases – DFFITS, Cook's Distance, and DFBETAS Measures: The researchers now wish to ascertain whether previously identified outlying cases were also influential cases. In Insert 10.9 we begin by specifying *proc reg* and with the *model* statement regressing Y against $X1$ and $X2$, and requesting *influence* statistics as an option in the *model* statement. We also save the *DFFITS* values in a variable named *dffits* and the *Cook's distance measure* in a variable named *cookd*. *DFBETAS* values can be printed as in SAS Output 10.11. The researcher can visually inspect *DFBETAS*, such as those shown in SAS Output 10.11 for the

Body Fat example, to determine if any values are indicative of influential cases. For large data sets, we use the syntax in Insert 10.9 to search for cases that may be influential.

```
                               INSERT 10.9

    data bodyfat;
      infile 'c:\alrm\chapter  7 data sets\ch07ta01.txt';
      input x1 x2 x3 y;
        case+1;
    proc reg;
      model y = x1 x2 / influence;
        output out=result1 cookd=cookd dffits=dffits;
          ods output outputstatistics=result2;
      run;
    data result1;
      set result1;
      If (dffits > 2*sqrt(3/20))then dfflag=1;
      else dfflag=0;
      percent = 100*probf(cookd,3,17);
    data result2;
      set result2;
      if (abs(dfb_x1) > 1) then b1flag=1;
        else b1flag=0;
      if (abs(dfb_x2) > 1) then b2flag=1;
        else b2flag=0;
    data both;
      merge result1 result2;
      run;
    proc print data=both;
      where dfflag=1 or percent > 20 or b1flag=1 or b2flag=1;
      var case dffits cookd percent dfb_x1 b1flag dfb_x2 b2flag;
    run;
```

A *DFFITS* value is judged to indicate an influential case if it exceeds 1 for a small to medium sample and $2\sqrt{p/n}$ for a large sample. In Insert 10.9, we create a new variable, *dfflag*, to indicate if the value of *DFFITS* exceeds $2\sqrt{p/n}$. We set *dfflag* = 1 if *DFFITS* (*dffits*) $> 2\sqrt{p/n}$, otherwise *dfflag* = 0. For *Cook's Distance measure* (D_i), the text authors suggest that D_i be compared to the *F*-distribution with *df1* = *p* and *df2* = *n—p* degrees of freedom. A case is not considered influential if the corresponding percentile value is less than 10 to 20 percent. However, a case is judged to have a major influence on the fitted regression function if the corresponding percentile value is near 50 percent or greater. In Insert 10.9 we use the probability function for the *F*-distribution (*probf*) with *df1* = 3 (parameters) and *df2* = 17 (*n—p*) degrees of freedom to find the proportion of the *F*-distribution that is less than D_i. We multiply this proportion by 100 to express the value as a percent. The result is saved as a new variable, *percent*. A case that has an absolute *DFBETAS* value greater than 1 for small to medium samples or an absolute *DFBETAS* greater than $2/\sqrt{n}$ for large samples may be considered to be an influential case. In Insert 10.9, we create two new variables, *b1flag* and *b2flag*. We then set *b1flag* = 1 if the absolute value (*abs*) of *dfb_x1* is greater than 0, otherwise *b1flag*=0. Similarly, b2flag = 1 if the absolute value (*abs*) of *dfb_x2* is greater than 0, otherwise *b2flag*=0.

To clarify the program structure of Insert 10.9, we first saved the Cook's *Distance measure* (*cookd*) and *DFFITS* (*dffits*) values in a data set named *result1* (*output out=result1 cookd=cookd dffits=dffits;*). *DFBETAS*, however, cannot be saved from this output statement. We must use the SAS Output Delivery System (ODS) to save these statistics. The *"ods output outputstatistics=result2;"* statement results in the *DFBETAS* statistics being saved in a data set named *result2*. *DFBETAS* for *X1* and *X2* are saved by default as *dbf_x1* and *dbf_x2*. We then point to the *result1* data set to test the *dffits* and *cookd* values. To test the *DFBETAS* values (*dbf_x1* and *dbf_x2*), we point to the *result2* data set. Since the *where* option in the *proc print* statement requires data from both *result1* and *result2* data sets, we merge the two data sets into one data set named *both*. The *data=both* option on the *proc print* statement points SAS to the *both* data set to test the conditions of the *where* statement.

We use the *where* option on the *proc print* statement to print all variables for cases that have *dfflag*=1 or have a *percent* value greater than 20 or *betaflag*=1. If we were to print all *dffit* and *percent* values, we would find that all values of *dfflag* = 0, indicating that no *DFFIT* value was greater than $2\sqrt{p/n}$. As reported on text page 404, we also would find that the percent value (*percent*) for case 3 is 30.6 percent and that the next largest percent value is for case 13 at 11 percent. Using *DFBETAS*, we find that *b1flag* = 1 and *b2flag* = 1 for case 3, indicating that *DFBETAS* for both *X1* and *X2* had an absolute value of greater than 1.

Influence on Inferences: Case 3 was identified as an outlying *X* observation and all three influence measures identified case 3 as influential. Thus, the researcher fitted regression functions with and without case 3. The researcher then calculated the percent difference between the predicted value of *Y* ($\hat{Y}_i$) based on the regression function with all 20 cases and the predicted value of *Y* ($\hat{Y}_{i(3)}$) based on the regression function when case 3 was omitted. From Insert 10.10, we begin this analysis by regressing *Y* on *X1* and *X2* and saving the *unstandardized predicted values* as *pred1*. We then use the *where* option in the *proc reg* procedure to select cases other than *case* 3 and again regress *Y* on *X1* and *X2*. (We used *case+1* to create the case number.) We do not save the *unstandardized predicted values* for this regression analysis because case 3 will not have a fitted value and we need the fitted values for all 20 cases; this analysis is only to find the parameter estimates for the model without case 3 in the data set. To find $\hat{Y}_{i(3)}$ for all 20 cases, we use the parameter estimates (shown on text page 406) from the regression function with case 3 omitted to compute *pred2*. We then find *pcinf*, the absolute value (*abs*) of $((\hat{Y}_{i(3)} - \hat{Y}_i)/\hat{Y}_i)*100$.

Finally, we use the *proc means* procedure to find the mean of *pcinf* = 3.1039. We can also generate a frequency distribution of *pcinf* using *proc freq* to see that the mean difference for 17 of the 20 cases is less than 5 percent. Based on this direct evidence, the researcher concluded that case 3 did not exert undue influence on the fitted regression function.

116

INSERT 10.10	INSERT 10.10 continued
```data bodyfat; infile 'c:\alrm\chapter 7 data sets\ch07ta01.txt'; input x1 x2 x3 y; case+1; proc reg; model y = x1 x2 ; output out=results p=pred1; run; proc reg; where case ne 3; model y = x1 x2 ; run;```	```data results; set results; pred2=-12.428+.5641*(x1)+.3635*(x2); pcinf=(abs((pred2-pred1)/pred1))*100; proc means; var pcinf; run; proc freq; tables pcinf; run;```

Multicollinearity Diagnostics Variance Inflation Factor: The researchers now fitted a regression model with all three predictor variables. From text Chapter 7, we recall that triceps skinfold thickness and thigh circumference were highly correlated. We also found large changes in parameter estimates when a variable was added to the model and that one estimated coefficient was negative when a positive value was expected. Accordingly, the researchers were concerned that there may be *multicollinearity* among the predictor variables. They generated *Variance Inflation Factors* (*VIF*) as a formal test for the presence of *multicollinearity*. Insert 10.11 requests *collinearity diagnostics* (*vif*) and SAS Output 10.13 contains partial results. The *VIF*'s for *X1*, *X2*, and *X3* are identical to those reported on text page 409. SAS does not report the mean *VIF* value.

INSERT 10.11

```
data bodyfat;
  infile 'c:\alrm\chapter  7 data sets\ch07ta01.txt';
  input x1 x2 x3 y;
proc reg;
 model y = x1 x2 x3 / vif;
 run;
```

SAS Output 10.13

Parameter Estimates						
Variable	DF	Parameter Estimate	Standard Error	t Value	Pr > |t|	Variance Inflation
Intercept	1	117.08469	99.78240	1.17	0.2578	0
x1	1	4.33409	3.01551	1.44	0.1699	708.84291
x2	1	-2.85685	2.58202	-1.11	0.2849	564.34339
x3	1	-2.18606	1.59550	-1.37	0.1896	104.60601

117

Building the Regression
Model III: Remedial Measures

When we detect inadequacies in a regression model that we are building, we take steps to improve the model so that it is more appropriate for the problem at hand. In earlier chapters, the authors used *remedial measures* such as transformations of Y "to linearize the regression relation, to make the error distributions more nearly normal, or to make the variances of the error terms more nearly equal." Further remedial measures are presented in Chapter 11 to improve model adequacy in the presence of unequal error variances, multicollinearity, and overly influential observations. The use of *weighted least squares* in regression analysis is discussed as an alternative to transformations in building models where the variances of the error terms are unequal. *Ridge regression* is presented as a method for dealing with serious multicollinearity among the predictor variables. *Robust regression* techniques are used to dampen the influence of outlying observations. Two methods of *nonparametric regression* are described: *lowess* and *regression trees*. The method of *bootstrapping* is presented to deal with the problem of estimating the precision of some of the complex estimators discussed in this chapter.

Unequal Error Variances Remedial Measures – Weighted Least Squares: Section 11.1 defines the *generalized multiple regression model*:

$$Y_i = \beta_0 + \beta_1 X_{i1} + \dots + \beta_{p-1} X_{i,p-1} + \varepsilon_i$$

where

$\beta_0, \beta_1, \dots, \beta_{p-1}$ are unknown parameters

$X_{i1}, X_{i2}, \dots, X_{i,p-1}$ are known constants

ε_i are independent $N(0, \sigma_i^2)$

$i = 1, 2, \dots, n$

The method of *weighted least squares* is frequently used to find estimators of the parameters in a generalized multiple regression model because the resulting estimators are *minimum variance unbiased estimators*. Let

$$w_i = \frac{1}{\sigma_i^2}$$

and

$$\mathbf{W} = \begin{bmatrix} w_1 & 0 & \cdots & 0 \\ 0 & w_2 & \cdots & 0 \\ \vdots & \vdots & \ddots & \vdots \\ 0 & 0 & \cdots & w_n \end{bmatrix}$$

Assume the σ_i^2 and hence the w_i are known. Then the weighted least squares (and maximum likelihood) estimators of the coefficients in the generalized multiple regression model are:

$$\mathbf{b_W} = (\mathbf{X'WX})^{-1}\mathbf{X'WY}$$

and the variance-covariance matrix of these estimators is:

$$\sigma^2\{\mathbf{b_W}\} = (\mathbf{X'WX})^{-1}$$

The σ_i^2 are unknown in most applications and the authors present methods for dealing with this circumstance.

The Blood Pressure Example

Data File: ch11ta01.txt

A researcher studied the relationship between diastolic blood pressure (*dbp*) and age (*age*). A sketch of the data is given in Table 11.1, text page 427. See Chapter 0, "Working with SAS," for instructions on opening the Blood Pressure example data file and naming the variables *age* and *dbp*. After opening ch11ta01.txt, the first 5 lines of output should look like SAS Output 11.1.

SAS Output 11.1

Obs	age	dbp
1	27	73
2	21	66
3	22	63
4	24	75
5	25	71

We begin the preliminary analysis of the residuals by fitting an *unweighted least squares* regression model ($Y_i = \beta_0 + \beta_1 X_{i1} + \varepsilon_i$) and saving the *unstandardized residuals* (*residual*). (Insert 11.1) To produce the plots in Figure 11.1, text page 428, we use the SAS absolute value function to generate data values for a new variable, *absresid=abs(residual)*, which is the *absolute value of the unstandardized residual*. We produce Figure 11.1 using the *proc gplot* procedure and related syntax as given in Insert 11.1. The resulting plots are shown in SAS Output 11.2.

INSERT 11.1

```
data bp;
  infile 'c:\alrm\chapter 11 data sets\ch11ta01.txt';
  input age dbp;
proc reg;
 model dbp = age / r;
  output out=results r = residual;
data step2;
 set results;
  absresid=abs(residual);
goptions reset=all;
proc gplot;
  symbol color=black value=circle;
  plot dbp*age;
   run;
  plot residual*age;
   run;
  plot absresid*age;
run;
```

SAS Output 11.2

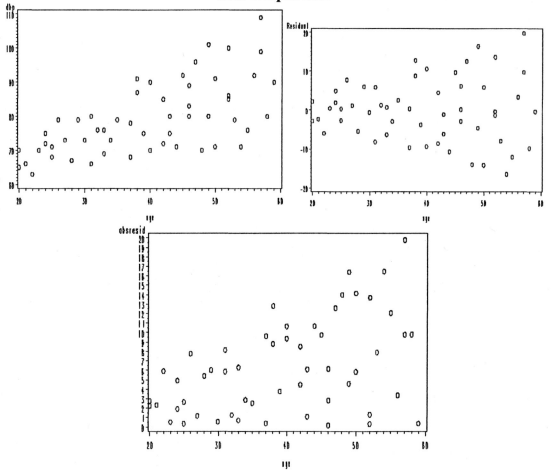

The plot of age against the *unstandardized residual* (*residual*) suggests that the error variance is not constant. From Figure 11.1c, text page 428, the researcher decided that a linear relation between the error standard deviation and *age* was reasonable. In *data step2* of Insert 11.2, we issue the statement *proc reg* with *model absresid = age / p* and obtain text equation 11.19, text page 428. The "*p*" following the slash requests the *fitted* (*unstandardized predicted*) *values*, denoted by $\hat{s}$ in the text. We use the statement *output out=results p = yhat* to specify that the predicted values are to be saved in an output file named *results* under the variable name *yhat*. In *data step3*, we specify *wt=1/(yhat**2)* to compute the weights following equation 11.16a, text page 425:

$$w_i = \frac{1}{(\hat{s}_i)^2} \ \text{ or } \ w_i = \frac{1}{(pre\_1)^2}.$$

The *proc print* statement results in a printout of the first five lines of the current data file (*step3*) as shown in SAS Output 11.3. This printout indicates that file *step2* is the same as the data presented in Table 11.1, text page 427.

INSERT 11.2	INSERT 11.2 continued

data bp;
  infile 'c:\alrm\chapter 11 data sets\ch11ta01.txt';
  input age dbp;
  line+1;
proc reg;
 model dbp = age / r;
  output out=results1 r = residual;
data step2;
 set results1;
  absresid=abs(residual);
proc reg;
 model absresid = age / p;
  output out=results2 p = yhat;
run;
	data step3;
	 set results2;
	  wt=1/(yhat**2);
	proc print;
	 where line < 6;
	proc reg;
	 model dbp = age / clb;
	  weight wt;
	run;

SAS Output 11.3

Obs	age	dbp	line	residual	absresid	yhat	wt
1	27	73	1	1.18224	1.18224	3.80117	0.06921
2	21	66	2	-2.33758	2.33758	2.61214	0.14656
3	22	63	3	-5.91761	5.91761	2.81031	0.12662
4	24	75	4	4.92233	4.92233	3.20666	0.09725
5	25	71	5	0.34230	0.34230	3.40483	0.08626

In *data step3* of Insert 11.2, we use weighted least squares to fit a regression model by regressing *dbp* on *age* using the statement *weight = wt* to specify *wt* as the weighting variable. We use the *clb* option in the model statement to get confidence limits for the model parameters. A partial output is shown in SAS Output 11.4 where we find:

$$\hat{Y} = 55.566 + 0.596\,X$$

and the 95% confidence interval is:

$$0.437 \le \beta_1 \le 0.755$$

consistent with text pages 428-429.

SAS Output 11.4

Parameter Estimates							
Variable	DF	Parameter Estimate	Standard Error	t Value	Pr > |t|	95% Confidence Limits	
Intercept	1	55.56577	2.52092	22.04	<.0001	50.50718	60.62436
age	1	0.59634	0.07924	7.53	<.0001	0.43734	0.75534

Multicollinearity Remedial Measures – Ridge Regression: One approach to estimating regression coefficients in the presence of multicollinearity among the predictor variables is to use a method known as *ridge regression*. The idea is to use estimators that have small bias but greater precision than their unbiased competitors. Let $\mathbf{r}_{XX}$ denote the correlation matrix of the X variables in the model of interest and let $\mathbf{r}_{XY}$ denote the vector of coefficients of simple correlation between Y and each X variable. The ridge standardized regression coefficients are:

$$\mathbf{b}^R = \left(\mathbf{r}_{XX} + c\mathbf{I}\right)^{-1}\mathbf{r}_{XY}$$

The constant c is the amount of bias in the estimators and $\mathbf{I}$ is the $(p-1)$ x $(p-1)$ *identity matrix.* When $c = 0$, $\mathbf{b}^R$ reduces to the standardized least squares estimators. When $c > 0$, the estimators are biased but more precise than ordinary least squares estimators. The strategy is to first obtain estimators based on $c = 0$. Then a small increment is added to c and another set of estimators is obtained. This process is continued for a large number of c values, usually values between 0 and 1. Then, we examine the *ridge trace* and *variance inflation factors* $(VIF)_k$ and try to choose the smallest value of c where the coefficients $\mathbf{b}^R$ first become stable and the $(VIF)_k$ are small. This requires judgement, but we take the value of $\mathbf{b}^R$ corresponding to the chosen c as the ridge estimate for the problem at hand. (See text page 434).

The Body Fat Example

Data File: ch07ta01.txt

On text page 434, a researcher studied the relationship between the amount of body fat (Y) and triceps skinfold thickness ($X1$), thigh circumference ($X2$), and midarm circumference ($X3$). See Chapter 0, "Working with SAS," for instructions on opening the Body Fat example data file and naming the variables. After opening the Body Fat example, the first 5 lines of output should look like SAS Output 11.5.

SAS Output 11.5

Obs	x1	x2	x3	y
1	19.5	43.1	29.1	11.9
2	24.7	49.8	28.2	22.8
3	30.7	51.9	37.0	18.7
4	29.8	54.3	31.1	20.1
5	19.1	42.2	30.9	12.9

In text Chapter 7, the researchers noted severe multicollinearity and decided that an appropriate alternative to ordinary least square regression was ridge regression. In Insert 11.3, we use *proc reg* and specify *outest = ridge* to request that variance inflation factors (*outvif*), standardized betas (*outstb*), and R^2 values be output to the *ridge* data set. The *ridge* = option requests a ridge regression analysis and specifies the values of the ridge constant k. (Here, k rather than c is used to represent the biasing constant.) Each value of k produces ridge regression estimates that are placed in the *outest = ridge* data set. The values of k are saved in the variable _RIDGE_. We write "*ridge = 0 to 1.0 by .002*" to specify the starting and ending constant value (biasing constant) and the value by which the constant will increment. We use the *plot* portion of the *proc reg* statement to produce SAS Output 11.6, which is consistent with Figure 11.3, text page 435. We use the two *proc print* statements to first print k and the standardized betas and then print k and the variance inflation factors. Partial results are shown in SAS Output 11.7 and are consistent with Tables 11.2 and 11.3, text page 434. Although we can print the variance inflation factors and standardized betas for each value of k, there is no readily available routine in SAS that will produce *r square* for a given biasing constant.

INSERT 11.3	INSERT 11.3 continued

data bodyfat;
  infile 'c:\alrm\chapter 7 data sets\ch07ta01.txt';
  input x1 x2 x3 y;
goptions reset=all;
proc reg outest = ridge outvif outstb rsquare ridge =
0 to 0.07 by .002;
  model y = x1 x2 x3 / influence vif;
  plot / ridgeplot;
  symbol1 value=dot color=black interpol=none;
  symbol2 value=circle color=black interpol=none;
 	
  symbol3 value=square color=black
	interpol=none;
	run;
	* print Table 11.2, text page 434;
	proc print data=ridge;
	  where _type_ = 'RIDGESTB';
	  var _ridge_ x1 x2 x3;
	run;
	* print Table 11.3, text page 434;
	proc print data=ridge;
	  where _type_ = 'RIDGEVIF'; var _ridge_ x1 x2 x3;
	
	  run;

123

SAS Output 11.6

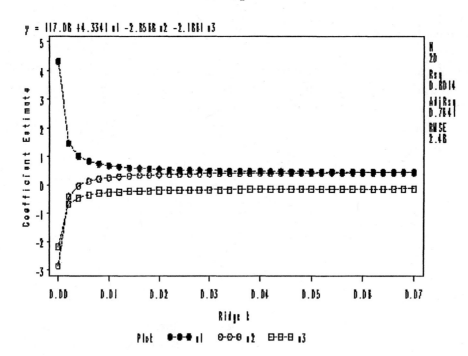

$y = 117.08 + 4.334 \, x1 - 2.856 \, x2 - 2.186 \, x3$

Plot   ●—●—● x1   ⊖—⊖—⊖ x2   ⊟—⊟—⊟ x3

SAS Output 11.7

Obs	_RIDGE_	x1	x2	x3
4	0.000	4.26370	-2.92870	-1.56142
7	0.002	1.44066	-0.41129	-0.48127
10	0.004	1.00632	-0.02484	-0.31487
13	0.006	0.83002	0.13142	-0.24716
16	0.008	0.73433	0.21576	-0.21030

We can use the *ridge = value* statement to specify a single constant value (biasing constant) with which to produce estimated regression coefficients. Insert 11.4, for instance, we request a *ridge regression* using a *biasing constant* = 0.02. Partial (truncated) results are shown in SAS Output 11.8 and are consistent with the fitted regression model on text page 435.

INSERT 11.4

```
data bodyfat;
  infile 'c:\alrm\chapter  7 data sets\ch07ta01.txt';
  input x1 x2 x3 y;
proc reg outest = ridge outvif outstb ridge = .02;
  model y = x1 x2 x3;
 run;
proc print;
 where _type_ = 'RIDGESTB';
 var _ridge_ x1 x2 x3 ;
 run;
```

SAS Output 11.8

Obs	_RIDGE_	x1	x2	x3
4	0.02	0.54633	0.37740	-0.13687

Remedial Measures for Influential Cases - Iteratively Reweighted Least Squares (IRLS) Robust Regression. The text authors state "For robust regression, weighted least squares is used to reduce the influence of outlying cases by employing weights that vary inversely with the size of the residual. Outlying cases that have large residuals are thereby given smaller weights. The weights are revised as each iteration yields new residuals until the estimation process stabilizes." Two widely used weight functions, the *Huber* and *bisquare weight functions*, are described. These functions are used to dampen the effect of outlying observations.

Mathematics Proficiency Example

Data File: ch11ta04.txt

On text page 441, a researcher studied the relationship between the home environment and aggregate school performance (mathematics proficiency) for 37 states, the District of Columbia, Guam, and the Virgin Islands. The data are given by state and states are the study units in this example. The average mathematics proficiency score, denoted *Y*, is the dependent variable of interest. The predictor variables are parents (*X1*), home library (*X2*), reading (*X3*), watch TV (*X4*), and absences (*X5*). This example considers only the predictor variable *X2* (home library) which represents the percentage of eighth-grade students who had three or more types of reading materials at home. See Chapter 0, "Working with SAS," for instructions on opening the Mathematics Proficiency example data file and naming the variables. After opening the Mathematics Proficiency data file, the first 5 lines of output should look like SAS Output 11.9.

SAS Output 11.9

Obs	state	y	x1	x2	x3	x4	x5
1	Alabama	252	75	78	34	18	18
2	Arizona	259	75	73	41	12	26
3	Arkansas	256	77	77	28	20	23
4	Califтом	256	78	68	42	11	28
5	Colorado	267	78	85	38	9	25

To produce the scatter plots (a) and (b) on text page 442 we first regress *Y* on *X2* and save the residuals (*residuals*). (Insert 11.5) We then use *proc gplot* to request a *scatter plot* of *Y* and *X2* (SAS Output 11.10) and *residual* and *X2* (SAS Output 11.11). From the plots, we notice that there are 3 states with outlying *Y* values (mathematics proficiency scores). Also, there are 6 states with low *X2* scores (resources) whose math proficiency scores are above the fitted regression line. We also issue *proc loess* to get the lowess regression fit as shown on Figure 11.5 (a), text page 442. (SAS Output not shown)

125

```
data math;
infile 'c:\alrm\chapter 11 data sets\ch11ta04.txt';
input state $ y x1 x2 x3 x4;
proc reg;
 model y = x2 / r;
  output out=results r = residual;
run;
goptions reset=all;
proc gplot;
 plot y*x2;
  symbol1 value=dot color=black interpol=rl;
 plot residual*x2;
  symbol1 value=dot color=black interpol=rl;
run;
proc loess data = math;
  model y = x2 / smooth = 0.5 to 0.8 by 0.1 residual;
  ods output OutputStatistics = Results;
run;
data step2;
  set results;
proc sort;
  by SmoothingParameter x2;
 run;
goptions reset=all;
proc gplot;
  by SmoothingParameter;
  symbol1 color=black value=none interpol=join ;
  symbol2 color=black value=circle;
   plot Pred*x2 DepVar*x2 / overlay ;
run;
```

SAS Output 11.10   **SAS Output 11.11**

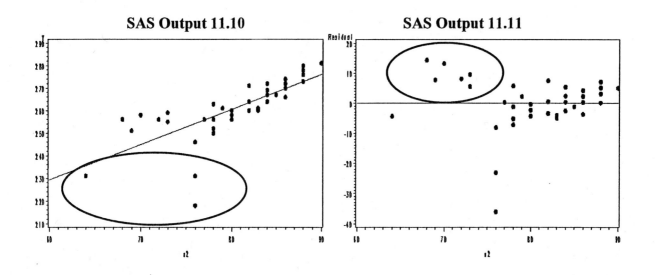

The researchers considered a second order model using *X2* (*homelib*) centered, and its square as the single predictor variable (equation 11.48, text page 443). To diminish the effect of outliers the researchers used *iteratively reweighted least squares* and the *Huber weight function*. We detail each step in the following and in Insert 11.7, although the output for each step is not shown. We begin in Insert 11.7 by centering *X2* (*x2cnt*) and finding the square of the centered variable (*x2cnt2*). Next, we issue *proc reg* and specify "*model y = x2cnt x2cnt2 / r*" to fit the second order model and compute the residuals. We then:

1) Save the unstandardized residuals in a data file named *step2* under the variable name *resid*.

2) Use *proc univariate* to find the median value ($median\{e_i\}$) of *resid* = 0.70629.

3) Compute (name arbitrary) *v1*, the absolute value of *resid* minus the median of the residuals ($|e_i - median\{e_i\}|$).

4) Find the median of $[median\{|e_i - median\{e_i\}|\}]$ *v1* = 3.1488.

5) Use *proc univariate* to find the median absolute deviation estimator (*MAD*) = *median of v1 − constant* = 0.6745 (3.1488 / 0.6745) = 4.6683.

6) Compute the scaled residual, *u1* = *resid* ÷ *mad* = *resid* ÷ 4.6683.

7) Use the scaled residual in the computation of the Huber weight (*w*), where *w* = 1 if the absolute value of *u1* ≤ 1.345 and *w* = (1.345 / *absolute value of u1*) if the *absolute value of u1* > 1.345.

8) Regress *y* on *x2cnt* and *x2cnt2* using *w* as a weighting variable and save the unstandardized residuals. This produces the regression model: $\hat{Y} = 259.39 + 1.670\,x_2 + 0.064\,x_2^2$. The *proc print* statement in Step 8 of Insert 11.7 requests the current Huber weights (*w*) and residual values (*resid2*) as shown in SAS Output 11.12. These values are identical to columns 3 and 4 of Table 11.5, text page 444.

9) Return to Step 2 for the second iteration and use the residuals saved in Step 8 for the next iteration.

INSERT 11.6	INSERT 11.6 continued

``
data math;
infile 'c:\alrm\chapter 11 data sets\ch11ta04.txt';
input state $ y x1 x2 x3 x4;
x2cnt=x2-80.4;
x2cnt2=x2cnt**2;
* STEP 1;
proc reg;
 model y = x2cnt x2cnt2 / r;
 output out=step2 residual = resid;
run;
* STEP 2;
proc univariate data = step2;
 var resid;
run;
* STEP 3;
data new;
 set step2;
  v1 = abs(resid - 0.70629);
run;
``
	``
	* STEP 4;
	proc univariate;
	 var v1;
	run;
	* STEPS 5, 6 & 7;
	data new;
	 set step2;
	  mad=3.1488 /.6745;
	  u = resid / mad;
	   if abs(u)<= 1.345 then wt=1;
	   if abs(u) > 1.345 then wt=(1.345/abs(u));
	run;
	* STEP 8;
	proc reg;
	 weight wt;
	 model y = x2cnt x2cnt2 / r;
	  output out=step3 residual = resid2;
	run;
	proc print data=step3;
	 var wt resid2;
	run;
	``

SAS Output 11.12

Obs	wt	resid2
1	1.00000	-3.7542
2	0.59390	8.4297
3	1.00000	1.5411
4	0.60899	7.3822
5	1.00000	-1.4402

Nonparametric Regression: SAS/STAT uses proc loess for performing nonparametric regression using the lowess method. To our knowledge, SAS/STAT does not currently have a readily available method for performing regression trees.

Remedial Measures for Evaluating precision in Nonstandard Situations – Bootstrapping
To our knowledge, SAS/STAT has no readily available procedure for performing a bootstrap analysis in this application.

Chapter 12 SAS®

Autocorrelation in Time Series Data

In the context of the generalized multiple regression model, a time series is a set of data collected by observing a random variable and one or more independent variables at each time in a sequence of times. For example, company sales (the dependent variable) and industry sales (the independent variable) could be observed quarterly over a five-year interval. Or, net income (the dependent variable), gross revenues, and labor costs (the independent variables) for a company could be observed monthly over a three-year period. The error terms for regression models for times series data are often correlated and are, therefore, referred to as being *autocorrelated* or *serially correlated*. A generalized multiple regression model in which the error terms are autocorrelated is called an *autoregressive error model*. In this chapter, the authors of the text introduce the concept of *autocorrelation* and associated *autoregressive error models*. They discuss the *Durbin-Watson test for autocorrelation* as a means for detecting *autocorrelation* among the error terms of a model. They then discuss remedial measures for dealing with data where *autocorrelation* has been detected.

The Blaisdell Company Example

Data File: ch12ta02.txt

The Blaisdell Company (text page 488) used a trade association record of industry sales to predict their own sales. See Chapter 0, "Working with SAS," for instructions on opening the Blaisdell Company data file and naming the variables *company* and *industry*. After opening ch12ta02.txt, the first 5 lines of output should look like SAS Output 12.1.

SAS Output 12.1

Obs	company	industry
1	20.96	127.3
2	21.40	130.0
3	21.96	132.7
4	21.52	129.4
5	22.39	135.0

In Insert 12.1, we begin the analysis by producing a scatter plot of the company and industry sales data. (SAS Output 12.2) Since the scatter plot suggested that a linear regression model might be appropriate, the analyst working with Blaisdell Company fitted an ordinary least squares regression model and saved the unstandardized residuals (*residual*). To plot the residuals against time, we used the *time+1* statement in Insert 12.1 to create a sequential record number for each line in the Blaisdell Company data file. A plot of the residuals against *time* is shown in SAS Output 12.3.

129

INSERT 12.1	INSERT 12.1 Continued

* Produce Scatter Plot Output 12.2;
data Blaisdell;
  infile 'c:\alrm\chapter 12 data sets\ch12ta02.txt';
  input company industry;
time+1;
proc gplot;
 plot company*industry;
 run;
	proc reg;
	 model company = industry / r;
	  output out=results r=residual;
	run;
	data results;
	 set results;
	
	* Produce Scatter Plot Output 12.3;
	proc gplot;
	 symbol value=dot color=black interpol=none
	vref=0;
	 plot residual*time;
	 run;

SAS Output 12.2   SAS Output 12.3

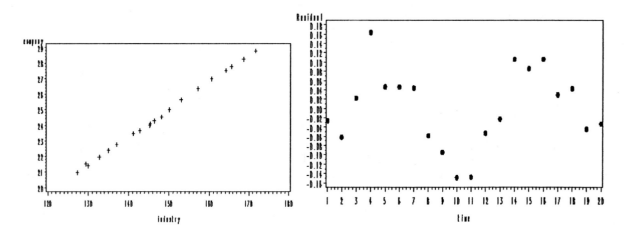

Durbin-Watson Test for Autocorrelation: Since SAS Output 12.3 indicates a positive autocorrelation in the error terms, the analyst produced the *Durbin-Watson Test for Autocorrelation*. To request the *Durbin-Watson* test, we use the first regression procedure in Insert 12.1 and modify the *model* statement to read, "*model company = industry / dw*;". Partial output is shown in SAS Output 12.4. The Durbin-Watson value = 0.735 as reported on text page 488. Using Table B.7, text page 675, we conclude that the error terms are positively autocorrelated.

SAS Output 12.4

Durbin-Watson D	0.735
Number of Observations	20
1st Order Autocorrelation	0.626

Cochrane-Orcutt Procedure: On text page 492, the analyst first used the *Cochrane-Orcutt procedure* to estimate the autocorrelation parameter ρ. Column 1 (e_t) of Table 12.3, text page 493, contains the residual values for the regression model fitted in Insert 12.1. Each line in text Table 12.3, column 2, is the residual value in column 1 for the preceding time period. In Insert 12.2, we use SAS' *Lag operation* that takes the form, *new=lag(old)* where *new* is the value of *old* for the previous case. That is, we use [*res_m1=lag(residual)*] to compute a new variable, *res_m1*, that is equal to the residual (*residual*) of the preceding time period (line) in the data file. This produces text Table 12.3, column 2, denoted by the symbol e_{t-1}. We also find *etm1et=residual*res_m1* which is the residual times the lagged residual. This produces text Table 12.3, column 3, denoted $e_{t-1}e_t$. We then square the lagged residual [*compute etm1et2=res_m1**2*] to produce text Table 12.3, column 4, denoted e_{t-1}^2. Finally, in SAS Output 12.5, we find the sum of *etm1et* = 0.0834478 and *etm1et2* = 0.1322127. We now obtain *r*, an estimate of ρ, by equation 12.22, text page 492:

$$r = 0.0834479 / 0.1333023 = 0.631166$$

INSERT 12.2	INSERT 12.2 continued

data Blaisdell;
  infile 'c:\alrm\chapter 12 data sets\ch12ta02.txt';
  input company industry;
time+1;
proc reg;
 model company = industry / r;
  output out=results r=residual;
run;
	*Produce Output 12.5;
	data results;
	 set results;
	res_m1=lag(residual);
	etm1et=residual*res_m1;
	etm1et2=res_m1**2;
	proc means sum;
	 var etm1et etm1et2;
	run;
	*Produce Output 12.6;
	data results;
	 set results;
	ypt=company-.631166*lag(company);
	xpt=industry-.631166*lag(industry);
	proc print;
	run;

SAS Output 12.5

Variable	Sum
etm1et	0.0834479
etm1et2	0.1322127

In Insert 12.2, we use equation 12.18, text page 491, to find Y_t' (ypt) and X_t' (xpt). This produces the 3rd and 4th columns of Table 12.4, text page 493. To complete the *Cochrane-Orcutt* procedure, we then regress *ytp* on *xpt* and request the Durbin-Watson statistic (syntax not show). Partial results are shown in SAS Output 12.6. The Durbin-Watson statistic = 1.65, consistent with text page 494. This leads to the conclusion that the autocorrelation of error terms is not significantly different from zero. From SAS Output 12.6 we note that $b_0' = -0.394$, $b_1' = 0.174$, $s\{b_0'\} = 0.16723$, and $s\{b_1'\} = 0.00296$. We use equation 12.20, text pages 491-492, to transform back to a fitted regression model in terms of the original variables and equation 12.21 to obtain estimated standard deviations of the back-transformed fitted model coefficients (syntax not shown).

SAS Output 12.6

Durbin-Watson D	1.650
Number of Observations	19
1st Order Autocorrelation	0.147

Parameter Estimates					
Variable	DF	Parameter Estimate	Standard Error	t Value	Pr > |t|
Intercept	1	-0.39411	0.16723	-2.36	0.0307
xpt	1	0.17376	0.00296	58.77	<.0001

Hildreth-Lu Procedure: The *Hildreth-Lu procedure* uses the value of ρ that minimizes the error sum of squares (*SSE*) for the transformed regression model. As noted in the text, we can repeatedly fit regression models while varying the value of ρ to find the value that minimizes *SSE*. An efficient method of producing Table 12.5, text page 495, is to use SAS' *IML* procedure to automatically perform the numerical search across a large range of ρ values as in Insert 12.3.

In Insert 12.3, use the *lag operation* to find Y_{t-1} (ylag) and X_{t-1} (xlag). Since there are 20 lines of data in the file with values for *company* and *industry*, the first line will have a missing value for the lagged variables *ylag* and *xlag*. If we read *ylag* and *xlag* into matrices and subsequently attempt to perform arithmetic using these variables, we will get an error due to the **xlag** and **ylag** vectors containing missing values. To circumvent the problem of missing values in the first row of these vectors, we read all variables into vectors that begin with a *t*, e.g., *txlag*. We then set a new vector equal to the 2nd through 20th row of the vector that begins with *t*, *xlag=txlag[2:20]* so that **xlag** now contains 19 rows that are equal to the last 19 rows of **txlag** 20x1 column vector.

To reproduce the *Hildreth-Lu* results shown in Table12.5, text page 495, we calculate *SSE* for a range of ρ values from 0.01 to 1.0 in increments of 0.01 (*do r=.01 to 1.0 by .01;*). We find X_t'

(**xstep1**) and concatenate **X0** (a column vectors of 1's) and **xstep1** to form matrix **xmat**. We find Y_t' (**ystep**) (also from equation 12.17, text page 491) and use this as our "Y" matrix. Finally, we use equations 5.60 and 5.85a, text pages 200 and 204, respectively, to find the **b** matrix and **SSE**. The numerical output from this insert (not shown) is consistent with the selected values of ρ and corresponding *SSE* values shown on Table 12.5, text page 495.

INSERT 12.3

```
data blaisdell;
 infile 'c:\alrm\chapter 12 data sets\ch12ta02.txt';
 input company industry;
x0=1;
ylag=lag(company);
xlag=lag(industry);
proc iml;
  use blaisdell;
  read all var {'x0'} into tx0;
  read all var {'xlag'} into txlag;
  read all var {'industry'} into tindustry;
  read all var {'ylag'} into tylag;
  read all var {'company'} into tcompany;
x0=tx0[2:20];
xlag=txlag[2:20];
industry=tindustry[2:20];
ylag=tylag[2:20];
company=tcompany[2:20];
do r=.01 to 1.0 by .01;
xstep1=industry-(r*xlag);
xmat=(x0||xstep1);
ystep1=company-(r*ylag);
* Find b matrix by text equation 5.56;
b=inv(t(xmat)*xmat)*t(xmat)*ystep1;
* Find sse by text equation 5.84a;
sse=t(ystep1)*ystep1-t(b)*t(xmat)*ystep1;
print r [format = 4.2] "   " sse [format = 6.5];
end;
quit;
```

First Difference Procedure: On text page 496, the analyst used the *First Difference procedure* where $\rho = 1.0$ for the transformed model (equation 12.17, text page 491), where $Y_t' = Y_t - Y_{t-1}$ and $X_t' = X_t - X_{t-1}$, and the regression model is based on "regression through the origin." In Insert 12.4, we compute Y_t' (*fdyp*) and X_t' (*fdxp*) using the *lag operation*. We then regress *fdyp* on *fdxp*, and specify that the fitted model has no intercept (*noint*). This results in SAS Output 12.7 and a fitted regression model $\hat{Y}' = 0.168X'$. To produce the Durbin-Watson statistic, we run the same regression model as above except we include the intercept term in the model. SAS Output 12.8 shows the Durbin-Watson statistic, which is consistent with text page 497.

```
data Blaisdell;
  infile 'c:\alrm\chapter 12 data sets\ch12ta02.txt';
  input company industry;
fdyp=company-lag(company);
fdxp=industry-lag(industry);
proc reg;
 model fdyp = fdxp / noint;
 run;
proc reg;
 model fdyp = fdxp / dw;
 run;
```

SAS Output 12.7

		Parameter Estimates			
Variable	DF	Parameter Estimate	Standard Error	t Value	Pr > |t|
fdxp	1	0.16849	0.00510	33.06	<.0001

SAS Output 12.8

Durbin-Watson D	1.749
Number of Observations	19
1st Order Autocorrelation	0.116

We use equation 12.31, text pages 496-497, to transform the coefficients of the fitted model back to coefficients of a model fitted in terms of the original variables. If in the first regression procedure in Insert 12.1, we had modified the *proc reg* statement to read, "*proc reg simple;*" we would have obtained SAS Output 12.9. We use equation 12.31a and 12.31b, respectively, and SAS Output 12.7 and 12.9 to find:

$$b_0 = \overline{Y} - b_1' \overline{X}$$

$$b_0 = 24.5690 - 0.168(147.6250) = -0.30349$$

and

$$b_1 = b_1'$$

$$b_1 = 0.16849$$

Thus, the fitted model in terms of the original variables is:

$$\hat{Y} = -.30349 + 0.16849X$$

SAS Output 12.9

Descriptive Statistics					
Variable	Sum	Mean	Uncorrected SS	Variance	Standard Deviation
Intercept	20.00000	1.00000	20.00000	0	0
industry	2952.50000	147.62500	439411	186.73776	13.66520
company	491.38000	24.56900	12183	5.81001	2.41040

Forecasting with Autocorrelated Error Terms

Forecasting Based on Cochrane-Orcutt Estimates: On text page 500, the analyst forecasts the Blaisdell Company sales for the 21st quarter. The trade association's estimate for this quarter is \$175.3 million. We use the *Cochrane-Orcutt* results presented in Insert 12.2 and SAS Output 12.6 where:

$$\hat{Y} = -0.3941 + 0.17376X'$$

We transform the model back to the original variables as shown on text page 497, which leads to:

$$\hat{Y} = -1.0685 + 0.17376X$$

In Insert 12.5, we find X'_t (*xpt*) and Y'_t (*ypt*) for subsequent use and set the estimate of ρ: (*r*) = 0.631166. We find (*e*) $e_{20} = Y_{20} - \hat{Y}_{20}$ where $Y_{20} = 28.78$ and $\hat{Y}_{20} = -1.0685 + 0.17376*171.7$. Using the above fitted regression model, the predicted value (*yhat*) $\hat{Y}_{21}$ when the independent variable is $X_{21} = 175.3$, is $-1.0685 + 0.17376*175.3 = 29.392$. We then compute F_{21} (*forcast*) = $\hat{Y}_{21} + re_{20} = 29.40$. We find X'_{n+1} where $X_{n+1} = 175.3$ (X_{21} above), $r = 0.631166$, and $X_{20} = 171.7$. Thus:

$$X'_{n+1} = 175.3 - 0.631166*171.7 = 66.629$$

Using the above numerical results, we find $s\{pred\}$ using equation 2.38, text page 59. Here, *MSE* = 0.0045097 (regress *ypt* on *xpt*), $(X_h - \bar{X})^2 = [(66.929 - 56.3186)**2]$, and $\sum(X_i - \bar{X})^2 = 515.8551$. Thus, text equation 2.38a:

$$s^2\{pred\} = MSE\left[1 + \frac{1}{n} + \frac{(X_h - \overline{X})^2}{\sum(X_i - \overline{X})^2}\right]$$

where $n = 19$, $X_h = X'_{n+1}$, $\overline{X}$ = mean of XPT, and $X_i = XPT_i$, becomes:

$$s\{pred\} = \sqrt{0.004509\left[1 + \frac{1}{19} + \frac{(66.929 - 56.3186)^2}{515.8551}\right]}$$

The value of *spred* = 0.07570 is consistent with text page 500. Finally, we find *tvalue* = *idf.t*(.975,17) = 2.11 and then compute the lower (*cilo*) and upper (*cihi*) prediction intervals as F_{21} (*forcast*) = $29.40 \pm 2.11 * 0.0757$ or $29.24 \leq Y_{21(new)} \leq 29.56$.

INSERT 12.5

```
data Blaisdell;
 infile 'c:\alrm\chapter 12 data sets\ch12ta02.txt';
 input company industry;
ypt=company-.631166*lag(company);
xpt=industry-.631166*lag(industry);
r = 0.631166;
e = company-(-1.0685+(.17376*industry));
yhat = -1.0685+(.17376*175.3);
forecast = yhat+(r*e);
* Use text equation 2.38 to find s{pred} (spred);
* Find mean of xpt=56.3186 using proc means;
proc means;
 var xpt;
run;
data blaisdell;
 set blaisdell;
xptdif2=(xpt-56.3186)**2;
* Find sum of xptdif2 = 515.8551 using Analyze > Descriptives and Sum Option;
proc means sum;
 var xptdif2;
data blaisdell;
 set blaisdell;
spred=sqrt(0.0045097+ (0.0045097*((1/19)+(((66.929- 56.3186)**2)/515.8551))));
tvalue=tinv(.975,17);
cilo=29.4-(tvalue*spred);
cihi=29.4+(tvalue*spred);
proc print data=blaisdell (obs=1);
 var cilo cihi;
 run;
```

Forecasting Based on First Difference Estimates: On text page 501, the authors use the first differences fitted regression function to forecast sales for quarter 21. To find F_{21} we recall that the first part of the expression [(−0.30349 + 0.16849*(175.3)] comes from equation 12.33, text page 497, and from X_{21} = 175.3 above. The latter part of the expression {1.0[28.78 + 0.30349 − 0.16849*171.70)]} comes from text equation 12.33, the fact that ρ = 1.0 in the *First Difference* model, and X_{20} = 171.70. The forecasted value for the 21st quarter is F_{21} = 29.39 as shown on text page 501. We find the prediction intervals using equation 4.20, text page 162. We first find X_h^2 (*xh2*), which is the forecasted value (F_{21}) squared. We then find X_i^2 (*fdxp2*) and, subsequently, use *Descriptives* to find the sum of *fdxp2* = 185.42. We find X_{n+1}' (*xnp1*) = X_{21} − ($r* X_{20}$) where X_{20} = *industry*$_{20}$. The MSE = 0.00482 is from regressing *fdyp* on *fdxp* and specifying the fitted model with no intercept term. The statistic *tvalue* [*tinv*(.975,18)] is based on 18 *df*. First, 19 of the original 20 records were used in the analysis because we "lost" a record when we lagged to get *xpt* and *ypt*. (This was also true for the *Cochrane-Orcutt* procedure above where we used 17 *df* (*n*−2) to find *tvalue*.) We have 18 *df* (*n*−1) here because we estimated only 1 parameter (β_1) (this model has no intercept term.) We then find the 95% prediction intervals for F_{21} = 29.39 as shown on text page 501: $29.24 \le Y_{21(new)} \le 29.52$

INSERT 12.6

```
data Blaisdell;
  infile 'c:\alrm\chapter 12 data sets\ch12ta02.txt';
  input company industry;
fdyp=company-lag(company);
fdxp=industry-lag(industry);
xh2=29.39**2;
fdxp2=fdxp**2;
xnp1=175.3-(1.0*171.7);
xnp12=xnp1**2;
* find sum of fdxp2=185.42;
proc means sum;
 var fdxp2;
  run;
data blaisdell;
 set blaisdell;
spred=sqrt(.00482*(1+(12.96/185.42)));
tvalue=tinv(.975,18);
cilo=29.39-(tvalue*spred);
cihi=29.39+(tvalue*spred);
proc print data=blaisdell (obs=1);
 var cilo cihi;
run;
```

Forecasting Based on Hildreth-Lu Estimates: On text page 501, the authors find the 95 percent prediction interval based on the *Hildreth-Lu* procedure. Similar to the *Cochrane-Orcutt* procedure above, we first find Y_t' (*ytp*) and X_t' (*xpt*) using text equation 12.17, and set *r* = 0.96 as shown in Table 12.5, text page 495. (Insert 12.7) We transform the fitted regression model (equation 12.28, text page 496) back to a fitted model in terms of the original variables to obtain: $\hat{Y} = 1.7793 + .16045X$. We use this model to calculate e and $yhat$. Since X'_{n+1} is not given, we compute $xnp1 = X_{21} - (r * X_{20}) = X'_{n+1}$ (note that $X_{20} = industry_{20}$). We use equation 2.38, text page 59, to find $s^2\{pred\}$ ($spred$) where MSE is the mean square error from regressing ypt on xpt. The lower ($cilo$) and upper ($cihi$) prediction intervals are obtained by printing the 20th line in the current data set.

INSERT 12.7

```
data Blaisdell;
  infile 'c:\alrm\chapter 12 data sets\ch12ta02.txt';
  input company industry;
  line+1;
r = 0.96;
ypt=company-r*lag(company);
xpt=industry-r*lag(industry);
e = company-(1.7793+(.16045*industry));
yhat = 1.7793+(.16045*175.3);
forecast = yhat+(r*e);
xnp1=175.3-(.96*171.7);
* Use text equation 2.38 to find s{pred} (spred);
* Find mean of xpt=8.1912 using proc means;
proc means;
 var xpt;
 run;
data blaisdell;
 set blaisdell;
xptdif2=(xpt-8.1912)**2;
* Find sum of xptdif2 = 90.1088 using proc menas Sum Option;
proc means sum;
 var xptdif2;
 run;
data blaisdell;
 set blaisdell;
spred=sqrt(0.00422+ (0.00422*((1/19)+(((10.468- 8.1911)**2)/90.1088))));
tvalue=tinv(.975,17);
cilo=forecast-(tvalue*spred);
cihi=forecast+(tvalue*spred);
proc print;
 where line=20;
 var cilo cihi;
run;
```

Chapter 13 SAS®

Non Linear Regression

The general linear regression model is:

$$Y_i = \beta_0 + \beta_1 X_{i1} + \beta_2 X_{i2} + ... + \beta_{p-1} X_{i,p-1} + \varepsilon_i$$

In this model, Y_i is a linear function of the $\beta's$ (the parameters). The general linear regression model can also be expressed as:

$$Y_i = f(\mathbf{X}_i, \boldsymbol{\beta}) + \varepsilon_i$$

where $\mathbf{X}_i$ is the vector of predictor variables for the ith subject, β is the vector of model parameters, and

$$f(\mathbf{X}_i, \boldsymbol{\beta}) = \beta_0 + \beta_1 X_{i1} + \beta_2 X_{i2} + ... + \beta_{p-1} X_{i,p-1}$$

The nonlinear regression model can be expressed as (text page 511):

$$Y_i = f(\mathbf{X}_i, \boldsymbol{\gamma}) + \varepsilon_i$$

In this representation of the nonlinear regression model, the authors use γ to denote the vector of parameters rather than β as a reminder that in this model $f(\mathbf{X}_i, \boldsymbol{\gamma})$ is a nonlinear function of the parameters.

Estimation of the parameters of a nonlinear regression model is usually based on the method of least squares or the method of maximum likelihood. However, it is usually not possible to find closed form "formulas" for the estimators of the parameters of nonlinear regression models. Instead, iterative search procedures are required. Thus, although the approach to finding estimators is the same as that used for linear regression models, the computations are much more tedious.

The problem reduces to one of finding values of the parameters that maximize (minimize) a nonlinear function of the parameters. Many search procedures have been developed for finding the maxima (minima) of nonlinear functions. SAS offers five methods for estimating regression parameters and statistics for models that are not linear in their parameters. The default method is Gauss-Newton. Other available options that may be specified (e.g., *proc nlin method= marquardt;*) are Marquardt, Newton, Gradient, and DUD.

Severely Injured Patients Example

Data File: ch13ta01.txt.

A hospital administrator wanted to predict long-term recovery among a group of severely injured patients (text page 514). The predictor variable used in the regression model was number of days of hospitalization and the response variable was a long-term recovery index measure. See Chapter 0, "Working with SAS," for instructions on opening the Severely Injured Patient data file and naming the variables *index* and *days*. After opening ch13ta01.txt, the first 5 lines of output should look like SAS Output 13.1.

SAS Output 13.1

Obs	index	days
1	54	2
2	50	5
3	45	7
4	37	10
5	35	14

The administrator began by producing a scatter plot (Insert 13.1 – SAS Output 13.2) of the data, which suggested an exponential relationship between number of days hospitalized and long-term recovery. A review of related literature also suggested an exponential relationship. Thus, the administrator explored the appropriateness of a two-parameter nonlinear exponential regression model (text page 514).

INSERT 13.1

```
data Injured;
   infile 'c:\alrm\chapter 13 data sets\ch13ta01.txt';
   input index days;
goptions reset=all;
proc gplot;
   symbol color=black value=circle;
   plot index*days;
run;
```

SAS Output 13.2

140

To fit this nonlinear regression model, we first find the initial starting values $g_0^{(0)}$ and $g_1^{(0)}$ (text page 521). We create a new variable, *logy*, the logarithmic transformation of *index*, and perform an ordinary least squares regression using *logy* as the response variable and *days* as the predictor variable using the syntax shown in Insert 13.2. From SAS Output 13.3, we find the parameter estimates for this model to be $b_0 = 4.0371$ and $b_1 = -0.03797$. Using a hand-held calculator, we find $g_0^{(0)} =$ exponential of $b_0 = \exp(4.0371) = 56.6646$ and $g_1^{(0)} = b_1 = -0.03797$ (text page 521). We now use SAS *proc nlin* as shown in Insert 13.3 to produce estimates of the parameters of the nonlinear model and related results. The final least square estimates, including the *MSE*, shown in SAS Output 14.4 are consistent with Table 13.3, text page 524. The 95% confidence intervals for β_0 and β_1 are also shown in SAS Output 13.4.

INSERT 13.2

```
data Injured;
  infile 'c:\alrm\chapter 13 data sets\ch13ta01.txt';
  input index days;
logy=log(index);
goptions reset=all;
proc reg;
 model logy = days;
run;
```

INSERT 13.3

```
data Injured;
  infile 'c:\alrm\chapter 13 data sets\ch13ta01.txt';
  input index days;
logy=log(index);
proc nlin;
 parms g0 = 56.6646 g1 = -0.03797;
 model index = g0*exp(g1*days);
run;
```

SAS Output 13.3

Parameter Estimates					
Variable	**DF**	**Parameter Estimate**	**Standard Error**	**t Value**	**Pr > |t|**
Intercept	1	4.03716	0.08410	48.00	<.0001
days	1	-0.03797	0.00228	-16.62	<.0001

SAS Output 13.4

Iter	g0	g1	Sum of Squares
0	56.6646	-0.0380	56.0867
1	58.5578	-0.0395	49.4638
2	58.6055	-0.0396	49.4593
3	58.6065	-0.0396	49.4593

NOTE: Convergence criterion met.

Estimation Summary	
Method	Gauss-Newton
Iterations	3
R	9.015E-6
PPC(g1)	1.405E-6
RPC(g1)	0.000043
Object	7.896E-8
Objective	49.4593
Observations Read	15
Observations Used	15
Observations Missing	0

NOTE: An intercept was not specified for this model.

Source	DF	Sum of Squares	Mean Square	F Value	Approx Pr > F
Regression	2	12060.5	6030.3	1585.01	<.0001
Residual	13	49.4593	3.8046		
Uncorrected Total	15	12110.0			
Corrected Total	14	3943.3			

Parameter	Estimate	Approx Std Error	Approximate 95% Confidence Limits	
g0	58.6065	1.4722	55.4261	61.7869
g1	-0.0396	0.00171	-0.0433	-0.0359

Approximate Correlation Matrix		
	g0	g1
g0	1.0000000	-0.7071473
g1	-0.7071473	1.0000000

On text page 530, the authors generated bootstrap samples to check the appropriateness of large-sample inferences. To our knowledge, SAS/STAT has no readily available procedure for generating bootstrap samples in this application.

Learning Curve Example

Data File: ch13ta04.txt

An electronic products manufacturer produces a new product at two different locations: location A ($X1$=1) and location B ($X1$=0). (Text page 533) The response variable was relative efficiency, a measure related to production costs at each location. See Chapter 0, "Working with SAS," for instructions on opening the Learning Curve Example data file and naming the variables $X1$, $X2$, and Y. After opening ch13ta04.txt, the first 5 lines of output should look like SAS Output 13.5.

Obs	x1	x2	y
1	1	1	0.483
2	1	2	0.539
3	1	3	0.618
4	1	5	0.707
5	1	7	0.762

It was well known that production efficiency of a new product initially increases with time and then levels off. We use the syntax in Insert 13.4 to produce the scatter plot of relative efficiency (Y) and time ($X2$) with separate symbols to be used for Locations A and B (SAS Output 13.6). The engineers decided to fit an exponential model that used location ($X1$) and the relationship between relative efficiency (Y) and time ($X2$). The selected nonlinear model (equation 13.38, text page 534) was:

$$Y_i = \gamma_0 + \gamma_1 X_{i1} + \gamma_3 \exp(\gamma_2 X_{i2}) + \varepsilon_i$$

On text page 535, the authors detailed the rationale for determining the starting values for estimates of the parameters: $g_0^{(0)} = 1.025$, $g_1^{(0)} = -0.0459$, $g_2^{(0)} - 0.122$, and $g_3^{(0)} = -0.5$. Continuing in Insert 13.4, we specify *proc nonlin* and define the above non-linear regression model to be used in the nonlinear regression analysis. Results are shown in Output 13.8 and are consistent with Table 13.5, text page 535.

INSERT 13.4

```
data learning;
  infile 'c:\alrm\chapter 13 data sets\ch13ta04.txt';
  input x1 x2 y;
*Produce scatter plot shown in Figure 13.5, text page 534
proc gplot;
  symbol1 color=black value=circle interpol=join;
  symbol2 color=black value=dot i=join;
 plot y*x2=x1;
 run;
*Perform nonlinear regression analysis
proc nlin;
 parms g0 = 1.025 g1 = -0.0459 g2 = -.122 g3 = -0.5 ;
  model y = g0+g1*x1+g3*(exp(g2*x2));
 run;
```

SAS Output 13.6

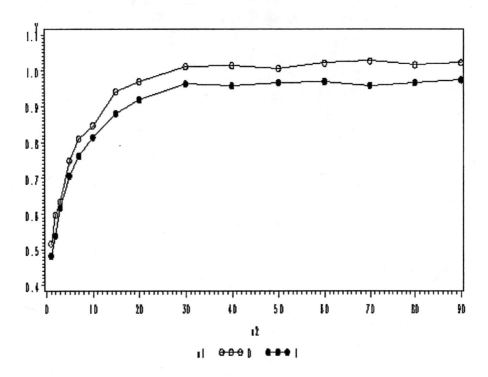

SAS Output 13.8

Iter	g0	g1	g2	g3	Sum of Squares
0	1.0250	-0.0459	-0.1220	-0.5000	0.0160
1	1.0157	-0.0473	-0.1343	-0.5505	0.00330
2	1.0156	-0.0473	-0.1348	-0.5524	0.00329
3	1.0156	-0.0473	-0.1348	-0.5524	0.00329
4	1.0156	-0.0473	-0.1348	-0.5524	0.00329

NOTE: Convergence criterion met.

Estimation Summary	
Method	Gauss-Newton
Iterations	4
R	6.969E-6
PPC(g2)	1.149E-6
RPC(g2)	0.000017

145

Estimation Summary	
Object	1.08E-8
Objective	0.003293
Observations Read	30
Observations Used	30
Observations Missing	0

Source	DF	Sum of Squares	Mean Square	F Value	Approx Pr > F
Regression	4	22.8541	5.7135	2272.64	<.0001
Residual	26	0.00329	0.000127		
Uncorrected Total	30	22.8574			
Corrected Total	29	0.8667			

Parameter	Estimate	Approx Std Error	Approximate 95% Confidence Limits	
g0	1.0156	0.00367	1.0081	1.0231
g1	-0.0473	0.00411	-0.0557	-0.0388
g2	-0.1348	0.00436	-0.1438	-0.1258
g3	-0.5524	0.00816	-0.5692	-0.5357

Approximate Correlation Matrix				
	g0	g1	g2	g3
g0	1.0000000	-0.5595744	0.4112806	-0.1441959
g1	-0.5595744	1.0000000	-0.0000000	-0.0000000
g2	0.4112806	-0.0000000	1.0000000	0.5603318
g3	-0.1441959	-0.0000000	0.5603318	1.0000000

Neural Networks

On text page 537, the authors introduce Neural Network Modeling. To our knowledge, SAS/STAT has no readily available procedure for generating neural networks.

Logistic Regression, Poisson Regression, And Generalized Linear Models

In this chapter, the authors consider nonlinear regression models for dependent variables that are categorical. In the simplest case, the dependent variable is binary and has two possible outcomes. Examples of binary outcome categories are success versus failure, disease absent versus disease present, and good performance versus bad performance. At the next level, the dependent variable is multinomial and has more than two possible outcomes. Examples of multinomial outcome categories are complete success, partial success, failure; no disease, minimum disease, moderate disease, severe disease; and performance on an increasing scale of 1 to 5. Lastly, the dependent variable may be a count of the number of events where large counts are rare and thought to be distributed as a Poisson random variable. Examples include counts of the number of accidents at a particular intersection, the number of people with a rare disease, and the number of sales.

One type of nonlinear regression model for categorical random variables is known as the logistic regression model. If the dependent variable is binary, we refer to the model as a logistic regression model whereas, if it is multinomial, we refer to the model as a polynomial or polytomous logistic regression model. If the dependent variable has a Poisson distribution, we refer to the nonlinear regression model as the Poisson regression model. This chapter of the text focuses on the statistical analysis of data based on the logistic model but also gives the essentials for the polytomous logistic regression model and the Poisson regression model.

Linear and nonlinear models considered in the text, including the three models mentioned above, belong to a family of models called *generalized linear models*. The salient feature of generalized linear models is that the initial model formulation is the same for each application with the exception of the specification of a link function. Thus, a unified approach to fitting regression models is provided. A concise description of generalized linear models is given on text pages 623-624.

Simple Logistic Regression

The Programming Task Example

Data File: ch14ta01.txt

An analyst studied the relationship between computer programming experience and success in completing a complex programming task within a specified time period (text page 565). That is, the response variable was coded $Y=1$ if the task was successfully completed and $Y=0$ if the task was not successfully completed. The predictor variable was the number of months of previous programming experience. See Chapter 0, "Working with SAS," for instructions on opening the Programming Task example data file (Table 14.1, text page 566) and naming the variables X, Y, and *FITTED*. After opening ch14ta01.txt, the first 5 lines of output should look like SAS Output 11.1.

SAS Output 11.1

Obs	x	y	fitted
1	14	0	0.31026
2	29	0	0.83526
3	6	0	0.11000
4	25	1	0.72660
5	18	1	0.46184

We begin the preliminary analysis by plotting Y against X values and overlaying a curve representing the fitted values plotted against X. (Insert 14.1 - SAS 14.2a) In a second graph, we plot the Y against X values and overlay a Lowess curve. (Insert 14.1 - SAS Output 14.2b) The plots suggest the underlying model is a sigmoidal S-shaped function. Accordingly, the analyst decided to use logistic regression to analyze the data.

Continuing in Insert 14.1, we issue the *proc logistic* statement with "*model y = x / cl*" to perform a logistic regression analysis using Y as the response variable and X as the single predictor variable. The *cl* option requests confidence intervals for the parameters of the model. We specify "*output out = estimates p = pie*" to save the predicted values (fitted values) to an output file named *estimates* under a variable named *pie*. Finally, we issue a *proc print data = estimates* to print the fitted estimates as shown in Table 14.1, column (3), text page 566. From the results shown in SAS Output 14.3, we see that the estimate of the model intercept is $b_0 = -3.0597$ and the estimate of the coefficient for X is $b_1 = 0.1615$. Further, the estimate of the odds ratio is 1.175. Thus, we estimate that the odds of successfully completing the task increase by 17.5% with each additional month of experience. These results are consistent with those given on text page 567. The printout of the fitted values (not shown) indicates that they are also consistent those reported in Table 14.1 of the text.

In some applications, the response variable is coded as 0 for non-event and 1 for event, and in other applications the coding is reversed. This coding is important because, by default, *proc logistic* models the probability of response levels of the lowest ordered value, i.e., response = 0, the non-event, relative to the highest ordered value, i.e., response = 1, the event. There are several ways to reverse this order of modeling. One is to code the response 0 for an event and 1 for a non-event but this is counterintuitive. Another option is to choose either the *ascending* or *descending* option on the *proc logistic* statement. The appropriate choice of the *ascending* or *descending* option depends on how the response variable (Y) is coded in the data file. We must be careful in using *proc logistic* to correctly use the ordering option and interpret the results accordingly.

In Insert 14.1, we specified the *descending* option for the logistic regression procedure and obtained 1.175 as an estimate of the odds ratio. If we executed the procedure without the *descending* option, the resulting odds ratio would be 0.851. Given what we now know about the *ascending* and *descending* options with *proc reg*, we know that $1/0.851 = 1.175$ so we can use these options to estimate odds ratios according to our interest and the problem at hand. Since the event (task completed successfully) is coded internally as 0, the odds ratio obtained from *proc reg* estimates the odds that the task is completed successfully. Although it does not matter mathematically how the event or non-event is coded, it is less confusing to have the event we want to model (task completed successfully rather than unsuccessfully) in the numerator. Thus, in Insert 14.1, we include the *descending* option on the *proc logistic* statement to override the default.

INSERT 14.1

```
data program;
  infile 'c:\alrm\chapter 14 data sets\ch14ta01.txt';
  input x y fitted;
run;
*Produce plot of y vs x and fitted vs x shown in SAS Output 14.2 (a);
data graphs;
  set program;
proc sort;
  by fitted;
proc gplot;
  symbol1 color=black value=none interpol=join ;
  symbol2 color=black value=circle;
  title 'Scatter Plot fitted vs months';
  plot fitted*x y*x / overlay haxis=0 to 40 by 10 vaxis=0 to 1 by .5;
run;
*Produce Lowess Curve;
proc loess;
  model y = x / smooth = 0.4 to 0.8 by 0.1 residual;
  ods output OutputStatistics = Results;
run;
*Produce plot of y vs x and Lowess pred vs x shown in SAS Output 14.2 (b);
proc sort data = results;
 by smoothingparameter pred;
 run;
goptions reset=all;
proc gplot;
  by SmoothingParameter;
  symbol1 color=black value=none interpol=join ;
  symbol2 color=black value=circle;
   title 'Scatter Plot and Lowess Curve';
  plot Pred*x DepVar*x / overlay haxis=0 to 40 by 10 vaxis=0 to 1 by .5;
run;
*Run logistic regression analysis;
proc logistic  data = program descending;
 model y = x /  cl ;
 output out = estimates p = pie;
proc print data = estimates;
run;
```

SAS Output 14.2

(a)

(a)   **(b)**

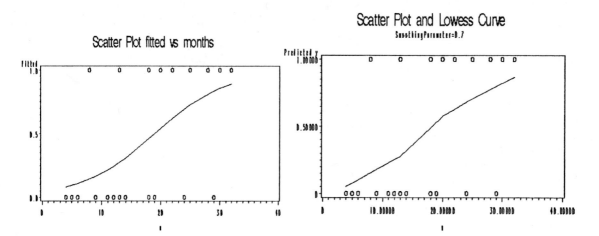

SAS Output 14.3

Analysis of Maximum Likelihood Estimates					
Parameter	DF	Estimate	Standard Error	Chi-Square	Pr > ChiSq
Intercept	1	-3.0597	1.2594	5.9029	0.0151
x	1	0.1615	0.0650	6.1760	0.0129

Odds Ratio Estimates			
Effect	Point Estimate	95% Wald Confidence Limits	
x	1.175	1.035	1.335

Wald Confidence Interval for Parameters			
Parameter	Estimate	95% Confidence Limits	
Intercept	-3.0597	-5.5280	-0.5914
x	0.1615	0.0341	0.2888

The Coupon Effectiveness Example

Data File: ch14ta02.txt

Marketing researchers mailed packets containing promotional material and related coupons to 1000 selected homes. Two hundred homes each received one of 5 different coupons that offered price reductions of 5, 10, 15, 20, or 25 dollars. Thus, the predictor variable was the amount of price reduction and the response variable (Y) was coded to indicate whether or not the home redeemed the coupon within a six-month period. See Chapter 0, "Working with SAS," for instructions on opening the Coupon Effectiveness example data file (Table 14.2, text page 569) and naming the variables X, n, Y, and *pro*. After opening ch14ta02.txt, the 5 lines of output should look like SAS Output 14.4.

SAS Output 14.4

Obs	x	n	y	pro
1	5	200	30	0.150
2	10	200	55	0.275
3	15	200	70	0.350
4	20	200	100	0.500
5	30	200	137	0.685

In Insert 14.2, we input the data in the form of Table 14.2, text page 569 (SAS Output 14.4). Next, we issue the *proc logistic* statement and specify *model y/n = x*. Here, Y is the number of coupons redeemed and n is the number of households that received a coupon with price reduction X. Thus, y/n is the proportion of coupons redeemed at each price reduction x. Execution of the logistic procedure with this model produces SAS Output 14.5 with $b_0 = -2.0443$, $b_1 = 0.0968$, and odds ratio = 1.102 as shown on text page 570.

Following the model statement in Insert 14.2, we specify "*output out = estimates p = pie*." This requests estimates of the predicted values to be stored in a file named *estimates* under the variable name *pie*. We then specify "*proc print data=estimates*" to print (SAS Output 14.5) the model based estimates in Column (5) of Table 14.2, text page 569.

INSERT 14.2

```
data Program;
  infile 'c:\alrm\chapter 14 data sets\ch14ta02.txt';
  input x n y pro;
proc logistic;
  model y/n = x;
output out = estimates p = pie;
run;
proc print data = estimates;
run;
```

Analysis of Maximum Likelihood Estimates					
Parameter	DF	Estimate	Standard Error	Chi-Square	Pr > ChiSq
Intercept	1	-2.0443	0.1610	161.2794	<.0001
x	1	0.0968	0.00855	128.2924	<.0001

Odds Ratio Estimates		
Effect	Point Estimate	95% Wald Confidence Limits
x	1.102	1.083 1.120

SAS Output 14.5 continued – Model Based Estimates

Obs	x	n	y	pro	pie
1	5	200	30	0.150	0.17362
2	10	200	55	0.275	0.25426
3	15	200	70	0.350	0.35621
4	20	200	100	0.500	0.47311
5	30	200	137	0.685	0.70280

Multiple Logistic Regression

The Disease Outbreak Example – Model Building

Data File: ch14ta03.txt and appenc10.txt

Note: Data for the Disease Outbreak example are available in two files: ch14ta03.txt and appenc10.txt. There are differences in the files that we will identify. For all analyses involving the Disease Outbreak example, we must pay close attention to the data file we use, i.e., ch14ta03.txt versus appenc10.txt. We will be unable to match the chapter results if we use the wrong file.

Using ch14ta03.txt, a researcher studied the relationship between disease outbreak ($Y=1$ if disease was present and $Y=0$ if disease was not present) and the individual's age ($X1$), socioeconomic status (represented by indicator variables $X2$ and $X3$), and the sector in which the individual lived ($X4$ coded 0 or 1). See Chapter 0, "Working with SAS," for instructions on opening the Disease Outbreak example data file (Table 14.3, text page 574) and naming the variables *case, X1, X2, X3, X4, and Y*. After opening ch14ta03.txt, the first 5 lines of output should look like SAS Output 14.6a.

SAS Output 14.6 (a)

Obs	case	x1	x2	x3	x4	y
1	1	33	0	0	0	0
2	2	35	0	0	0	0
3	3	6	0	0	0	0
4	4	60	0	0	0	0
5	5	18	0	1	0	1

Similar to the steps for requesting the logistic regression analysis in the previous example (Insert 14.1), for the Disease Outbreak example we enter *Y* as the dependent variable and *X1*, *X2*, *X3*, and *X4* as covariates. Partial results are shown in SAS Output 14.7(a). The regression function is consistent with equation 14.46, text page 574. The odds ratio (listed under *Point Estimate*) for *X1*=1.03, indicating that, because age is coded in units of years, the risk of disease increases by 1.03 or approximately 3% for each additional year of age. The odds ratio for *X4*=4.83, indicating that individuals living in sector 2 are 4.829 times more likely to have contracted the disease compared to individuals living in sector 1.

As noted above, data file ch14ta03.txt comprises the first 98 cases of a sample of 196 cases used in the Disease Outbreak example. The entire 196 cases are available in the data file appenc10.txt. After opening appenc10.txt and naming the variables *case, age, status, sector, y,* and *other*, the first 10 lines of output should look like SAS Output 14.6(b). In this file, socioeconomic status is represented by one variable (*status*) where *status* = 1 if status is "upper," *status* = 2 if status is "middle," and *status* = 3 if status is "lower." The sector is coded *sector* = 1 if the individual lives in sector 1 and *sector* = 2 if the individual lives in sector 2.

SAS Output 14.6 (b)

Obs	case	age	status	sector	y	other
1	1	33	1	1	0	1
2	2	35	1	1	0	1
3	3	6	1	1	0	0
4	4	60	1	1	0	1
5	5	18	3	1	1	0
6	6	26	3	1	0	0
7	7	6	3	1	0	0
8	8	31	2	1	1	1
9	9	26	2	1	1	0
10	10	37	2	1	0	0

In ch14ta03.txt, two indicator variables (*X2*, *X3*) were used to represent the 3 levels of socioeconomic status. We illustrate the use *status* as a *class* variable to represent socioeconomic status. However, SAS uses the larger of the coded values as the reference category, and our desired reference category (upper) is coded 1. Accordingly, we compute a new variable named *newstat* where we simply reverse the coding of *status* so that upper socioeconomic status (*newstat*=3) will now be the reference category. Using this coding scheme we can enter *age*, *newstat*, and *sector*, as predictor variables and specify *newstat* as a *class* variable in the *proc logistic* procedure (Insert 14.3).

For each variable specified as a *class* variable, SAS will automatically code a set of indicator variables. Thus, we replaced *X2* and *X3* in the analysis that produced the partial output shown in SAS Output 14.7(a) with *newstat*. We also replaced *X1* with *age*, although these two variables are equal, and replaced *X4* with *sector*. This results in SAS Output 14.7(b). So, SAS Output 14.7(a) is a result of using *X1, X2, X3,* and *X4* as predictors from the ch14ta03.txt data set. On the other hand, the partial output shown in SAS Output 14.7(b) is a result of using *age*, *newstat*, and *sector* as predictors from the appenc10.txt data set (and using *where case* < 99). Although the variables used in the two models were coded differently, they contain the same "information." Thus, the two analyses produce identical the odd ratios. The coding scheme used to produce SAS Output 14.7(b) also yields an overall test statistic that is indicative of lack of significance for the *newstat* variable, $p = 0.547$.

SAS Output 14.7

(a)

	Odds Ratio Estimates		
Effect	Point Estimate	95% Wald Confidence Limits	
x1	1.030	1.003	1.058
x2	1.505	0.465	4.868
x3	0.737	0.226	2.408
x4	4.829	1.807	12.907

(b)

	Odds Ratio Estimates		
Effect	Point Estimate	95% Wald Confidence Limits	
age	1.030	1.003	1.058
newstat 1 vs 3	0.737	0.226	2.408
newstat 2 vs 3	1.505	0.465	4.868
sector	4.829	1.807	12.907

154

The Initial Public Offerings Example

Data File: appenc11.txt

A researcher studies the relationship between whether or not a company was financed by venture capital (*VC*) and the face value (*facval*) of the company (text page 575). See Chapter 0, "Working with SAS," for instructions on opening the IPO example data file and naming the variables *case, VC, FACEVAL, SHARES,* and *X3*. After opening appenc11.txt, the first 5 lines of output should look like SAS Output 14.8.

SAS Output 14.8

Obs	case	vc	faceval	shares	x3
1	1	0	1200000	3000000	0
2	2	0	1454000	1454000	1
3	3	0	1500000	300000	0
4	4	0	1530000	510000	0
5	5	0	2000000	800000	0

In Insert 14.4, we find *LNFACE*, the natural logarithm of *FACEVAL*. We use *proc loess* and *proc gplot* to generate a scatter plot of *VC* and *LNFACE* and overlay a Lowess curve [SAS Output 14.9(a)]. We then regress *VC* against *LNVACE* and save the fitted values (*linpie*) to a file named *linear*. In the next *proc plot* statement we use *linpie* and other values as shown to generate the first order logistic curve [SAS Output 14.9 (b)]. SAS Output 14.9 (a) and (b) display the Lowess curve and the first order logistic curve, respectively, as shown on Figure 14.9 (a), text page 575. The mound shaped relationship of SAS Output 14.9 (a) was not adequately depicted by the first order regression fit of SAS Output 14.9 (b). This indicated that a first-order logistic regression would not be appropriate; the authors decided to explore a second-order model using *LNFACE* (centered) and *LNFACE* (centered) squared as predictors. Continuing in Insert 14.4, we center *LNFACE* (*xcnt*), find its square (*xcnt2*), and regress *VC* on *xcnt* and *xcnt2*. Partial results of this model are shown in SAS Output 14.10 and are consistent with Table 14.5, text page 576. Finally, we use *proc gplot* to generate a scatter plot of the estimated probabilities (*pie*) against *xcnt* with an overlay of *VC* against *xcnt* [SAS Output 14.9 (c)]. This second order logistic curve is consistent with Figure 14.9 (b), text page 575.

INSERT 14.4	INSERT 14.4 continued

```data ipo;```
``` infile 'c:\alrm\appendix c data sets\appenc11.txt';```
``` input case vc faceval shares x3;```
```lnface = LOG(faceval);```
```*Run lowess analysis;```
```proc loess;```
``` model vc = lnface / smooth = 0.5 residual;```
``` ods output OutputStatistics = Results;```
```run;```
```*Produce scatter plot and lowess curve;```
```data graph1;```
``` set results;```
```proc sort;```
``` by lnface;```
``` run;```
```goptions reset=all;```
```proc gplot;```
``` by SmoothingParameter;```
``` symbol1 color=black value=none interpol=join;```
``` symbol2 color=black value=circle;```
```   title 'Scatter Plot and Lowess Curve';```
``` plot Pred*lnface DepVar*lnface / overlay ;```
```run;```
```*Run 1st order logistic regression analysis;```
```proc logistic data = ipo descending;```
``` model vc = lnface;```
``` output out = linear p = linpie;```
```run;```
```*Produce scatterplot and fitted 1st order logistic;```
```data graph2;```
``` set linear;```
```proc sort;```
``` by lnface;```
	```proc gplot;```
	``` symbol1 color=black value=none interpol=join ;```
	``` symbol2 color=black value=circle;```
	``` title 'Scatter Plot and 1st Order Logit Curve';```
	``` plot linpie*lnface vc*lnface / overlay;```
	```run;```
	```* Find mean of lnface=16.7088 ;```
	```proc means;```
	``` var lnface;```
	```run;```
	```*Run 2nd order logistic regression analysis;```
	```data step2;```
	``` set linear;```
	```xcnt = 16.7088 - lnface ;```
	```xcnt2 = xcnt ** 2 ;```
	```proc logistic descending;```
	``` model vc = xcnt xcnt2/ cl;```
	``` output out = estimates p = pie;```
	```run;```
	```*Produce scatter plot and fitted 2nd order logistic;```
	```data graph;```
	``` set estimates;```
	```proc sort;```
	``` by xcnt;```
	``` proc gplot;```
	``` symbol1 color=black value=none interpol=join ;```
	``` symbol2 color=black value=circle;```
	``` title 'Scatter Plot and 2nd Order Logit Curve';```
	``` plot pie*xcnt vc*xcnt / overlay ;```
	```run;```

SAS Output 14.9

(a)   **(b)**

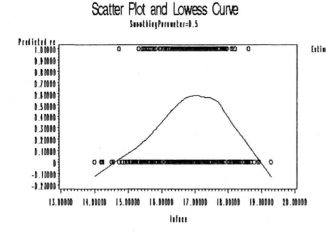

(c)

Scatter Plot and 2nd Order Logit Curve

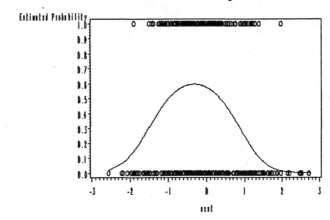

SAS Output 14.10

Analysis of Maximum Likelihood Estimates					
Parameter	DF	Estimate	Standard Error	Chi-Square	Pr > ChiSq
Intercept	1	0.3004	0.1240	5.8741	0.0154
xcnt	1	-0.5517	0.1385	15.8718	<.0001
xcnt2	1	-0.8615	0.1404	37.6504	<.0001

Inferences about Regression Parameters: **The Programming Task Example**

On text page 578, the authors wish to test $H_0 : \beta_1 \leq 0$ versus $H_a : \beta_1 > 0$. The test for β_1 in SAS Output 14.3 is a two-sided test. For a one-sided test, we simply divide the probability level (0.0129 listed under *Pr > ChiSq* and across from *x*) by 2. This yields the one-sided probability level of 0.0065 as noted on text page 579. The text authors report $z^* = 2.485$ and the square of *z* is equal to the Wald statistic, which is distributed approximately as chi-square distribution with 1 degree of freedom. Thus, the Wald statistic is $X^2 = 2.485^2 = 6.176$ as shown in SAS Output 14.3.

The confidence interval for β_1 shown in SAS Output 14.3 is consistent with text page 579. This is a result of the *cl* option on the *proc logistic* statement in Insert 14.1. We know from SAS Output 14.3 that $b_1 = 0.161$ and $s\{b_1\} = 0.065$. For a 95% confidence interval, we require $z = 1.96$. We compute the 95% confidence interval directly from these results as follows:

$$0.161 \pm 1.96(0.065)$$

or

$$0.034 \leq \beta_1 \leq 0.288.$$

The Disease Outbreak Example

Data File: ch14ta03.txt

On text page 581, the researchers now wish to determine if *X1* can be dropped from the model in the Disease Outbreak Example. We use *proc logistic* (syntax not shown) to regress *Y* on *X2*, *X3*, and *X4* and refer to this as the reduced model (reduced because *X1* is no longer in the model). In SAS Output 14.11, we see that –2 Log Likelihood statistic = 106.204. We now regress *Y* on *X1*, *X2*, *X3*, and *X4* and refer to this as the full model. From SAS Output 14.11,the –2 Log Likelihood statistic = 101.054. Using equation 14.60, text page 581, we find $G^2 = 106.204 - 101.054 = 5.15$. For $\alpha = 0.05$ we require $X^2(.95;1) = 3.84$. Since our computed G^2 value (5.15) is greater than the critical value (3.84) we conclude H_a, that *X1* should not be dropped from the model.

SAS Output 14.11

Reduced Model

Model Fit Statistics		
Criterion	Intercept Only	Intercept and Covariates
AIC	124.318	114.204
SC	126.903	124.544
-2 Log L	122.318	106.204

Full Model

Model Fit Statistics		
Criterion	Intercept Only	Intercept and Covariates
AIC	124.318	111.054
SC	126.903	123.979
-2 Log L	122.318	101.054

Automatic Model Selection Methods: The Disease Outbreak Example

Data File: appenc10.txt

Best Subsets Procedure. SAS' *proc logistic* offers a best subset procedure. We specify on the *proc logistic* statement that *n* models with the highest score chi-square statistics for each model

158

size. The *best* option is used exclusively with *selection=score*. For example, we can specify that the 2 models with the highest score chi-square statistics be displayed for models with one, two, three, and four predictor variables, e.g., "*model y = x1 x2 x3 x4 / selection=score best=2;*". Additionally, the $-2\log_e L(\mathbf{b})$, AIC_p and SBC_p statistics are available for each *proc logistic* analysis but are not displayed as part of the best subset procedure.

Stepwise Model Selection. SAS' *proc logistic* offers stepwise selection, forward selection, and backward selection. The methods generally follow the description of selection procedures described for multiple linear regression in Chapter 9.

From the appenc10.txt data file, we create four new variables to match the coding scheme of ch14ta03.txt and we select the first 98 cases (Insert 14.5). That is, socioeconomic status will be represented by two indicator variables: $X2 = 1$ if class is middle, otherwise $X2 = 0$, and $X3=1$ if class is lower, otherwise $X3 = 0$. Thus, upper socioeconomic status is the reference class. Also, $X4 = 0$ if the individual lives in sector 1 and $X4=1$ if the individual lives in sector 2. Finally, to use the same variable names as the text, we set $X1=age$.

We present the syntax to perform forward stepwise selection in Insert 14.5. Partial results are shown in SAS Output 14.12. At Step 0, before any variables are entered into the model, $X4$ has the lowest probability level of the 4 variables under consideration and, thus, is entered into the model at Step 1. Of the variables not in the model after Step 1, $X1$ has the lowest probability level of the three candidate variables and is entered into the model at Step 2. Of the variables not in the model after Step 2, $X2$ and $X3$ both have probability levels greater than the probability level required for entering a variable into the model and the stepping process stops.

Although we arrive at the same conclusion as the text, the *constant* values in SAS Output 14.12 (Analysis of Maximum Likelihood Estimates) do not match those displayed in Table 14.11, text page 585. In the model specified in Insert 14.5, we used $X1$, $X2$, $X3$, and $X4$ as predictor variables. The text example used *age*, *status*, and *sector* as predictors. Both sets of predictors contain the same "information" but socioeconomic status and city sector are parameterized differently in the two models. Thus, although we arrive at the same conclusion regarding the regression coefficients, the intercepts differ due to the different parameterization.

INSERT 14.5

```
data disease;
  infile 'c:\alrm\appendix c data sets\appenc10.txt';
  input case age status sector y other;
x1=age;
if status eq 2 then x2=1; else x2=0;
if status eq 3 then x3=1; else x3=0;
if sector eq 2 then x4=1; else x4=0;
proc logistic descending;
  where case < 99;
  model y = x1 x2 x3 x4 / selection=forward detail;
run;
```

SAS Output 14.12 Step 0 – Intercept only

Analysis of Effects Not in the Model			
Effect	DF	Score Chi-Square	Pr > ChiSq
x2	1	1.4797	0.2238
x3	1	3.9087	0.0480
x4	1	14.7805	0.0001
x1	1	7.5802	0.0059

After Step 1

Analysis of Effects Not in the Model			
Effect	DF	Score Chi-Square	Pr > ChiSq
x2	1	0.6638	0.4152
x3	1	1.1549	0.2825
x1	1	5.2385	0.0221

After Step 2

Analysis of Effects Not in the Model			
Effect	DF	Score Chi-Square	Pr > ChiSq
x2	1	0.9681	0.3252
x3	1	0.7332	0.3919

Summary of Variables in Final Model

Analysis of Maximum Likelihood Estimates					
Parameter	DF	Estimate	Standard Error	Chi-Square	Pr > ChiSq
Intercept	1	-2.3350	0.5111	20.8713	<.0001
x4	1	1.6734	0.4873	11.7906	0.0006
x1	1	0.0293	0.0132	4.9455	0.0262

Tests for Goodness of Fit

Pearson Chi-Square Goodness of Fit Test. Using the Coupon Effectiveness example (ch14ta02.txt), the authors present 3 goodness of fit measures (text page 586) including the *Pearson Chi-Square Goodness of Fit*. In Insert 14.6, we open the Coupon Effectiveness data file. To request the *Pearson Chi-Square Goodness of Fit* test, we use the *aggregate* and *scale* options on the *model* statement, i.e.," *model y/n = x / aggregate=(x) scale=pearson;*". This produces SAS Output 14.13. The Pearson Chi-Square test statistic = 2.1486, consistent with the value (X^2=2.15) reported on text page 588.

Deviance Goodness of Fit Test. SAS will compute the *Deviance Goodness of Fit* statistic as described above. Notice from SAS Output 14.13 that the Deviance test statistic = 2.1668, consistent with the value [$DEV(X_0, X_1)$=2.16] reported on text page 586. This statistic can also be obtained as described in the text. As noted on text page 588, the *deviance goodness of fit test* for logistic regression is analogous to the *F lack of fit test* discussed in previous chapters. We present the steps without the specific syntax. Find –2 log likelihood (1191.973) for the reduced model by regressing Y on X using the Coupon Effectiveness example described above. To find –2 log likelihood for the saturated model, we first compute j=1 if X=5, j=2 if X=10, … , j=5 if X=30. Now use *proc logistic* to regress Y on X and j, and specify j as a *class* variable. This results in –2 log likelihood for the saturated model = 1189.806. The difference in the 2 values is 2.167, consistent with text page 589 where $DEV(X_0, X_1) = 2.16$.

Hosmer-Lemeshow Goodness of Fit Test. On text page 590, the authors calculated the Hosmer-Lemeshow goodness of fit test for the Disease Outbreak example. They based the calculation on 5 classes. SAS, however, uses approximately 10 classes based on the percentiles of the estimated probabilities. To generate the Hosmer-Lemeshow goodness of fit statistic, we use the *lackfit* option on the *proc logistic* model statement, e.g., *model y = x1 x2 x3 x4 / lackfit;*. Partial results are found in SAS Output 14.14. Note that $X^2 = 9.180$ ($p = 0.327$) and the text reports $X^2 = 1.98$ ($p = 0.58$). Although the difference in magnitudes of the test statistics is large, the p-values lead to the same conclusion. The discrepancy is due to the test statistics being based on different contingency tables, i.e., 2x5 versus 2x10. The use of 20 cells (2x10) decreases the expected frequencies per cell relative to the use of 10 cells (2x5). However, Hosmer and Lemeshow take a liberal approach to expected values and argue that their method is an accurate assessment of goodness of fit.

INSERT 14.6

```
data Program;
  infile 'c:\alrm\chapter 14 data sets\ch14ta02.txt';
  input x n y pro;
proc logistic;
  model y/n = x / aggregate=(x) scale=pearson;
run;
```

Deviance and Pearson Goodness-of-Fit Statistics				
Criterion	**DF**	**Value**	**Value/DF**	**Pr > ChiSq**
Deviance	3	2.1668	0.7223	0.5385
Pearson	3	2.1486	0.7162	0.5421

SAS Output 14.14

Hosmer and Lemeshow Goodness-of-Fit Test		
Chi-Square	**DF**	**Pr > ChiSq**
9.1871	8	0.3268

Logistic Regression Diagnostics and Detection of Influential Observations

Logistic Regression Residuals and Influence on Pearson Chi-Square and the Deviance Statistics. On text page 591, the authors demonstrate the calculation of various residuals and other statistics for the Disease Outbreak example. We illustrate the steps needed to produce the statistics presented in Tables 14.9 and 14.11 (text pages 593 and 599, respectively). In Insert 14.7, we use the first 98 cases of the Disease Outbreak data file as given in appenc10.txt and compute *X1*, *X2*, *X3*, and *X4* as before. In *proc logistic*, the diagonal elements of the hat matrix (leverage values) are available only through the *ods* output system. Thus, we specify *influence* on the *model* statement. This creates an *ods* output table named *influence*. In the *ods output* statement we put the contents of the *influence* table into a file named *results2*. By default, the diagonal elements of the hat matrix are stored in a variable named *hatdiag*. In the *output* statement immediately following the *model* statement we set *yhat* = *p* (estimated probability, $\hat{\pi}_i$), *chires* = *reschi* (Pearson residual, r_{Pi}), *devres* = *resdev* (deviance residual, dev_i), *difchsq* = *difchisq* (delta chi-square, ΔX_i^2), and *difdev* = *difdev* (delta deviance, Δdev_i). (Here, *p*, *reschi*, *resdev*, *difchisq*, and *difdev* are SAS key words for the respective variables.) We then *merge results1* and *results2* into one file named *both* so that the data for all variables will be assessable in one data file. We then use equation 14.81 (text page 592) to find *rsp* (*studentized Pearson residual*, r_{SPi}) and equation 14.87 (text page 600) to find *cookd* (*Cook's Distance*, D_i). Printed output (first 5 lines) is shown in SAS Output 14.15. These values are consistent with Tables 14.9 and 14.11 (text pages 593 and 599, respectively). These values can be used to generate various plots as displayed in the text. For example, we can generate Figure 14.15 (d), text page 600, by using *proc gplot* to *plot difdev*yhat;*.

```
data disease;
  infile 'c:\alrm\appendix c data sets\appenc10.txt';
  input case age status sector y other;
x1=age;
if status eq 2 then x2=1; else x2=0;
if status eq 3 then x3=1; else x3=0;
if sector eq 2 then x4=1; else x4=0;
proc logistic descending;
 where case < 99;
 model y = x1 x2 x3 x4 / influence ;
 output out = results1 p = yhat reschi=chires resdev=devres
   difchisq=difchisq difdev=difdev;
 ods output influence=results2 ;
 run;
data both;
 merge results1 results2;
 rsp=chires/(sqrt(1-hatdiag));
 cookd=((chires**2)*hatdiag)/(5*(1-hatdiag)**2);
proc print;
 var yhat chires rsp hatdiag devres difchisq difdev cookd;
run;
```

SAS Output 14.15

Obs	yhat	chires	rsp	HatDiag	devres	difchisq	difdev	cookd
1	0.20898	-0.51399	-0.52425	0.0387	-0.68473	0.2748	0.4795	0.002215
2	0.21898	-0.52951	-0.54054	0.0404	-0.70308	0.2922	0.5061	0.002461
3	0.10581	-0.34400	-0.34985	0.0332	-0.47295	0.1224	0.2277	0.000841
4	0.37100	-0.76800	-0.80486	0.0895	-0.96293	0.6478	0.9852	0.012737
5	0.11082	2.83262	2.86899	0.0252	2.09755	8.2311	4.6070	0.042535

Inferences about Mean Response

On text page 603, the authors wish to calculate the 95% confidence interval for a new observation. Completion of this task requires that we find $s^2\{\mathbf{b}\}$ for use in equation 14.92, text page 602. To generate this variance-covariance matrix, we use the *covb* option on the *proc logistic* model statement, e.g., *model y = x1 x2 x3 x4 / covb;*. We could now use SAS *iml* matrix language to find $\mathbf{X}_h' s^2\{\mathbf{b}\} \mathbf{X}_h$, where $\mathbf{X}_h$ is given on text page 603.

Prediction of a New Observation: The Disease Outbreak Example

Data File: appenc10.txt

We use the first 96 cases of appenc10.txt for this analysis. The authors wish to obtain the best cutoff point for predicting whether the outcome of a new observation will be "1" or "0". If 0.5 is used as the cutoff point, for example, the prediction rule would be: if $\hat{\pi}_h \geq 0.5$, predict the outcome will be "1"; otherwise predict it will be "0". Table 14.12, text page 605, displays the classification table based on text equation 14.95. That is, because 0.316 is the proportion of the diseased individuals in the sample, we use 0.316 as the starting cutoff point.

In Insert 14.8, we save the predicted probability in a variable named *yhat* (*p=yhat*) and then set *newgp*=1 if *yhat* is greater than or equal to 0.316, otherwise *newgp*=0. We then select the first 98 cases from the Disease Outbreak data file and form a 2x2 table of *Y* values (diseased versus non-diseased) by *newgp* values (predicted diseased versus predicted non-diseased). This produces SAS Output 14.16. We could produce text Table 14.12 (b) by changing the value in the *if* statement from 0.316 to 0.325 in Insert 14.8.

```
INSERT 14.8


data disease;
  infile 'c:\alrm\appendix c data sets\appenc10.txt';
  input case age status sector y other;
x1=age;
if status eq 2 then x2=1; else x2=0;
if status eq 3 then x3=1; else x3=0;
if sector eq 2 then x4=1; else x4=0;
proc logistic descending;
 where case < 99;
 model y = x1 x2 x3 x4 / outroc=roc ;
 output out = results p = yhat;
 run;
data results;
 set results;
 if yhat ge .316 then newgp=1; else newgp=0;
proc freq;
 where case < 99;
 table y*newgp / nopercent norow nocol;
 run;
proc gplot data=roc;
  symbol1 color=black value=none interpol=join ;
 plot _SENSIT_ * _1MSPEC_;
run;
```

SAS Output 14.16

Table of y by newgp			
y	**newgp**		
Frequency	**0**	**1**	**Total**
0	49	18	67
1	8	23	31
Total	57	41	98

There were 8+23=31 (false negatives + true positives) diseased individuals and 49+18=67 (true negatives + false positives) non-diseased individuals. The fitted regression function correctly predicted 23 of the 31 diseased individuals. Thus, the sensitivity is calculated as true positives /

(false negatives + true positives) = 23 / 8 + 23 = 0.7419 as reported on text page 606. The logistic regression function's specificity is defined as true negatives / (false positives + true negatives) = 48 / (18 + 48) = 0.727. This is not consistent with text page 607 where the specificity is reported as 47 / 67 = 0.7014. The difference comes from numerator used to calculate specificity. Table 14.12 (a), text page 605, uses 47 and we determined the number of true negatives to be 49. We assumed that the regression model was to be based on using *age, status,* and *sector* as predictors. If we use *age* and *sector* as suggested by the stepwise regression above, the classification table will match Table 14.12, text page 605.

To approximate Figure 14.17, text page 606, we use the *outroc* option on the *model* statement. This saves the data necessary to produce the ROC curve in a data file named *roc* (*outroc=roc*). We can view the contents of the *roc* data file by specifying *proc contents data=roc; run;*. Two of the variables stored in the *roc* data set are *_sensit_* (sensitivity) and *_1mspec_* (1-specificity). In the *proc gplot* we plot *_sensit_* against *_1mspec_* to form the ROC curve (SAS Output 14.17) presented in Figure 14.17, text page 606.

SAS Output 14.17

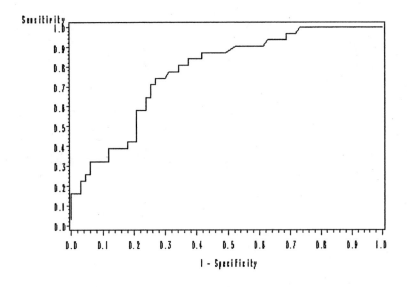

Validation of Prediction Error Rate:

The Disease Outbreak Example – Validation Data Set

Data File: appenc10.txt

On text page 607, the researchers now wish to use the regression function based on the Model-Building data set to classify observations in the Validation data set. After opening appenc10.txt and naming the variables *case, age, status, sector, y,* and *other,* the first 10 lines of output should look like SAS Output 14.18.

Obs	case	age	status	sector	y	other
1	1	33	1	1	0	1
2	2	35	1	1	0	1
3	3	6	1	1	0	0
4	4	60	1	1	0	1
5	5	18	3	1	1	0
6	6	26	3	1	0	0
7	7	6	3	1	0	0
8	8	31	2	1	1	1
9	9	26	2	1	1	0
10	10	37	2	1	0	0

In Inert 14.9 we first compute *X1, X2, X3*, and *X4* as before. We compute the estimated probabilities (*yhat*) for each case using the regression function (equation 14.46, text page 574) based on the model-building data set. We then use prediction rule 14.96 (text page 606) to establish group membership (*newgp*) where *newgp* = 1 if *yhat* ≥ 0.325 and *newgp* = 0 if *yhat* < 0.325. We also select on cases 99 to 196. Finally, we use the *proc freq* procedure to form a 2x2 table of *Y* values (diseased versus non-diseased) by *newgp* values (predicted diseased versus predicted non-diseased). Results are shown in SAS Output 14.19 and are consistent with text page 608. The off-diagonal percentages indicate that 12 (46.2%) of the 26 diseased individuals were incorrectly classified (prediction error) as non-diseased. Twenty-eight (38.9%) of the 72 non-diseased individuals were incorrectly classified as diseased.

INSERT 14.9

```
data disease;
  infile 'c:\alrm\appendix c data sets\appenc10.txt';
  input case age status sector y other;
x1=age;
if status eq 2 then x2=1; else x2=0;
if status eq 3 then x3=1; else x3=0;
if sector eq 2 then x4=1; else x4=0;
yhat=(1+exp(2.3129-(.02975*x1)-(.4088*x2)+(.30525*x3)-(1.5747*x4)))**-1;
if yhat ge .325 then newgp=1; else newgp=0;
proc freq;
  where case > 98;
  table y * newgp / nocol nopct;
run;
```

Table of y by newgp			
y	newgp		
Frequency Row Pct	0	1	Total
0	44 61.11	28 38.89	72
1	12 46.15	14 53.85	26
Total	56	42	98

Polyotomous Logistic Regression for Nominal and Ordinal Response:

Pregnancy Duration Example

Data File: ch14ta13.txt

On text page 609, a researcher studied the relationship between gestation period and several predictor variables. Gestation period was classified as less than 36 weeks ($Y=1$), 36 to 37 weeks ($Y=2$), and greater than 37 weeks ($Y=3$). After opening ch14ta13.txt and naming the variables *case, Y, RC1, RC2, RC3, X1, X2, X3, X4*, and *X5*, the first 5 lines of output should look like SAS Output 14.20. In Insert 14.10, we specify *proc logistic* with *X1, X2, X3, X4*, and *X5* as predictor variables; *X2, X3, X4*, and *X5* as *class* variables; and *Y* as the dependent variable. Since there are 3 levels of *Y*, SAS will assume that the variable is ordinal and perform Ordinal Logistic Regression. Results are shown in SAS Output 14.21 and are consistent with Figure 14.19, text page 618.

SAS Output 14.20

Obs	case	y	rc1	rc2	rc3	x1	x2	x3	x4	x5
1	1	1	1	0	0	150	0	0	0	1
2	2	1	1	0	0	124	1	0	0	0
3	3	1	1	0	0	128	0	0	0	1
4	4	1	1	0	0	128	1	0	0	1
5	5	1	1	0	0	133	0	0	1	1

INSERT 14.10

```
data Pregnancy;
 infile 'c:\alrm\chapter 14 data sets\ch14ta13.txt';
 input case y rc1 rc2 rc3 x1 x2 x3 x4 x5;
 x2=1-x2;
 x3=1-x3;
 x4=1-x4;
 x5=1-x5;
proc logistic ;
 class  x2 x3 x4 x5;
 model y = x1 x2 x3 x4 x5;
 run;
```

SAS Output 14.21

Testing Global Null Hypothesis: BETA=0			
Test	Chi-Square	DF	Pr > ChiSq
Likelihood Ratio	47.1740	5	<.0001
Score	36.1844	5	<.0001
Wald	32.3954	5	<.0001

Type III Analysis of Effects			
Effect	DF	Wald Chi-Square	Pr > ChiSq
x1	1	17.4958	<.0001
x2	1	11.3137	0.0008
x3	1	6.0426	0.0140
x4	1	12.4767	0.0004
x5	1	12.3683	0.0004

Analysis of Maximum Likelihood Estimates						
Parameter		DF	Estimate	Standard Error	Chi-Square	Pr > ChiSq
Intercept		1	6.2303	1.5826	15.4982	<.0001
Intercept2		1	8.3251	1.6838	24.4446	<.0001
x1		1	-0.0489	0.0117	17.4958	<.0001
x2	0	1	0.9880	0.2937	11.3137	0.0008
x3	0	1	0.6817	0.2773	6.0426	0.0140
x4	0	1	0.8349	0.2364	12.4767	0.0004
x5	0	1	0.7958	0.2263	12.3683	0.0004

Odds Ratio Estimates			
Effect	Point Estimate	95% Wald Confidence Limits	
x1	0.952	0.931	0.974
x2 0 vs 1	7.214	2.281	22.815
x3 0 vs 1	3.910	1.318	11.595
x4 0 vs 1	5.311	2.103	13.415
x5 0 vs 1	4.911	2.023	11.924

The steps to perform polyotomous logistic regression for an ordinal response completely generalize to the case where there is a nominal response. Use Insert 14.10 and add the *link=glogit* option, i.e., "*model y = x1 x2 x3 x4 x5 / link=glogit;*". The resulting syntax will produce the appropriate analysis for a multinomial response (output not shown). Note that this *link* option is available in Release 8.2 or higher. For users of earlier releases, SAS does not offer a direct solution if the response variable is polyotomous nominal and any of the predictor variables is continuous. If all predictor variables are nominal, *proc catmod* may be used.

Poisson Regression: Miller Lumber Company Example

Data File: ch14ta14.txt

The Miller Lumber Company conducted an in-store customer survey. The researcher counted the number of customers who visited the store from each nearby census tract. The researcher also collected and subsequently retained five (quantitative) predictor variables for use in the Poisson Regression. See Chapter 0, "Working with SAS," for instructions on opening the Miller Lumber Company data file and naming the *Y*, *X1*, *X2*, *X3*, *X4*, and *X5*. The first 5 lines of output should look like SAS Output 14.22. Table 14.14, text page 622, appears to indicate that the variables are arranged in the data file in the order *X1*, *X2*, *X3*, *X4*, *X5*, and *Y*. By printing the data set, we sometimes find that the order of the variables listed in the SAS *input* statement is not consistent with the order in data file.

SAS Output 14.22

Obs	y	x1	x2	x3	x4	x5
1	9	606	41393	3	3.04	6.32
2	6	641	23635	18	1.95	8.89
3	28	505	55475	27	6.54	2.05
4	11	866	64646	31	1.67	5.81
5	4	599	31972	7	0.72	8.11

Poisson regression is available in the *proc genmod* procedure (Insert 14.11). *Proc genmod* fits a generalized linear model to the data. There are a number of link functions and probability distributions that can be specified by the user. We complete the Miller Lumber analysis by specifying the Poisson distribution on the *model* statement. Results are shown in SAS Output 14.23 below and are consistent with Table 14.15, text page 622.

```
                          INSERT 14.11
data Miller;
  infile 'c:\alrm\chapter 14 data sets\ch14ta14.txt';
  input y x1 x2 x3 x4 x5;
proc genmod;
 model y = x1 x2 x3 x4 x5 /
  dist = poisson ;
run;
```

SAS Output 14.23

Analysis Of Parameter Estimates							
Parameter	DF	Estimate	Standard Error	Wald 95% Confidence Limits		Chi-Square	Pr > ChiSq
Intercept	1	2.9424	0.2072	2.5362	3.3486	201.57	<.0001
x1	1	0.0006	0.0001	0.0003	0.0009	18.17	<.0001
x2	1	-0.0000	0.0000	-0.0000	-0.0000	30.63	<.0001
x3	1	-0.0037	0.0018	-0.0072	-0.0002	4.37	0.0365
x4	1	0.1684	0.0258	0.1179	0.2189	42.70	<.0001
x5	1	-0.1288	0.0162	-0.1605	-0.0970	63.17	<.0001
Scale	0	1.0000	0.0000	1.0000	1.0000		

SPSS® Program Solutions

Chapter 0 SPSS®

Working with SPSS®

The purpose of this chapter is to introduce the requisite methods for entering, creating, and opening the data files provided with *Applied Linear Regression Models, 4th Edition* by Kutner, Nachtsheim and Neter (2003). Although we will work with many SPSS programs as we progress through *Applied Linear Regression Models,* the present chapter is not intended to be a comprehensive review of the SPSS capabilities.

SPSS is a modular, integrated, product line for the analytical process. The SPSS products are available using a simple point-and-click graphical interface and can be used for:

- Planning
- Data collection
- Data access and management
- Report writing
- Statistical analysis
- File handling

SPSS is available for most mainframe and personal computers running under the Windows® or Macintosh® operating systems. The modules required to perform the analyses in the forthcoming chapters are SPSS Base®, SPSS Advanced Models®, and SPSS Regression Models®.

Starting SPSS: Most compute systems have an SPSS icon on the desktop. Double click the icon to start SPSS and, if needed, close the window that may be superimposed over SPSS Output 0.1. The SPSS Data Editor screen will appear as shown in SPSS Output 0.1.

SPSS Output 0.1

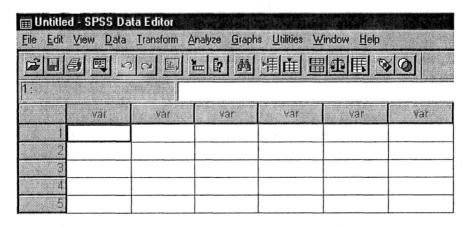

SPSS Data Editor: SPSS Data Editor is versatile spreadsheet-like system that can be used for entering, editing, defining, and displaying data. It is from this window that most of the procedures used in this book will be performed.

Entering data in the Data Editor: On text page 3, the authors present a data set containing the number of units sold (*X*) and the dollar sales amount (*Y*) for 3 time periods. To enter this data, we first name the variables to be entered. Click (✔) the Variable View tab in the lower left hand corner of the Data Editor. Name the three variables *period*, *units*, and *sales*, respectively, as shown in SPSS Output 0.2. These names are arbitrary but follow conventions for variable names within SPSS. In general, the name must be no more than 8 characters, cannot contain a special character, and must begin with one of the alphabetic letters *a* through *z*. So, instead of using *units* as the name for the variable "Number of Units Sold," we could have used *unit*, *unitsold*, *sold*, *x*, or any other name that was within the naming convention for a variable. From the screen shown in SPSS Output 0.2, we can also define the type of variable, the width of the display for each variable as well as define missing values, variable labels, and value labels. For instance, each of the 3 variables defined on text page 3 contains only numeric values and, thus, should be defined as *numeric* variables under the *Type* column. If we were to enter the name of a city, such as "New York," we would have to define the variable as a *String* variable in the *Type* column so that non-numeric characters could be entered to represent the city name. But, once a variable is defined as a *String*, mathematical operations cannot be performed on the variable. So, we can take the square root of *units* if it is defined as a *Numeric* variable but not if it is defined as a *String* variable. A variable label for *units* such as "Number of Units Sold" can also be defined. The variable label will be included in addition to or in place of the variable name in program output. We may also define *Value Labels* for any variable. For instance, we could assign *Value Labels* for *period* where 1="January thru April," 2="May thru August," and 3="Sept thru Dec." Finally, the column labeled *Decimals* in SPSS Output 0.2 indicates the number of decimal places displayed in the Data Editor. It does not affect the number of decimals that can be entered and stored in SPSS. A column Width=8 and Decimal=2 is the default settings in SPSS and throughout this book we will use these default values.

After naming the variables *period*, *units*, and *sales* in the Data Editor, ✔ the Data View tab in the lower left hand corner of the Data Editor and enter the 3 columns of data shown on text page 3. The Data Editor will now appear as SPSS Output 0.3.

SPSS Output 0.2

	Name	Type	Width	Decimals	Label	Values	Missing	Columns
1	period	Numeric	8	2		None	None	8
2	units	Numeric	8	2		None	None	8
3	sales	Numeric	8	2		None	None	8
4								
5								

SPSS Output 0.3

	period	units	sales	var	var

Untitled - SPSS Data Editor

File Edit View Data Transform Analyze Graphs Utilities Window Help

3 : sales   260

	period	units	sales	var	var
1	1.00	75.00	150		
2	2.00	25.00	50		
3	3.00	130.00	260		
4					
5					

Reading Data files into the Data Editor: A SPSS data file that has been saved by SPSS will have a "*sav*" extension. To read a SPSS data file, simply ✔ Open from the Data Editor and point SPSS to the appropriate file. However, the data files provided with *Applied Linear Regression Models* are "*txt*" files and must be read using SPSS's Text Import Wizard. To open the Toluca Company example from text Chapter 1 (Table 1.1, text page 19), ✔ Open ✔ Data and point SPSS to ch01ta01.txt in the appropriate directory. (We recommend that you transfer your data files from the disk provided with your text to a directory on a permanent disk drive, e.g., C:\alrm*.) SPSS will recognize that the requested file is not a SPSS file and open the Import Wizard. (SPSS Output 0.4) For most files, we simply need to ✔ Next 5 times and then ✔ Finish. The Data Editor will now look like SPSS Output 0.5. Note that the default variable names are *v1* and *v2*. We can rename the variables *lotsize* and *workhrs* as described above and add value labels, variable labels, etc. as appropriate. We can now save 🖫 the data file, by default, as a SPSS .sav file type.

SPSS Output 0.4

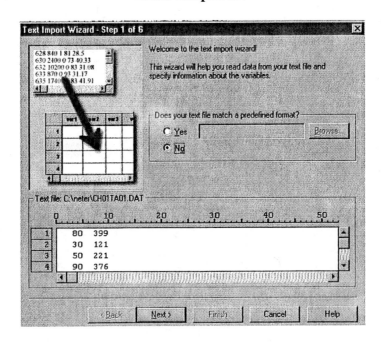

Note that the default file type to be opened when using the ⌻ icon is determined by the current Window type in SPSS. That is, while in the Data Editor, pressing the ⌻ icon will only display files that are SPSS Data Files (*.sav* extension). While in the Syntax Editor, pressing the ⌻ icon will only display files that are syntax files (*.sps* extension). Thus, if we are in an SPSS Output file (*.spo* extension), for example, and want to open a syntax file, we must ✔ File, ✔ Open, ✔ Syntax to view files in the current directory that are syntax files (*.sps* extension).

Creating a new variable: Suppose in the Toluca Company example above we wish to find the square of *lotsize* for use in polynomial regression. From the Data Editor ✔ Transform ✔ Compute. This will open a Compute Variable dialogue box. (SPSS Output 0.6) Insert 0.1 details the steps to create a new variable, *lotsqr*, the square of *lotsize*. In Insert 0.1 and throughout this book, an * in an insert box represents a comment statement or instructions needed to complete the current task. Following the steps in Insert 0.1 up to but not including ✔ OK will result in a Compute Variable dialogue box identical to SPSS Output 0.6. When we ✔ OK, *lotsqr* will be computed and placed in the Data Editor with *lotsize* and *workhrs*. Use of Insert 0.1 presupposes that the Toluca Company data file is open in the Data Editor and the variables are named *lotsize* and *workhrs*.

SPSS Output 0.6

```
                         INSERT 0.1

    ✔ Transform
    ✔ Compute
    * Type "lotsqr" in the Target Variable" box.
    ✔ lotsize
    ✔ ▶
    ✔ **
    ✔ 2
    ✔ OK
```

There may be times when we need to begin writing SPSS code for a data analysis before the complete data set is available. For instance, in the Toluca Company example, the analyst may have only a portion of a data set available but because of time constraints needs to begin writing the code for the final data analysis. Suppose that the analyst knows from past experience with the Toluca Company that a polynomial regression will be required. The analyst can develop and save the computational steps necessary for finding *lotsqr* prior to having the complete data set. Repeat the steps in Insert 0.1 up to but not including ✔ OK. Instead of ✔ OK, ✔ Paste. SPSS will paste the syntax into a SPSS Syntax Editor. If we now ✔ Run, and ✔ All in the SPSS Syntax Editor, *lotsqr* will be computed and placed in the Data Editor with *lotsize* and *workhrs*. However, since the complete data set is not available, we simply click save 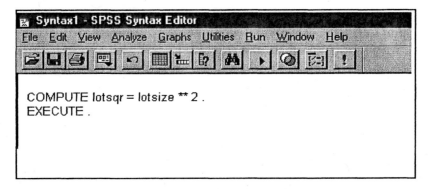 in the Syntax Editor (SPSS Output 0.7) to save the syntax file for later use.

SPSS Output 0.7

We can also create a syntax file by ✔ File, ✔ New, ✔ Syntax from the Data Editor. Once in the Syntax Editor, we can type the desired syntax so that the Syntax Editor looks like SPSS Output 0.7. For many applications the use of the point-and-click Compute Variable dialogue box is convenient and practical. For other applications, it is easier to open the Syntax Editor and type the appropriate syntax. For example, in SPSS Chapter 3 we find the expected value under normality,

$$ev = \text{sqrt}(2383.716) * \text{probit} ((rres\_1 - .375)/(25 + .25)).$$

Although this calculation can be done in a Compute Variable dialogue box, it is simpler and less prone to error to type a *Compute* statement in the Syntax Editor for this computation.

Many SPSS procedures allow the user to save various statistics. For example, we can specify that unstandardized residuals for a regression model be saved to the Data Editor. By default, the residuals will be saved in a variable named *res_1*. If we save the residuals for a second model, they will be saved as *res_2* if *res_1* has not been deleted from the Data Editor. In this book, we may instruct you to save residuals for use in subsequent calculations. For each set of instructions we assume that all previous residuals have been deleted from the Data Editor and the current residuals will be saved as *res_1*. If they will be saved as *res_3*, for example, you should use *res_3* in the calculation instead of *res_1*.

Default settings in SPSS: In the SPSS Data Editor, the user has available a number of options for customizing the software. These options can be edited by ✔ Edit, ✔ Options. (SPSS Output 0.8) If the system is set to Display Names, SPSS will display the Variable Name in a dialogue box. If the system is set to Display Labels, SPSS will display the Variable Label followed by the Variable Name in parentheses in a dialogue box. For example, if we save the unstandardized residuals from a regression procedure, they are saved under the Variable Name *res_1* with the Variable Label "Unstandardized Residuals." Selecting Display Names will result in SPSS Output 0.9 and selecting Display Labels will result in SPSS Output 0.10. We will use the Display Names option under Variable List throughout this text. The other option that may cause a difference in appearance of data files is the number of decimals displayed. We use Decimal Places=2 throughout this book unless otherwise specified. This option is found under the Data tab in the Options dialogue box.

SPSS Output 0.8

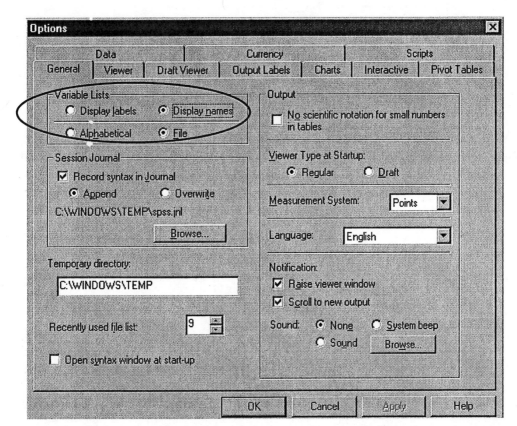

SPSS Output 0.9

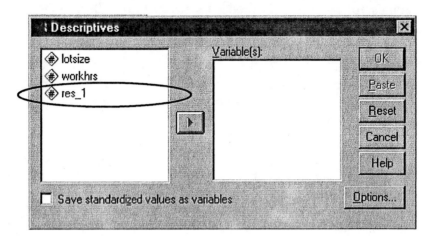

SPSS Output 0.10

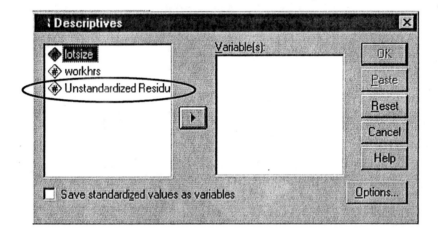

SPSS provides an extensive help system available from the Help dropdown menu. Additional help can be obtained at www.spss.com/tech using AnserNet and other technical support options. We are grateful to SPSS Technical Support for their assistance. SPSS® software is licensed by:

SPSS Inc. Headquarters
233 S. Wacker Drive
11th floor
Chicago, Illinois 60606
www.SPSS.com

Chapter **1** SPSS®

Linear Regression with One Predictor Variable

This chapter of the text introduces the simple linear regression model. The idea is to use statistical methods to find a model that allows us to predict the value of a dependent variable Y given the value of an independent variable X. We are given Y_i and X_i for a sample of n subjects and we want to find a "best fitting" model for predicting Y given X. The usual approach is to begin by assuming the model is of the form:

$$Y_i = \beta_0 + \beta_1 X_i + \varepsilon_i$$

for the ith subject, i = 1, ... , n. This is often referred to as a *simple linear regression model*. The first step is to find estimates of β_0 and β_1 so that the "fitted" model is best, as defined in the text, among all models of this form. The statistical analysis that involves estimating, constructing confidence intervals, and testing hypotheses about β_0 and β_1 is called *regression analysis*. The model is generalized in subsequent chapters to include more than one independent variable. The text introduces the basic concepts and gives easy to follow numerical examples that illustrate the ideas. Also, the text presents excerpts of a large data set in Table 1.1 followed by a description of the problem and details of a regression analysis. The text refers to this example as the Toluca Company example. A step-by-step set of SPSS solutions is given below.

The Toluca Company Example

Data File: ch01ta01.txt

The Toluca Company manufactures refrigeration equipment and replacement parts (text page 19). The parts are manufactured in "runs" and groups of runs comprise lots of varying sizes. The data disk provided with the text contains data on 25 manufacturing runs under the file name ch01ta01.txt. For each run, this file contains data on lot size and work hours, defined as the number of labor hours required to produce the lot. The aim of the example is to investigate the relationship between work hours and lot size. We wish to fit a simple linear regression model that allows us to predict work hours from lot size. We let the variable names *workhrs* denote work hours and *lotsize* denote lot size, and take these to be the dependent variable Y and the independent variable X, respectively, in the simple linear regression model. See Chapter 0, "Working with SPSS," for instructions on opening the Toluca Company data file and renaming the variables *lotsize* and *workhrs*. After opening ch01ta01.txt, the first 5 lines in the SPSS Data Editor should look like SPSS Output 1.1.

SPSS Output 1.1

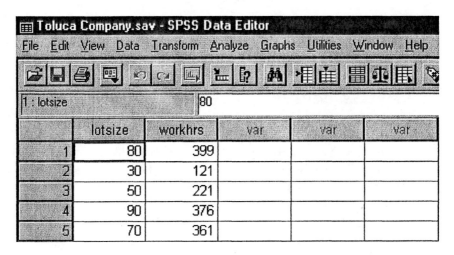

Scatter plots: The text begins a regression analysis of the Toluca Company data with a scatter plot showing work hours plotted against lot size in Figure 1.10 (a & b), text page 20. We provide SPSS syntax for constructing this scatter plot in Insert 1.1 and display the results in SPSS Output 1.2 (a & b) below.

INSERT 1.1

* Produce Scatter Plot.
✔ Graphs
✔ Scatter
✔ Simple
✔ Define
✔ *lotsize*
✔ ►(to the X axis)
✔ *workhrs*
✔ ►(to the Y axis)
✔ OK

* **Release 11.5 and lower**: To draw the regression line as shown in text Figure 1.10(b),
* double click on the graph in the SPSS Viewer. This brings up a SPSS Chart Editor window.
✔ Chart
✔ Options
✔ Total (under Fit Line)
✔ OK
* Close the Chart Editor.

* **Release 12.0 and higher**: To draw the regression line as shown in text Figure 1.10(b), .
* double click on the graph in the SPSS Viewer. This brings up a SPSS Chart Editor.
* window. Double click on any data point on the graph.
✔ Chart
✔ Add Chart Element
✔ Fit Line at Total
✔ Linear
✔ OK
* Close the Chart Editor.

(a)   **SPSS Output 1.2**   **(b)**

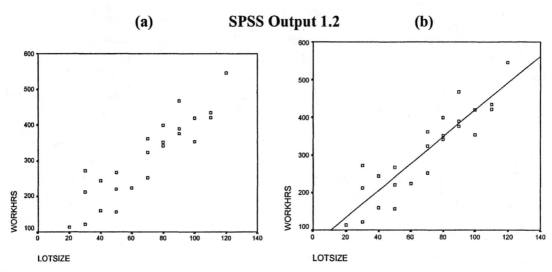

Estimation of β_0 and β_1: To find the least squares estimates b_0 and b_1 of β_0 and β_1, respectively, we follow the overview of the required computations that are outlined in Table 1.1, text page 19. In Insert 1.2, we show the SPSS syntax that performs these computations. We first find the means for *workhrs* ($\overline{X} = 70$) and *lotsize* ($\overline{Y} = 312.28$). Next we find $X_i - \overline{X}$ (*xdev*), $Y_i - \overline{Y}$ (*ydev*), $(X_i - \overline{X})^2$ (*xdev2*), $(Y_i - \overline{Y})^2$ (*ydev2*), $(X_i - \overline{X})(Y_i - \overline{Y})$ (*xdev * ydev = cross*). The last step in Insert 1.2 is to find the sum of $(X_i - \overline{X})(Y_i - \overline{Y})$ ($\sum$ *cross*=70,690) and $(X_i - \overline{X})^2$ ($\sum$ *xdev2*=19,800). Using equation 1.10, text page 17, we calculate:

$$b_1 = \frac{\sum(X_i - \overline{X})(Y_i - \overline{Y})}{\sum(X_i - \overline{X})} = \frac{70,690}{19,800} = 3.5702$$

and

$$b_0 = \overline{Y} - b_1\overline{X} = 312.28 - 3.5702 * 70 = 62.365$$

INSERT 1.2	INSERT 1.2 continued
* Find mean of *lotsize* and *workhrs*.	compute ydev2=(workhrs-312.28)**2.
✔ Analyze	compute cross=xdev*ydev.
✔ Descriptive Statistics	execute.
✔ Descriptives	✔ Run
✔ *lotsize*	✔ All
✔ ▶	* Find sum of *cross* and *xdev2*.
✔ *workhrs*	✔ Analyze
✔ ▶	✔ Descriptive Statistics
✔ OK	✔ Descriptives
* Calculate least squares estimates.	✔ *cross*
✔ File	✔ ▶
✔ New	✔ *xdev2*
✔ Syntax	✔ ▶
* Type the following syntax.	✔ Options
compute xdev=(lotsize-70).	✔ Sum
compute ydev=workhrs-312.28.	✔ Continue
compute xdev2=(lotsize-70)**2.	✔ OK

181

Estimated Regression Function: We used the SPSS syntax shown in Insert 1.2 to demonstrate how SPSS code can be written to provide intermediate results when we compute the parameter estimates b_0 and b_1. We now use SPSS' *Regression* procedure to find b_0 and b_1. This procedure is shorter and easier to use but it does not provide the results of the intermediate computations shown on text page 19. In Insert 1.3, we request a regression analysis where *workhrs* is regressed on *lotsize*. Partial results are shown in SPSS Output 1.3. Notice that the "Unstandardized Coefficients" shown under column heading B give the results $b_0 = 62.366$ and $b_1 = 3.570$. Here b_0 is given in the row labeled (Constant) and b_1 is given in the row labeled LOTSIZE. These results are consistent with those shown on text page 20 and in the calculations of Insert 1.2 above. We can easily obtain more precision in the SPSS output for parameter estimates. For example, after executing Insert 1.3, the output file will appear as SPSS Output 1.3. Double click on the Coefficients table and then double click on the "62.366." The output will display the estimate $b_0 = 62.3658585$.

INSERT 1.3

- ✔ Analyze
- ✔ Regression
- ✔ Linear
- ✔ *workhrs*
- ✔ ▶ (to the Dependent box)
- ✔ *lotsize*
- ✔ ▶ (to the Independent box)
- ✔ OK

SPSS Output 1.3

ANOVA[b]

Model		Sum of Squares	df	Mean Square	F	Sig.
1	Regression	252377.6	1	252377.581	105.876	.000[a]
	Residual	54825.459	23	2383.716		
	Total	307203.0	24			

a. Predictors: (Constant), LOTSIZE

b. Dependent Variable: WORKHRS

Coefficients[a]

Model		Unstandardized Coefficients		Standardized Coefficients	t	Sig.
		B	Std. Error	Beta		
1	(Constant)	62.366	26.177		2.382	.026
	LOTSIZE	3.570	.347	.906	10.290	.000

a. Dependent Variable: WORKHRS

182

From the above regression analysis, we estimate the regression function to be: $\hat{Y}= 62.377 + 3.570(X)$. To estimate the mean response (the mean number of work hours required) or predicted value for each lot size specified in the data as in column 3 of Table 1.2, text page 22, we instruct SPSS to save the unstandardized predicted values for the regression analysis (Insert 1.4). Similarly, to obtain the residuals as in column 4 of text Table 1.2, we instruct SPSS to save the unstandardized residuals. SPSS saves the unstandardized predicted values and the unstandarized residuals in the active Data Editor as *pre_1* and *res_1*, respectively. (SPSS Output 1.4). The partial results shown in SPSS Output 1.4 are consistent with those shown in text Table 1.2 with the exception of the Squared Residual. To produce the squared residual, simply open a Syntax File or use a dialogue box as shown in Insert 1.5. We compute the square of the individual residuals as *res_1*res_1* and store the result as a new variable that we have arbitrarily called *sqres*. *Sqres* will be saved in the active Data Editor.

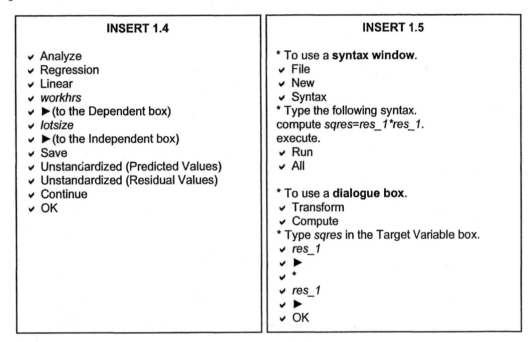

INSERT 1.4	INSERT 1.5
✔ Analyze	* To use a **syntax window**.
✔ Regression	✔ File
✔ Linear	✔ New
✔ *workhrs*	✔ Syntax
✔ ► (to the Dependent box)	* Type the following syntax.
✔ *lotsize*	compute *sqres=res_1*res_1*.
✔ ► (to the Independent box)	execute.
✔ Save	✔ Run
✔ Unstandardized (Predicted Values)	✔ All
✔ Unstandardized (Residual Values)	
✔ Continue	* To use a **dialogue box**.
✔ OK	✔ Transform
	✔ Compute
	* Type *sqres* in the Target Variable box.
	✔ *res_1*
	✔ ►
	✔ *
	✔ *res_1*
	✔ ►
	✔ OK

SPSS Output 1.4

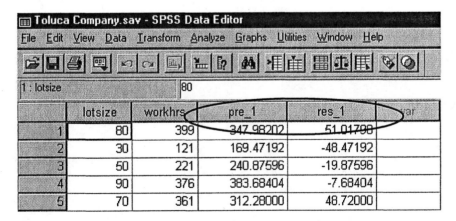

Chapter 2 SPSS®

Inferences in Regression and Correlation Analysis

In this chapter, the authors introduce the fundamentals for making statistical inferences about the parameters of the regression model. The model of interest is:

$$Y_i = \beta_0 + \beta_1 X_i + \varepsilon_i$$

where:

β_0 and β_1 are unknown parameters

X_i are known constants for $i = 1, 2, \dots, n$.

ε_i are unknown errors associated with the true model.

We assume the ε_i are independent $N(0, \sigma^2)$ for $i = 1, 2, \dots, n$. Note that the assumptions made here about the errors differ from those made in Chapter 1. The normality assumption is required by theoretical considerations used to develop confidence intervals and test statistics for making inferences involving the model parameters. Inferences were not considered in Chapter 1 and, therefore, normally distributed error terms were not required. The inferences of specific interest pertain to confidence intervals and tests of hypotheses for model parameters and functions of model parameters. The statistical analysis associated with estimation and inference involving regression models is called *regression analysis*.

Chapter 1 of the text presents formulas for the estimators b_0 and b_1 of the parameters β_0 and β_1, respectively. These estimators yield estimates that vary from sample to sample and, therefore, have a (sampling) distribution of their own. It is desirable then to make inferences about the true values of the parameters based on their sample estimates. Let β_j represent either β_0 or β_1 depending on whether $j = 0$ or 1 and let b_j denote the corresponding estimator. We can estimate the variance of b_j from a single sample of Y's and X's and the square root of this variance estimate is an estimate of the standard error of b_j. The text denotes the variance and standard error estimators by $s^2\{b_j\}$ and $s\{b_j\}$, respectively, and gives their computational formulas (see equation 2.9 page 43 and equation 2.23 page 49 of the text).

Consider the null hypothesis:

$$H_0: \beta_j = 0$$

A statistic for testing this hypothesis against either a one or two-sided alternative is:

$$t = \frac{b_j}{s\{b_j\}}$$

Under the assumptions of the model this statistic is distributed as Student's t with $n - 2$ degrees of freedom.

Let $t(1 - \alpha/2; n - 2)$ denote the $(\alpha/2)100$ percentile of the t distribution with $n - 2$ degrees of freedom. The $1 - \alpha$ confidence limits for β_j are:

$$b_j \pm t(1 - \alpha/2; n - 2)\, s\{b_j\}$$

Chapter 2 of the text presents these formulas and others of inferential interest in the simple linear regression model.

Further Aspects of the Toluca Company Example

Data File: ch01ta01.txt

We return to the Toluca Company example first presented in Chapter 1 (text page 19). We let the variable names *workhrs* denote work hours and *lotsize* denote lot size, and take these to be the dependent variable Y and the independent variable X, respectively, in the simple linear regression model. See Chapter 0, "Working with SPSS," for instructions on opening the Toluca Company data file and renaming the variables *lotsize* and *workhrs*. After opening ch01ta01.txt, the first 5 lines in the SPSS Data Editor should look like SPSS Output 2.1.

SPSS Output 2.1

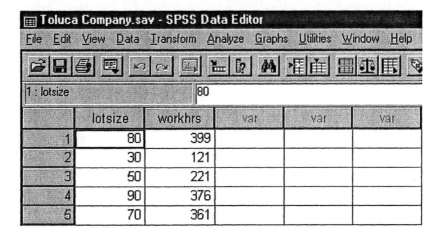

We wish to perform a regression analysis using the model:

$$\text{workhrs} = \beta_0 + \beta_1(\text{lotsize}) + \varepsilon$$

Confidence Interval for β_1: On text page 46 the management team wants an estimate of β_1 with a 95% confidence interval. The $1 - \alpha$ confidence limits for β_1 are (equation 2.15, text page 45):

$$b_1 \pm t(1 - \alpha/2; n-2)\, s\{b_1\}$$

We request confidence intervals for the regression coefficients using SPSS by the instructions given in Insert 2.1. From SPSS Output 2.2 we see that the 95% confidence interval is $2.852 \le \beta_1 \le 4.288$, consistent with text page 46.

Tests of Hypothesis Concerning β_1: The t-test for H_0: $\beta_1 = 0$ is:

$$t^* = \frac{b_1}{s\{b_1\}}$$

with $n - 2$ degrees of freedom (equation 2.17, text page 47). We get the numerical value for this test statistic as part of the output from the SPSS instructions given in Insert 2.1. In SPSS Output 2.2 we see $(t^*) = 10.290$ for *lotsize* and the associated p-value $= 0.000$ (Sig.). Since $p < 0.05$ we conclude $\beta_1 \ne 0$.

To test that $\beta_1 > 0$ (a one-sided test), we require that $|t^*| > t(1-\alpha; n-2)$. The standard output in SPSS displays the p-value for a two-sided test. To get the p-value for a one-sided test, first determine whether the numerical value of the t-test is negative or positive and hence whether it tends to favor the specified one-sided alternative. A negative t-test favors H_a: $\beta_1 < 0$ whereas a positive t-test favors H_a: $\beta_1 > 0$. If the t-test favors the specified one-sided alternative, just divide the p-value displayed in the SPSS output by 2. In the example, $(t^*) = 10.290$ is positive and this favors H_a: $\beta_1 > 0$ with p-value $= 0.000/2 = 0.000$. If the numerical value of the t-test does not favor the specified alternative, it is probably sufficient to quote $p > 0.50$ as the p-value. More precisely, the p-value for the one-sided t-test in this situation is 1 minus the p-value displayed in the SPSS output divided by 2. In the present example, if we had specified H_a: $\beta_1 < 0$ and found $(t^*) = 10.290$, the p-value would be $1 - 0.000/2 = 1.000$. In this circumstance, we would not reject H_0.

INSERT 2.1

- ✔ Analyze
- ✔ Regression
- ✔ Linear
- ✔ *workhrs*
- ✔ ▶(to the Dependent box)
- ✔ *lotsize*
- ✔ ▶(to the Independent box)
- ✔ Statistics
- ✔ Confidence Intervals (under Regression coefficients)
- ✔ Continue
- ✔ OK

Model Summary[b]

Model	R	R Square	Adjusted R Square	Std. Error of the Estimate
1	.906[a]	.822	.814	48.823

a. Predictors: (Constant), LOTSIZE

b. Dependent Variable: WORKHRS

ANOVA[b]

Model		Sum of Squares	df	Mean Square	F	Sig.
1	Regression	252377.6	1	252377.581	105.876	.000[a]
	Residual	54825.459	23	2383.716		
	Total	307203.0	24			

a. Predictors: (Constant), LOTSIZE

b. Dependent Variable: WORKHRS

Coefficients[a]

Model		Unstandardized Coefficients		Standardized Coefficients	t	Sig.	95% Confidence Interval for B	
		B	Std. Error	Beta			Lower Bound	Upper Bound
1	(Constant)	62.366	26.177		2.382	.026	8.214	116.518
	LOTSIZE	3.570	.347	.906	10.290	.000	2.852	4.288

a. Dependent Variable: WORKHRS

Sometimes it is desirable to find the critical value $t(1-\alpha; n-2)$ corresponding to a t-test. To find $t(1-\alpha; n-2)$ in SPSS, use a Compute Variable (✔ Transform ✔ Compute) dialogue box (SPSS Output 2.3) to request a new variable named *tvalue*. Let T denote a random variable that has a t-distribution with degrees of freedom = df. Suppose we want to find the value t such that $P(T > t) = \alpha$ for a value of α that we specify. Equivalently, suppose we want to find the value t such that $P(T \leq t) = 1-\alpha$. We can use the SPSS *inverse distribution function IDF.T$(1-\alpha, df)$* for this purpose. When we specify $1-\alpha$ and df, this function returns the associated t-value. To find the critical value for testing H_a: $\beta_1 > 0$ controlling the level of significance at $\alpha = 0.05$, we must find $t(1-\alpha; n-2)$ or $t(0.95, 23)$. This critical value ($t = 1.71$) is obtained from the Compute Variable dialogue box (SPSS Output 2.3) and is written into each of the 25 data-lines in the Toluca Data Editor. Since the obtained t-value ($t = 10.29$) is greater than the critical value $t = 1.71$, we conclude $\beta_1 > 0$. If we want to obtain the critical value for a two-sided test where $\alpha = 0.05$, we find $t(1-\alpha/2; n-2)$ or $t(0.975, 23)$ using the above *inverse distribution function*.

SPSS Output 2.3

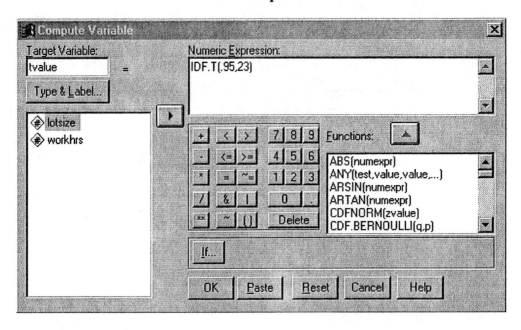

Confidence Interval for β_0: To obtain the 90% confidence interval for β_0 (text page 49), we must find $t(1-\alpha/2; n-2)$. (Note that Output 2.2 presents the 95% confidence interval.) Since we desire a 90% confidence interval, $\alpha = 0.10$ and $t(0.95, 23) = 1.71$. From Output 2.2, we find the standard deviation of b_0 (Constant) is $s\{b_0\} = 26.17$. From equation 2.25, text page 49, the 90% confidence interval is:

$$b_0 \pm t(1-\alpha/2; \ n-2) \ s\{b_0\}$$

or

$$62.366 \pm 1.71 * 26.17$$

Simplifying this expression, we find the 90% confidence interval:

$$17.5 \le \beta_0 \le 107.2$$

Confidence Interval for $E\{Y_h\}$: Once specific sample data have been used to produce parameter estimates b_0 and b_1, we denote the fitted model as:

$$\hat{Y}_i = b_0 + b_1 X_i \qquad i = 1, 2, \ldots, n$$

We call $\hat{Y}_i$ the *predicted value* of Y when the predictor variable is X_i. Let X_h denote a specific value of the independent variable in the simple linear regression model. As noted by the authors, "X_h may be a value which occurred in the sample, or it may be some other value of the predictor variable within the scope of the model." The corresponding predicted value is denoted $\hat{Y}_h$. We can select many random samples of data and different samples may produce different estimates

b_0 and b_1 and hence different $\hat{Y}_h$. Hence, for any such X_h there is a distribution corresponding to the random variable $\hat{Y}_h$. This random variable has a mean and variance denoted $E\{\hat{Y}_h\} = E\{Y_h\}$ and $\sigma^2\{\hat{Y}_h\}$, respectively. Estimators of the mean and variance are $\hat{Y}_h$ and $s^2\{\hat{Y}_h\}$ (see equation 2.30, text page 53 for the formula for $s^2\{\hat{Y}_h\}$).

On text page 54, the authors use Toluca Company Example to find the 90% confidence interval for $E\{Y_h\}$ when lot size (X_h) = 65 units. The $1 - \alpha$ confidence limits are given in equation 2.33, text page 54:

$$\hat{Y}_h \pm t(1 - \alpha/2; n-2)s\{\hat{Y}_h\}$$

To find $s^2\{\hat{Y}_h\}$ (equation 2.30, text page 53) we need the mean squared error (*MSE*), which is given in SPSS Output 2.2 above (*MSE* = 2383.716). To find $\overline{X}$, we use ✔ Analyze ✔ Descriptive Statistics ✔ Descriptives to find the mean for *lotsize* = 70. To find $(X_i - \overline{X})^2$ we use a Compute Variable dialogue box (✔ Transform ✔ Compute) and enter the syntax as shown in SPSS Output 2.4. This results in a new variable named *xdev2* being placed in your active Data Editor. To find the sum of this variable $(\sum(X_i - \overline{X})^2)$ we use ✔ Analyze ✔ Descriptive Statistics ✔ Descriptives ✔ Options ✔ Sum and move *xdev2* to the Variables box. This produces SPSS Output 2.5, which contains the sum of *xdev2* = 19,800. We can now find:

$$s^2\{\hat{Y}_h\} = MSE\left[\frac{1}{n} + \frac{(X_h - \overline{X})^2}{\sum(X_i - \overline{X})^2}\right]$$

Thus,

$$s^2\{\hat{Y}_h\} = 2383.7\left[\frac{1}{25} + \frac{(65 - 70)^2}{19,800}\right] = 98.37$$

$$s\{\hat{Y}_h\} = \sqrt{s^2\{\hat{Y}_h\}} = 9.918$$

To find the 90 percent confidence interval, we also need $t(1-\alpha/2; n-2)$. Using the *inverse distribution function* introduced above (SPSS Output 2.3) where $t(1-\alpha/2; n-2) = t(0.95; 23)$, *IDF.T*(0.95, 23) returns $t(0.95, 23) = 1.71$. Using $X_h = 65$ units and the parameter estimates in Output 2.2 above, we find:

$$\hat{Y}_h = b_0 + b_1 X_h$$

or

$$\hat{Y}_h = 62.366 + 3.570(65) = 294.4$$

Substituting these numerical values into equation 2.33, text page 54, we obtain:

$$294.4 \pm 1.71(9.918)$$

or

$$277.44 \leq E\{Y_h\} \leq 311.36$$

SPSS Output 2.4

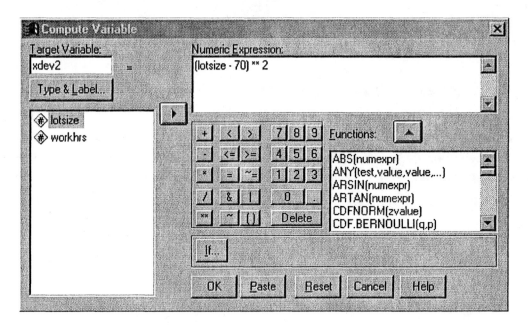

SPSS Output 2.5

Descriptive Statistics

	N	Minimum	Maximum	Sum	Mean	Std. Deviation
XDEVSQ	25	.00	2500.00	19800.00	792.0000	764.26435
Valid N (listwise)	25					

Confidence Band for Regression Line: To find the confidence band for the regression line as shown in equations 2.40 and 2.40a, text page 62, we use the Working-Hotelling $1-\alpha$ confidence band to establish boundary values for each *lotsize* value:

$$\hat{Y}_h \pm W s\{\hat{Y}_h\}$$

where:

$$W^2 = 2F(1-\alpha;\ 2, n-2)$$

190

In Insert 2.2, we use the SPSS *inverse distribution function IDF.F*$(1-\alpha, df_1, df_2)$ to find the critical F-value for the F-distribution with $1-\alpha = 0.90$ and $df_1 = 2$ and $df_2 = 23$ degrees of freedom and place the result in a new variable named *fvalue*. This results in SPSS writing the F-value associated with *IDF.F*$(0.9, 2, 23) = 2.55$ into every line in the Data Editor.

We now use text equations 2.28, 2.30, 2.40, and 2.40a to find the confidence band for the Toluca Company regression example. Insert 2.2 contains the necessary syntax to compute the upper (*bandhi*) and lower (*bandlo*) confidence bands. Notice that we have presented the syntax to be used in a Syntax Editor. Dialogue boxes could have been used in place of the Syntax Editor. We can now use ✔ Graphs ✔ Scatter ✔ Overlay ✔ Define to request (SPSS Output 2.6) the plot as shown in SPSS Output 2.7. This plot is equivalent to Figure 2.6, text page 63.

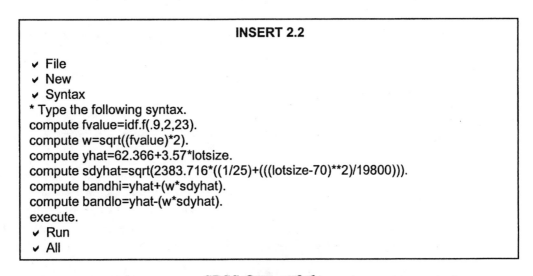

INSERT 2.2

✔ File
✔ New
✔ Syntax
* Type the following syntax.
compute fvalue=idf.f(.9,2,23).
compute w=sqrt((fvalue)*2).
compute yhat=62.366+3.57*lotsize.
compute sdyhat=sqrt(2383.716*((1/25)+(((lotsize-70)**2)/19800))).
compute bandhi=yhat+(w*sdyhat).
compute bandlo=yhat-(w*sdyhat).
execute.
✔ Run
✔ All

SPSS Output 2.6

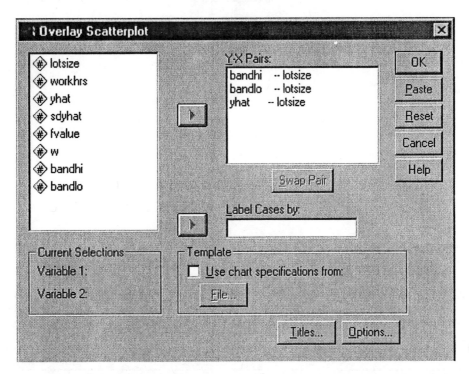

SPSS Output 2.7

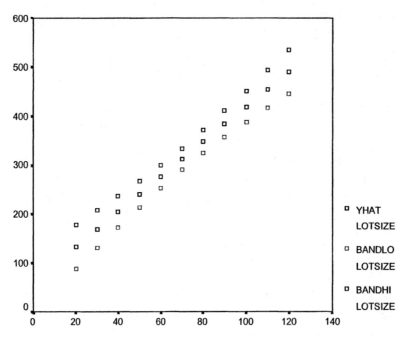

Now that we have a grasp of the computation of the confidence band for a regression line, we use SPSS' charting capabilities in Insert 2.3 to produce SPSS Output 2.8, which is equivalent to Output 2.7 above and to text Figure 2.6. For both SPSS Output 2.7 and 2.8, the range of values for the x-axis and y-axis is not identical to text Figure 2.6. The SPSS Chart Editor can additionally be used to modify a chart to conform to the values shown in text Figure 2.6.

INSERT 2.3

* Produce Scatter Plot.
✔ Graphs
✔ Scatter
✔ Simple
✔ Define
✔ *lotsize*
✔ ▶ (to the X axis)
✔ *workhrs*
✔ ▶ (to the Y axis)
✔ OK
* To draw the confidence bandsas shown in text Figure 2.6,.
* double click on the graph in the SPSS Viewer. This brings.
* up a SPSS Chart Editor window.
✔ Chart
✔ Options
✔ Total (under Fit Line)
✔ Fit Options
✔ Mean (under Regression Prediction Lines)
* Type "90" in Confidence Interval box.
✔ Continue
✔ OK
* Close the Chart Editor.

SPSS Output 2.8

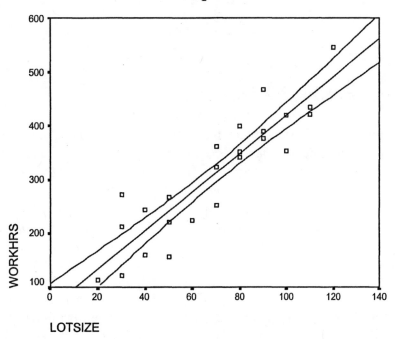

LOTSIZE

The coefficient of determination (R^2): From equation 2.72, text page 74, the coefficient of determination (R^2) is defined as Sum of Squares Regression / Sum of Squares Total; i.e.

$$R^2 = \frac{SSR}{SSTO} = 1 - \frac{SSE}{SSTO}$$

From SPSS Output 2.2 given above, we find $SSR = 252{,}377.6$ and $SSTO = 307{,}203.0$. Thus $R^2 = 0.822$ and the correlation coefficient (r) $= +\sqrt{R^2} = 0.907$.

Chapter 3 SPSS®

Diagnostics and Remedial Measures

In Chapter 1 we learned that regression analysis involves using a sample of data values to find a linear model that will predict (the *dependent variable*) Y when the values of the (*independent variable*) X's are known. We also referred to the X's as *predictor variables*. We began with a sample of data where each sample unit provides an observation on Y (the dependent variable) and on one or more X's (the independent or predictor variables). Next, we formulated a model that we thought might specify the relationship between Y and the predictor variables. Once we decided on a model that we thought might be appropriate, we assumed the model was in fact an appropriate model for our purpose. We used the sample data to estimate parameters of the model and this estimation provided us with a fitted model. The adequacy of the estimates and of related statistical inferences depends on whether the specified model and the assumptions made in specifying the model are at least approximately true. In most applications, we do not know for certain in advance that the model we decide to use is appropriate for the specific analysis.

In Chapter 3, the authors discuss methods for using the observed sample data to investigate the adequacy of a fitted regression model. A group of methods known as *regression diagnostics* are used for this purpose. When regression diagnostics identify inadequacies in the model, *remedial measures* can sometimes be used to correct or offset the inadequacies.

Many regression diagnostics are performed in terms of the residuals from the fitted model. Assuming the simple linear regression model is the specified model, the fitted model is denoted (equation 1.13, text page 21):

$$\hat{Y}_i = b_0 + b_1 X_i \qquad i = 1, \ldots, n$$

The estimates b_0 and b_1 are obtained from the sample data. Once they have been calculated and their numerical values obtained, we can substitute X_i into the fitted model and calculate the predicted value $\hat{Y}_i$ for the ith unit in the sample, $i = 1, 2, \ldots, n$. (See the example on text page 21). Ideally, the fitted model would predict the observed Y_i exactly for all $i = 1, 2, \ldots, n$ but this rarely happens in practice. Thus, there is usually a difference between the observed Y_i and its predicted value $\hat{Y}_i$. This difference is called the *residual*. The ith residual, denoted e_i, is defined as:

$$e_i = Y_i - \hat{Y}_i$$

In the discussion below, we use SPSS to find the predicted values and the residuals for the sample data. We illustrate how to obtain graphs to perform graphical diagnostics discussed in the text. We then demonstrate how to produce regression diagnostics in terms of regression model residuals.

The Toluca Company Example Continued

Data File: ch01ta01.txt

We return to the Toluca Company example from Chapters 1 and 2 (text page 19, 46). See Chapter 0, "Working with SPSS," for instructions on opening the Toluca Company data file and renaming the variables *lotsize* and *workhrs*. After opening ch01ta01.txt, the first 5 lines in your SPSS Data Editor should look like SPSS Output 3.1.

SPSS Output 3.1

	lotsize	workhrs	var	var	var
1	80	399			
2	30	121			
3	50	221			
4	90	376			
5	70	361			

Diagnostics for Predictor Variable: The authors begin by discussing graphical methods that can be used as diagnostic tools to see if there are any outlying X values in the sample. This is important because a few outlying X values could have too much influence on estimates of the model parameters and other aspects of the regression analysis. If the influence of a few outlying X values is dampened or eliminated altogether, we sometimes get a drastically different regression analysis.

We use SPSS (Insert 3.1) to produce Figure 3.1 a-d shown on text page 101. A dot plot identical to text Figure 3.1a is not available in SPSS but a bar chart (SPSS Output 3.2) approximates that display of information. In the Toluca example, there are 25 runs represented in the data file. Text Figure 3.1b is a plot of *lotsize* against the run number but the data file does not contain the run number. SPSS has a system variable named *$casenum* that is simply a sequential count of the data file records. That is, *$casenum* equals 1 for the first record, 2 for the second record, etc. However, *$casenum* is a "dynamic" variable in that it can change values as the data file is restructured by various procedures such as sorting the file. Thus, we need to convert *$casenum* to a variable in the active Toluca Data Editor for use in subsequent analyses. This is accomplished by the *compute run=$casenum* command in Insert 3.1. The active Data Editor will now contain a variable named *run*, whose values are a sequential numbering of the data-lines from 1 to 25. (Here, we assume the original Toluca Company data-lines are in order by run number.) We can now produce text Figure 3.1b as shown in SPSS Output 3.2. Next, we use SPSS' Explore to produce a stem-and-leaf and box plots of *lotsize*. (SPSS Output 3.2)

INSERT 3.1	INSERT 3.1 continued
* text Figure 3.1a. ✔ Graph ✔ Bar ✔ Simple ✔ Define ✔ *lotsize* ✔ ► (to the Category axis) ✔ OK * text Figure 3.1b. ✔ File ✔ New ✔ Syntax * Type the following syntax. compute run=$casenum. execute. ✔ Run ✔ All	✔ Graph ✔ Line ✔ Simple ✔ Define ✔ *lotsize* ✔ ► (to the Variable) ✔ *run* ✔ ► (to the Category axis) ✔ OK * text Figure 3.1c and d. ✔ Analyze ✔ Descriptive Statistics ✔ Explore ✔ *lotsize* ✔ ► (to the Dependent List) ✔ OK

SPSS Output 3.2 a - d

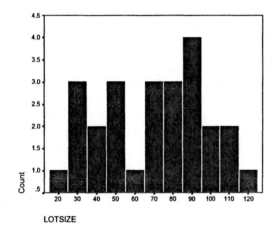

LOTSIZE

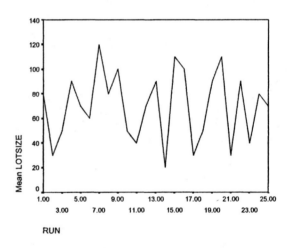

RUN

LOTSIZE Stem-and-Leaf Plot

```
    Frequency     Stem &   Leaf

         .00        0  .
        4.00        0  .  2333
        5.00        0  .  44555
        4.00        0  .  6777
        7.00        0  .  8889999
        4.00        1  .  0011
        1.00        1  .  2

    Stem width:    100
    Each leaf:         1 case(s)
```

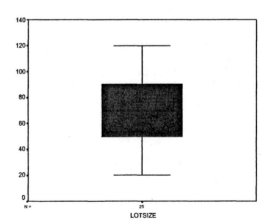

LOTSIZE

196

Diagnostics for Residuals: In text Section 3.3, the authors present informal diagnostic plots of *residuals*. Note that the authors make the distinction between *studentized* and *semistudentized* residuals. The authors point out that "both semistudentized residuals and studentized residuals can be very helpful in identifying outlying observations." SPSS does not have an option to produce semistudentized residuals but it does produce studentized residuals. Studentized residuals are probably the more important of the two types of residuals. Therefore, we use studentized residuals in place of semistudentized residuals to present solutions to text chapter examples.

As before, we perform a regression analysis using the model:

$$workhrs = \beta_0 + \beta_1(lotsize) + \varepsilon$$

In Insert 3.2, we specify this regression model and request that unstandardized residuals be saved to the active Data Editor. By default the residuals are saved as *res_1*. To produce text Figures 3.2 a, b, and c, simply use the syntax in Insert 3.1 above, replacing *lotsize* with *res_1*. To produce text Figure 3.2d, request a normality plot from the Explore dialogue box. (Insert 3.2) Notice that the standard SPSS output has the observed residual values on the *x*-axis, opposite of text Figure 3.2d. Using SPSS Chart Editor, we can switch the axis positions so the plot appears as text Figure 3.2d as shown in SPSS Output 3.3. The text presents numerous other diagnostic plots. For brevity, we do not present these plots but you should be able to produce them easily by generalizing the syntax shown in Inserts 3.1 and 3.2.

INSERT 3.2	INSERT 3.2 continued
✔ Analyze	* text Figure 3.1d.
✔ Regression	✔ Analyze
✔ Linear	✔ Descriptive Statistics
✔ *workhrs*	✔ Explore
✔ ▶(to the Dependent box)	✔ *res_1*
✔ *lotsize*	✔ ▶(to the Dependent List)
✔ ▶(to the Independent box)	✔ Plots
✔ Save	✔ Normality plots with tests
✔ Unstandardized (Residual Values)	✔ Continue
✔ Continue	✔ OK
✔ OK	* Double click the chart you wish to edit.
	✔ Series
	✔ Displayed
	* Use the Simple Scatterplot Displayed Data dialogue.
	* to reverse the plot axises.
	✔ OK
	* Close the Chart Editor.

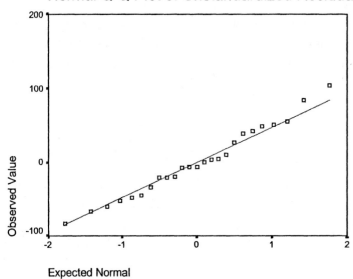

Normal Q-Q Plot of Unstandardized Residual

Correlation Test for Normality: On text page 111, the authors calculate the coefficient of correlation between the ordered residuals and their expected value under normality for the Toluca Company Example. The expected value under normality of the kth smallest observation from a sample of size n is given by:

$$\sqrt{MSE}\left[z\left(\frac{k-0.375}{n+0.25}\right)\right]$$

The first step in obtaining the expected value of the ordered residuals under normality is to calculate $(k-0.375)/(n+0.25)$. The syntax for this is shown in Insert 3.3. In the Toluca Company example, $n = 25$ and k is the rank of the ordered residuals. We rank the saved residuals (*res_1*) using the Rank Case function. SPSS creates another new variable, *rres_1*, that is the rank based on the value of *res_1*, the unstandardized residual based on a model with *lotsize* as the predictor variable. Thus, $(k-0.375)/(n+0.25)$ becomes $(rres\_1 - 0.375)/(25 + 0.25)$. This quantity represents the cumulative probability $(0 \leq$ cumulative probability $\leq 1.0)$ of the standard normal distribution. We now obtain a standard score (z) associated with the cumulative probability calculated above. In SPSS, the Probit function is an *inverse distribution function*. It takes the form *Probit(p)*, where p = cumulative probability. For instance, if cumulative probability = 0.50 [*Probit*(0.50)] then $z = 0$. If cumulative probability = 0.8413 (*Probit*(0.8413)) then $z = +1$. Thus, to obtain z we use *Probit* $((rres\_1 - 0.375)/(25 + 0.25))$. Finally, we multiply this quantity by the square root of the mean squared error, $MSE = 2383.716$ of the first order model (Chapter 2 – SPSS Output 2.2). From the above, we compute the expected value as:

$$ev = sqrt(2383.716) * Probit ((rres\_1 - 0.375)/(25 + 0.25)).$$

We can now obtain the coefficient of correlation between the residual (*res_1*) and the expected value under normality (*ev*). The result of the correlation between *ev* and *res_1* is 0.992 and is within rounding error of the correlation (0.991) reported on text page 115.

INSERT 3.3

- ✔ Analyze
- ✔ Regression
- ✔ Linear
- ✔ *workhrs*
- ✔ ▶(to the Dependent box)
- ✔ *lotsize*
- ✔ ▶(to the Independent box)
- ✔ Save
- ✔ Unstandardized (Residual Values)
- ✔ Continue
- ✔ OK
- ✔ Transform
- ✔ Rank Cases
- ✔ *res_1*
- ✔ ▶(to the Variables box)
- ✔ *OK*

- ✔ Transform
- ✔ Compute
- * Type "ev" in Target Variable box
- * Scroll down the list of functions
- ✔ SQRT(numexpr)
- ✔ ▲
- Type "2383.716"
- ✔ to the right of the closing) in the numeric expression
- ✔ *
- Scroll through the list of functions
- ✔ PROBIT(prob)
- ✔ ▲
- * Type "("
- ✔ *rres_1*
- ✔ ▶
- Type "-.375)/(25+.25)"
- ✔ OK

- ✔ Analyze
- ✔ Correlate
- ✔ Bivariate
- ✔ *ev*
- ✔ ▶
- ✔ *res_1*
- ✔ ▶
- ✔ OK

Note that the Transform>Compute (bracketed) portion of Insert 3.3 is cumbersome and prone to error. As we discussed in Chapter 0, "Working with SPSS," it is sometimes easier to work in a Syntax Editor window than to use a dialogue box (point-and-click). In doing so, we can type the formula for *ev* in a Syntax Editor in lieu of using a dialogue box. This is shown in Insert 3.4. Thus, the steps setoff with a single bracket in Insert 3.3 can be replaced by the steps in Insert 3.4. Both approaches accomplish the same outcome but the latter is much simpler.

INSERT 3.4

✔ File
✔ New
✔ Syntax
compute *ev* = sqrt(2383.716) * Probit(((*rres_1* - .375)/(25 + .25)).
execute.
✔ Run
✔ All

Test for Constancy of Error Variance - Brown-Forsythe: The authors present two formal tests for constancy of error variance. We begin with the *Brown-Forsythe* test using the Toluca Company example (text page 117). For simplicity, we have divided the process into steps:

- Step 1: (Insert 3.5) Divide the 25 data-lines into two subsets (*group*): those with *lotsize* between 20 and 70 (*lotsize*<=70) and those with *lotsize* > 70. Insert 3.5 contains syntax for using the Recode procedure and for using the If – Then procedure. Both procedures yield the same result.
- Step 2: Use *Explore* to obtain the median residual values $\tilde{e}_1$ (group 1) = −19.87 and $\tilde{e}_2$ (group 2) = −2.68.
- Step 3: Find *absdevmd*, the absolute value (*abs*) of the residual minus the median residual value for the respective group, d_{i1} and d_{i2}.
- Step 4: Find the means of *absdevmd* for the two groups, $\bar{d}_1 = 44.81$ and $\bar{d}_2 = 28.4$.
- Step 5: Find *sqrdev*, the square of *absdevmd* minus the respective group mean of *absdevmd*.
- Step 6: Find the sum of *sqrdev* for the two groups, $\sum(d_{11} - \bar{d}_1)^2 = 12{,}567.86$ and $\sum(d_{12} - \bar{d}_2)^2 = 9{,}621.54$.

We now have the required values to calculate the *Brown-Forsythe* statistic:

$$s^2 = \frac{12{,}567.86 + 9{,}621.54}{25 - 2} = 964.75$$

$$s = \sqrt{964.75} = 31.06$$

$$t^*_{BF} = \frac{44.81 - 28.4}{31.06\sqrt{\dfrac{1}{13} + \dfrac{1}{12}}} = 1.319$$

INSERT 3.5	INSERT 3.5 continued

* Step 1.
✔ File
✔ New
✔ Syntax
recode lotsize (0 thru 70=1) (else=2) into group .
execute .
* OR.
if (lotsize le 70)group=1.
if (lotsize gt 70)group=2.
execute.
✔ Run
✔ All
 	
* Step 2.
✔ Analyze
✔ Descriptive Statistics
✔ Explore
✔ *res_1*
✔ ▶(to the Dependent List)
✔ *group*
✔ ▶(to the Factor List)
✔ OK
 	
*Step 3.
✔ File
✔ New
✔ Syntax
* Type the following syntax.
if (group=1) absdevmd=abs(res_1-(-19.88)).
if (group=2) absdevmd=abs(res_1-(-2.68)).
execute.
✔ Run
✔ All
	* Step 4.
	✔ Analyze
	✔ Descriptive Statistics
	✔ Explore
	✔ *absdevmd*
	✔ ▶(to the Dependent List)
	✔ *group*
	✔ ▶(to the Factor List)
	✔ OK
	
	* Step 5.
	✔ File
	✔ New
	✔ Syntax
	* Type the following syntax.
	if (group=1) sqrdev=(absdevmd-44.81)**2.
	if (group=2) sqrdev=(absdevmd-29.45)**2.
	execute.
	✔ Run
	✔ All
	
	* Step 6.
	✔ Analyze
	✔ Compare Means
	✔ Means
	✔ *sqrdev*
	✔ ▶(to the Dependent List)
	✔ *group*
	✔ ▶(to the Independent List)
	✔ Options
	✔ Sum
	✔ ▶(to the Cell Statistics)
	✔ Continue
	✔ OK

Test for Constancy of Error Variance - Breusch-Pagan Test: The second Test for *Constancy of Error Variance* considered by the authors is the *Breusch-Pagan Test* (text page 118). To compute the *Breusch-Pagan* test we first save the residual values of the following model as shown in Insert 3.6:

$$workhrs = \beta_0 + \beta_1(lotsize) + \varepsilon$$

The residual values are saved in the active Data Editor as *res_1*. If we have *res_1* saved from a previous run, we do not need to resave the residuals. We now square *res_1* and place the result in a variable named *ressqr* (e^2). The last step is to regress *ressqr* against *lotsize* in the usual manner. The last regression procedure in Insert 3.6 results in SPSS Output 3.4 and contains SSR*=7,896,128. SSE (54,825.459) is shown in SPSS Output 3.5 and results from the first regression run in Insert 3.6. We now can find:

$$X^2_{BP} = \frac{SSR^*}{2} \div \left(\frac{SSE}{n}\right)^2$$

where SSR^* from regressing e^2 (*ressqr*) on X and SSE from regressing Y on X:

$$X^2_{BP} = \frac{7,896,128}{2} \div \left(\frac{54,825}{25}\right)^2 = 0.821$$

INSERT 3.6

- ✔ Analyze
- ✔ Regression
- ✔ Linear
- ✔ *workhrs*
- ✔ ▶(to the Dependent box)
- ✔ *lotsize*
- ✔ ▶(to the Independent box)
- ✔ Save
- ✔ Unstandardized (Residual Values)
- ✔ Continue
- ✔ OK

- ✔ File
- ✔ New
- ✔ Syntax
* Type the following syntax.
compute *ressqr=res_1**2*.
execute.
- ✔ Run
- ✔ All

- ✔ Analyze
- ✔ Regression
- ✔ Linear
- ✔ *ressqr*
- ✔ ▶(to the Dependent List)
- ✔ *lotsize*
- ✔ ▶(to the Independent List)
- ✔ OK

ANOVA[b]

Model		Sum of Squares	df	Mean Square	F	Sig.
1	Regression	7896142	1	7896141.961	1.091	.307[a]
	Residual	1.66E+08	23	7234604.179		
	Total	1.74E+08	24			

a. Predictors: (Constant), LOTSIZE

b. Dependent Variable: RESSQR

SPSS Output 3.5

ANOVA[b]

Model		Sum of Squares	df	Mean Square	F	Sig.
1	Regression	252377.6	1	252377.581	105.876	.000[a]
	Residual	54825.459	23	2383.716		
	Total	307203.0	24			

a. Predictors: (Constant), LOTSIZE

b. Dependent Variable: WORKHRS

The Bank Example

Data File: ch03ta04.txt

F Test for Lack of Fit: On text page 119, the authors present a formal test for determining if the regression function adequately fits the data. In an experiment involving 12 branch offices of a commercial bank, gifts were offered for customers who opened money market accounts. The gift was proportional to the minimum deposit and there were 6 levels of minimum deposit used in the experiment. The researchers studied the relationship between the minimum deposit and associated gift (*mindep*) and the number of new accounts (*newaccts*) opened at each of the 12 branches. After opening the Bank Example data (Table 3.4, text page 120), the first five lines of the Data Editor should look like SPSS Output 3.6.

To generate the Lack of Fit F* presented on text page 124, we switch to SPSS' *General Linear Model* (*GLM*) procedure. We demonstrate two ways of obtaining the lack of fit test in SPSS. The first is a longer method and is presented to parallel the solution presented in the text. We begin by fitting the full model to the data. This requires that we create a new variable, *J*. As noted in Insert 3.7, $j = 1$ if $X_1 = 75$, $j = 2$ if $X_2 = 100$ and so on. We now regress *Y* (*newaccts*) on *X* (*mindep*) and *J*. This is accomplished by the syntax shown in Insert 3.7 – Method 1. Note that we enter *J* as a fixed factor. Thus, *J* identifies 6 discrete classifications based on the values of *X*.

SPSS Output 3.6

	mindep	newaccts	var	var
1	125.00	160.00		
2	100.00	112.00		
3	200.00	124.00		
4	75.00	28.00		
5	150.00	152.00		

lof table 3.4.sav - SPSS Data Editor

File Edit View Data Transform Analyze Graphs Utilities W

11 : newaccts   136

Insert 3.7 - Method 1 produces SPSS Output 3.7a. Notice that $SSE = 1148.0$ and from text equation 3.16, $SSE(F) = SSPE$; therefore, $SSPE = 1148.0$. From text equation 3.24, we see that $SSLF = SSE(R) - SSPE$. If we were to regress *newaccts* on *mindep*, (reduced model) we would find that $SSE(R) = 14,741.6$. This is accomplished by the syntax shown in Insert 3.7 – Method 2, and the results are shown in SPSS Output 3.7b. Thus, $SSLF = 14,741.6 - 1148 = 13,593.6$. Notice, however, that this quantity is given in SPSS Output 3.7a under *Type III Sum of Squares* for *J*. Thus, regressing *Y* on *X* and *J* in SPSS' General Linear Model procedure gives us the quantities to find:

$$F^* = \frac{SSLF}{c-2} \div \frac{SSPE}{n-c}$$

or

$$F^* = \frac{13,593.6}{4} \div \frac{1,148}{5} = 14.8$$

(see text page 124). Now that we understand the computation of the lack of fit test, we note that SPSS has an easy way to request the lack of fit test. We simply use the *Lack of Fit* option within SPSS' *GLM* to perform the same test. Insert 3.7 - Method 2 contains the necessary steps to produce SPSS Output 3.7b and we do so without creating *J*. Again, notice the lack of fit $F = 14.8$, consistent with text page 124.

SPSS Output 3.7a

Tests of Between-Subjects Effects

Dependent Variable: NEWACCTS

Source	Type III Sum of Squares	df	Mean Square	F	Sig.
Corrected Model	18734.909a	5	3746.982	16.320	.004
Intercept	2729.288	1	2729.288	11.887	.018
MINDEP	.000	0	.	.	.
J	13593.571	4	3398.393	14.801	.006
Error	1148.000	5	229.600		
Total	170696.000	11			
Corrected Total	19882.909	10			

a. R Squared = .942 (Adjusted R Squared = .885)

204

INSERT 3.7	INSERT 3.7 continued

* Full Model.
✔ File
✔ New
✔ Syntax
* Type the following.
if (mindep = 75)j=1.
if (mindep = 100)j=2.
if (mindep = 125)j=3.
if (mindep = 150)j=4.
if (mindep = 175)j=5.
if (mindep = 200)j=6.
execute.
✔ Run
✔ All
 	
* Method 1.
✔ Analyze
✔ General Linear Model
✔ Univariate
✔ *newaccts*
✔ ► (to the Dependent box)
	✔ *j*
	✔ ► (to the Fixed factors)
	✔ *mindep*
	✔ ► (to the Covariate box)
	✔ OK
	
	* Method 2.
	✔ Analyze
	✔ General Linear Model
	✔ Univariate
	✔ *newaccts*
	✔ ► (to the Dependent box)
	✔ *mindep*
	✔ ► (to the Covariate box)
	✔ Options
	✔ Lack of Fit
	✔ Continue
	✔ OK

SPSS Output 3.7b

Tests of Between-Subjects Effects

Dependent Variable: NEWACCTS

Source	Type III Sum of Squares	df	Mean Square	F	Sig.
Corrected Model	5141.338[a]	1	5141.338	3.139	.110
Intercept	2714.916	1	2714.916	1.658	.230
MINDEP	5141.338	1	5141.338	3.139	.110
Error	14741.571	9	1637.952		
Total	170696.000	11			
Corrected Total	19882.909	10			

a. R Squared = .259 (Adjusted R Squared = .176)

Lack of Fit Tests

Dependent Variable: NEWACCTS

Source	Sum of Squares	df	Mean Square	F	Sig.
Lack of Fit	13593.571	4	3398.393	14.801	.006
Pure Error	1148.000	5	229.600		

Box-Cox Transformations: SPSS does not have a readily available option to generate *Box-Cox transformations* (text page 134). Insert 3.8 contains the necessary syntax in matrix language to produce Table 3.9 and Figure 3.17, text page 136. (SPSS matrix language will be covered in

Chapter 5.) The syntax requires the data file presented in Table 3.8, text page 133, to be open in the active Data Editor and the variables to be named *age*, *plasma*, and *logy*. Note that the syntax in Insert 3.8 will overwrite the active Data Editor; save the contents of the Data Editor as needed. The first two "get statements" set vectors **x** and **y** equal to *age* and *plasma*, respectively. This procedure uses a range of λ values starting at λ (*start*) = -2.3 and incrementing (*incremen*) by a value of 0.1 for 34 iterations (*noiter*). A plot of SSE as a function of λ is shown in SPSS Output 3.8. The start, increment, and iterations values can be varied to specify choices suitable for different data sets.

```
                              INSERT 3.8

    ✔ File
    ✔ New
    ✔ Syntax
    matrix.
    get x / variables = age.
    get y / variables = plasma.

    compute start = -2.3.
    compute noiter = 34.
    compute incremen = 0.1.

    compute n=nrow(y).
    compute x = {make(n,1,1),x}.
    compute hat = x*inv(t(x)*x)*t(x).
    compute lambda = {make(noiter,1,1)}.
    compute lambda(1,1)=start.
    loop i = 2 to noiter.
      compute lambda(i,1)=lambda(i-1,1) + incremen.
    end loop.
      compute sse={make(noiter,1,0)}.
    loop i = 1 to noiter.
      compute lamb = lambda(i).
      compute k2 = exp((1/n)&*({make(1,n,1)}*ln(y))).
      compute k1 = 1/(lamb*k2**(lamb-1)).
      do if (lamb > -.00000001 and lamb < .00000001).
      compute w=k2*ln(y).
      else.
      compute w = k1*( y&**lamb-{make(n,1,1)}).
      end if.
      compute e=(ident(n) - hat)*w.
      compute sse(i)=t(e)*e.
    end loop.
    compute newm={lambda,sse}.
      print newm.
      save newm / outfile=*.
    end matrix.
    graph
     /scatterplot(bivar)=col1 with col2
     /missing=listwise
     /title= 'Plot of Box-Cox text Figure 3.17'.
      execute.
    ✔ Run
    ✔ All
```

SPSS Output 3.8

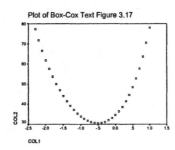

Plot of Box-Cox Text Figure 3.17

Exploration of Shape of Regression Function – Lowess Method: The authors state that "the name *lowess* stands for locally weighted regression scatter plot smoothing." The lowess procedure divides the independent variables into groups or *local neighborhoods* and then fits a regression model to each local neighborhood of data. The procedure uses weighted least squares to fit regression models for each neighborhood with the weights being a function of the distance of the X's for a given subject in a neighborhood from the center of the neighborhood. The final fitted model is a blending of these fitted models in each neighborhood. The proportion of the total number of observations to be included in each neighborhood (*percent of points to fit*) can be specified; the default value is 50%. SPSS also allows us to specify the *number of iterations*; the default value is 3. We can adjust the size of successive neighborhoods and iterations and examine a plot of the fitted loess curves to choose a reasonable fitting smooth curve.

In Insert 3.9, we present SPSS syntax to produce Figure 3.19a, text page 140. Using the Toluca Company example, we produce a scatter plot of *lotsize* and *workhrs*. In the Chart Editor we then request the lowess smoothed curve as shown in SPSS Output 3.9. SPSS does not produce the lowess curve and confidence bands on the same scatter plot.

INSERT 3.9

✔ Graphs
✔ Scatter
✔ Simple
✔ Define
✔ *lotsize*
✔ ► (to the X axis)
✔ *workhrs*
✔ ► (to the Y axis)
✔ OK
* To draw the regression line as shown in text Figure 1.10(b),.
* double click on the graph in the SPSS Viewer. This brings up a.
* SPSS Chart Editor window. For **Release 12.0** and higher see.
* Insert 1.1.
✔ Chart
✔ Options
✔ Total (under Fit Line)
✔ Fit Options
✔ Lowess
✔ Continue
✔ OK
* Close the Chart Editor.

SPSS Output 3.9

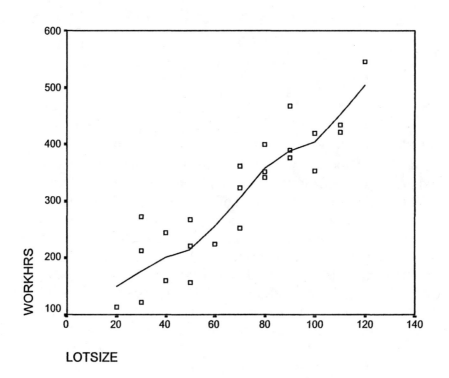

Chapter 4 SPSS®

Simultaneous Inferences and Other Topics in Regression Analysis

The level of significance of a statistical test is the probability the test rejects the null hypothesis given that the null hypothesis is true. This probability is usually denoted α and is referred to as the probability of a Type I error. These concepts are in the context of a single application of the test. If the test is applied to multiple hypotheses we often want to make a simultaneous or joint statement about the probability of a Type I error given that all the hypotheses being tested are true. We call a set of tests under consideration a family of tests. If each test in a family of g tests is tested at level of significance α, the probability of making one or more Type I errors if all g hypotheses are true is α^* where:

$$\alpha^* \leq 1 - (1-\alpha)^g$$

This inequality is a form of the Bonferroni Equality. In this statement, equality holds if the tests are mutually independent. It is apparent that as g increases α^* approaches 1 and, unless we take corrective action, the probability is close to one that we falsely reject at least one hypothesis even if they are all true. The right side of the statement is approximately equal to $g\alpha$ if α is small; i.e.,

$$1 - (1-\alpha)^g \approx g\alpha$$

Therefore, if we perform each of g tests at the α/g level we get:

$$1 - \left(1 - \frac{\alpha}{g}\right)^g \approx \alpha$$

This approach to multiple testing is referred to as the Bonferroni method or procedure. Using the Bonferroni procedure, we perform each of g tests at the α/g level and are assured that the probability is $\leq \alpha$ of making one or more Type I errors if all g hypotheses are true. For example, if we perform 4 tests and use the $0.05/4 = 0.0125$ level of significance for each test, the probability we falsely reject one or more of the 4 corresponding hypotheses is less than or equal to 0.05.

If we are conducting a single two-sided test, we find critical values from the t-distribution by finding $t(1 - \alpha/2; df)$ where df is the degrees of freedom of the test. Similarly, if we are conducting g two-sided multiple tests, we find critical values from the t-distribution by finding $t(1 - \alpha/(2g); df)$ where df is the degrees of freedom of the test.

Analogous to multiple testing, we often want to set g confidence intervals with the knowledge that the confidence level is $\geq 100(1 - \alpha)$ percent that all g intervals bracket their respective objective. In the simple linear regression model, simultaneous (Bonferroni) confidence limits for estimating β_0 and β_1 are (equation 4.3, text page 156):

$$b_0 \pm Bs\{b_0\}$$

$$b_1 \pm Bs\{b_1\}$$

where $B = t(1 - \alpha/4; n - 2)$. Here $g = 2$ so that $\alpha/(2g) = \alpha/4$.

Section 4.2, text page 157, discusses simultaneous confidence intervals for the mean responses at different values of X. In addition to the Bonferroni procedure, the authors discuss the *Working-Hotelling* procedure. Section 4.3 discusses simultaneous prediction intervals for new observations. Section 4.4 presents the simple model for regression through the origin while Section 4.5 discusses the effects of measurement errors while Section 4.6 presents inverse predictions.

More on The Toluca Company Example

Data File: ch01ta01.txt

We return to the Toluca Company example from Chapters 1, 2 and 3 (text page 19, 46, 156). See Chapter 0, "Working with SPSS," for instructions on opening the Toluca Company data file and renaming the variables *lotsize* and *workhrs*. After opening ch01ta01.dat, the first 5 lines in your SPSS Data Editor should look like SPSS Output 4.1.

SPSS Output 4.1

	lotsize	workhrs	var	var	var
1	80	399			
2	30	121			
3	50	221			
4	90	376			
5	70	361			

Bonferroni Joint Confidence Intervals for β_0 and β_1: On text page 156, the Toluca Company wants to estimate the 90 percent family confidence intervals for β_0 and β_1. To estimate the 90 percent confidence intervals for β_0 and β_1, the text reports $B = t(1 - 0.10/4; 23) = t(0.975; 23) =$

2.069. The *t*-value is actually based on $(1 - \alpha/(2g))$ where g is the number of joint intervals to be estimated. In the present example, we are estimating joint confidence intervals for β_0 and β_1 ($g=2$), so $B = t(1 - 0.10/(2*2); 23) = t(1 - 0.10/4; 23) = t(0.975; 23)$. Thus, we use SPSS' *inverse distribution function* to find $B = IDF.T(0.975, 23) = 2.068$. From SPSS Output 1.3, we know b_0, $s\{b_0\}$, b_1, and $s\{b_1\}$. In Insert 4.1 we use equation 4.3, text page 156, and the Bonferroni procedure for developing joint confidence intervals for b_0 (*b0lo* and *b0hi*) and b_1 (*b1lo* and *b1hi*).

INSERT 4.1

✔ File
✔ New
✔ Syntax
* Type the following syntax.
compute b=idf.t(.975,23).
compute b0lo=62.37-(b*26.18).
compute b0hi=62.37+(b*26.18).
compute b1lo=3.5702-(b*.3470).
compute b1hi=3.5702+(b*.3470).
execute.
✔ Run
✔ All

Simultaneous Estimation of Mean Responses: On text page 158, the Toluca Company wants to estimate the mean number of work hours required to produce lots of 30, 65, and 100 units where the family confidence coefficient is 0.90. In Chapter 2 we presented the steps to find the confidence interval for $E\{Y_h\}$ when lot size (X_h) = 65 units and the confidence band for the entire regression line. We repeat these steps here with fewer notations and more emphasis on SPSS computational statements.

a. Working-Hotelling Procedure: We first find the 90 percent confidence interval for work hours when lot size (X_h) = 65 units. We use the Working-Hotelling $1-\alpha$ band (text equations 2.40, 2.40a, 2.30) to establish boundary values for (X_h) = 65 units:

$$\hat{Y}_h \pm Ws\{\hat{Y}_h\}$$

where

$$W^2 = 2 F(1-\alpha; 2, n-2)$$

and

$$s^2\{\hat{Y}_h\} = MSE\left[\frac{1}{n} + \frac{(X_h - \bar{X})^2}{\sum(X_i - \bar{X})^2}\right]$$

In Insert 4.2 – Method 1, we first find the *F*-value (*fvalue*) associated with $\alpha = 0.90$ and 2, 23 degrees of freedom. We then find $W = \sqrt{2 * fvalue}$. Next, we find $\hat{Y}_h$ (*yhat*) when $X_h = 65$ units using previously found parameter estimates (SPSS Output 1.3), $b_0 = 62.366$ and $b_1 = 3.570$. We

211

find $s\{\hat{Y}_h\}$ (*sdyhat*), the square root of text equation 2.30. Finally, we compute $\hat{Y}_h \pm Ws\{\hat{Y}_h\}$ placing the lower boundary of the confidence interval in a variable named *bandlo* and the upper boundary of the confidence interval in a variable named *bandhi*. Execution of Insert 4.2 will place in the Data Editor the 90 percent confidence interval for *workhrs* when $X_h = 65$ units. After the syntax is working properly, as in Insert 4.2 – Method 1, we can rewrite the syntax in a more compact form if desired. Insert 4.2 Method 1 can be reduced to Method 2.

To find the confidence intervals for $X_h = 30$ and $X_h = 100$, simply change the computation of *yhat* and *sdyhat* in Insert 4.2 – Method 1 to reflect 30 or 100 units instead of 65 (bolded) units. Using this method, old values for *bandlo* and *bandhi* will be overwritten each time you change the value of X_h. If this is undesirable, the syntax in Insert 4.3 will find the 90 percent confidence interval for each value of X_h. These intervals will be placed in unique variables, i.e., *bandlo30*, *bandhi30*, *bandlo65*, *bandhi65*, *bandlo00*, and *bandhi00*.

INSERT 4.2

```
* Method 1.
✔ File
✔ New
✔ Syntax
* Type the following syntax.
compute fvalue=idf.f(.9,2,23).
compute w=sqrt(2*fvalue).
compute yhat=62.366+(3.570*65).
compute sdyhat=sqrt(2383.716*((1/25)+((65-70)**2/19800))).
compute bandlo=yhat-(w*sdyhat).
compute bandhi=yhat+(w*sdyhat).
execute.
✔ Run
✔ All
```

INSERT 4.2 continued

```
* Method 2.
✔ File
✔ New
✔ Syntax
* Type the following syntax.
compute fvalue=idf.f(.9,2,23).
compute w=sqrt(2*fvalue).
compute bandlo=(62.366+(3.570*65))-(sqrt(2*idf.f(.9,2,23))*sqrt(2383.716*((1/25)+((65-70)**2/19800)))).
compute bandhi=(62.366+(3.570*65))+(sqrt(2*idf.f(.9,2,23))*sqrt(2383.716*((1/25)+((65-70)**2/19800)))).
execute.
✔ Run
✔ All
```

✔ File
✔ New
✔ Syntax
* Type the following syntax.
compute fvalue=idf.f(.9,2,23).
compute w=sqrt(2*fvalue).
compute yhat30=62.366+(3.570***30**).
compute sdyhat30=sqrt(2383.716*((1/25)+((**30**-70)**2/19800))).
compute bandlo30=yhat30-(w*sdyhat30).
compute bandhi30=yhat30+(w*sdyhat30).
compute yhat65=62.366+(3.570***65**).
compute sdyhat65=sqrt(2383.716*((1/25)+((**65**-70)**2/19800))).
compute bandlo65=yhat65-(w*sdyhat65).
compute bandhi65=yhat65+(w*sdyhat65).
compute yhat100=62.366+(3.570***100**).
compute sdyhat00=sqrt(2383.716*((1/25)+((**100**-70)**2/19800))).
compute bandlo00=yhat100-(w*sdyhat00).
compute bandhi00=yhat100+(w*sdyhat00).
execute.
✔ Run
✔ All

b. Bonferroni Procedure: On text page 159, the authors use data from the Toluca Company example and the Bonferroni Procedure to estimate the mean number of work hours required to produce lots of 30, 65, and 100 units where the family confidence coefficient is 0.90. The Bonferroni confidence limits are:

$$\hat{Y}_h \pm Bs\{\hat{Y}_h\}$$

where:

$$B = t(1 - \alpha/2g; n-2)$$

and g is the number of joint confidence intervals to be estimated. In this example, there are three intervals in the family. Thus, $B = t(1 - 0.10/(2*\mathbf{3}); 23) = t(1 - 0.10/6; 23) = t(1 - 0.01666; 23) = t(0.9833, 23)$. We can use the *IDF.T* function as before to find $B = idf.t(0.98333, 23) = 2.263$. Computation of $s\{\hat{Y}_h\}$ is the same as for the Working-Hotelling procedure above including Insert 4.2. Thus, to use the Bonferroni Procedure to find the 90 percent confidence interval for work hours when $X_h = 65$ units, we use text equations 4.7 and 4.7a as in Insert 4.4 to find the upper (*bandhi*) and lower (*bandlo*) confidence limits. Confidence limits for $X_h = 30$ and $X_h = 100$ can be generalized from Inserts 4.2 and 4.3.

213

```
                          INSERT 4.4

  ✔ File
  ✔ New
  ✔ Syntax
  * Type the following syntax.
  compute b=idf.t(.98333,23).
  compute yhat=62.366+(3.570*65).
  compute sdyhat=sqrt(2383.716*((1/25)+((65-70)**2/19800))).
  compute bandlo=yhat-(b*sdyhat).
  compute bandhi=yhat+(b*sdyhat).
  execute.
  ✔ Run
  ✔ All
```

Prediction Intervals for New Observations: The Toluca Company now wants to predict the required work hours for lots of 80 and 100 units. The researchers first obtained S and B (text page160) to determine which prediction interval was smaller. Insert 4.5 uses the *IDF.T* and *IDF.F inverse distribution functions* to find $S = 2.616$ (equation 4.8a, text page 160) and $B = 2.398$ (equation 4.9a, text page 160). Because the Bonferroni procedure produces the tighter limits, we use B to establish the 95% family coefficients.

We now calculate $\hat{Y}_h$ for 80 (*yhat80*) and 100 (*yhat100*) units. We then use equation 2.38a, text page 59, to find $s\{pred\}$ for 80 (*sdpre80*) and 100 (*sdpre00*) units. Finally, we use equation 4.9, text page 160, to find the lower and upper 95% confidence limits for 80 (*bandlo80* and *bandhi80*, respectively) and 100 (*bandlo00* and *bandhi00*, respectively) units.

```
                          INSERT 4.5

  ✔ File
  ✔ New
  ✔ Syntax
  compute s=sqrt(2*idf.f(.95,2,23)).
  compute b=idf.t(1-(.05/(2*2)),23).
  compute yhat80=62.366+(3.570*80).
  compute sdpre80=sqrt(2383.716*(1+(1/25)+((80-70)**2/19800))).
  compute bandlo80=yhat80-(b*sdpre80).
  compute bandhi80=yhat80+(b*sdpre80).
  compute yhat100=62.366+(3.570*100).
  compute sdpre00=sqrt(2383.716*(1+(1/25)+((100-70)**2/19800))).
  compute bandlo00=yhat100-(b*sdpre00).
  compute bandhi00=yhat100+(b*sdpre00).
  execute.
  ✔ Run
  ✔ All
```

Inverse Predictions: The authors present an example of inverse prediction on text page 168. We do not have access to the source data file, however, and will use the Toluca Company for illustration. Suppose the auditing department of the Toluca Company's corporate home office requested that lot size $\hat{X}_{h(new)}$ be estimated with a 95% confidence interval when the work hours for a run is known to be $Y_{h(new)} = 300$. Although the accountants at the Toluca Company failed to appreciate the need for such a calculation, they complied with the request.

In Insert 4.6 we first find $\hat{X}_{h(new)} = Y_{h(new)} - 62.366 / 3.570 = 66.56$ using equation 4.31, text page 168, and parameter estimates from SPSS Output 1.3. For equation 4.32, text page 169, we find t using $IDF.T(1-0.05/2, 23) = 2.07$ and place the result in *tvalue*. Using equation 4.32a, text page 169, we find $s^2\{predX\}$ (*sdprex*) = 13.95. Finally, we use text equation 4.32 to find:

$$\hat{X}_{h(new)} \pm t(1-\alpha/2;\ n-2)s\{predX\}$$

$$66.56 \pm 2.07 * 13.95$$

$$37.70 < \hat{X}_{h(new)} < 95.42$$

The auditors at corporate the home office conclude with 95% confidence that the number of hours required to produce the lot size of 300 units is between 37.70 and 95.42 hours. The auditors calculate the error to be 66.56/((95.42-37.70)/2) or approximately $\pm 43\%$.

INSERT 4.6

✔ File
✔ New
✔ Syntax
* Inverse prediction.
compute xhnew=(300-62.366)/3.570.
compute tvalue=IDF.T(1-.05/2,23).
compute sdprex=sqrt((2383.716/(3.570**2)*(1+(1/25)+((**xhnew**-70)**2/19800)))).
compute prexlo=xhnew-(tvalue*sdprex).
compute prexhi=xhnew+(tvalue*sdprex).
execute.
✔ Run
✔ All

The Charles Plumbing Supplies Example

Data File: ch04ta02.txt

Regression through the Origin: From text page 162, the Charles Plumbing Supplies Company studied the relationship between the total labor costs (Y) and the number of work units performed (X). The authors use data from this example to illustrate fitting a regression model that has no intercept term and thus the fitted line passes through the origin. The explicit regression model used for this purpose is given in equation 4.10, page 161 of the text. After opening ch04ta02.txt and renaming the variables X and Y, the first 5 lines in the Data Editor should appear as SPSS Output 4.2. To perform regression through the origin, simply uncheck the Include Constant in Equation option (Insert 4.6). This results in SPSS Output 4.3 and is consistent with text pages 163-164.

SPSS Output 4.2

	x	y	var	var	var
1	20.00	114.00			
2	196.00	921.00			
3	115.00	560.00			
4	50.00	245.00			
5	122.00	575.00			

INSERT 4.6

- ✔ Analyze
- ✔ Regression
- ✔ Linear
- ✔ y
- ✔ ▶ (to the Dependent box)
- ✔ x
- ✔ ▶ (to the Independent box)
- ✔ Options
- ✔ Include Constant in Equation * Uncheck the box.
- ✔ Continue
- ✔ Statistics
- ✔ Confidence Intervals (under Regression coefficients)
- ✔ Continue
- ✔ OK

SPSS Output 4.3

ANOVA[c,d]

Model		Sum of Squares	df	Mean Square	F	Sig.
1	Regression	4191980	1	4191980.341	18762.480	.000[a]
	Residual	2457.659	11	223.424		
	Total	4194438[b]	12			

a. Predictors: V1

b. This total sum of squares is not corrected for the constant because the constant is zero for regression through the origin.

c. Dependent Variable: V2

d. Linear Regression through the Origin

Coefficients[a,b]

	Unstandardized Coefficients		Standardized Coefficients			95% Confidence Interval for B	
	B	Std. Error	Beta	t	Sig.	Lower Bound	Upper Bound
1	4.685	.034	1.000	136.976	.000	4.610	4.761

a. Dependent Variable: V2

b. Linear Regression through the Origin

217

Chapter 5 SPSS®

Matrix Approach to Simple Linear Regression Analysis

The text begins Chapter 5 by presenting an overview of the algebra for vectors and matrices. In Chapter 5 of this manual, we introduce *MATRIX - END MATRIX* commands, which are only available through the Syntax Editor. These commands allow the user to write SPSS statements for performing vector/matrix algebra. We follow examples in the text and show how to write SPSS programs to perform operations such as matrix addition, subtraction and multiplication. Following the overview of matrix algebra, the text expresses the simple linear regression model as a matrix equation and then repeats the steps followed in Chapter 1, only now these steps are described using matrix algebra.

Again, we are given Y_i and X_i for a sample of n subjects and we want to find a "best fitting" model for predicting Y given X. In Chapter 1, the simple linear regression model for the ith subject, $i = 1, \ldots, n$ was written as:

$$Y_i = \beta_0 + \beta_1 X_i + \varepsilon_i$$

To summarize the key results of the matrix formulation, let the capital bold-faced letter **Y** represent the nx1 column vector containing the data for the dependent variable and the bold-faced letter **X** represent the nx2 matrix with the first column being a column vector of ones and the second column being the column vector of data for the independent variable; i.e.,

$$\mathbf{Y} = \begin{bmatrix} Y_1 \\ Y_2 \\ \vdots \\ Y_n \end{bmatrix} \quad \text{and} \quad \mathbf{X} = \begin{bmatrix} 1 & X_1 \\ 1 & X_2 \\ \vdots & \vdots \\ 1 & X_n \end{bmatrix}$$

Further, let

$$\boldsymbol{\beta} = \begin{bmatrix} \beta_0 \\ \beta_1 \end{bmatrix} \quad \text{and} \quad \boldsymbol{\varepsilon} = \begin{bmatrix} \varepsilon_1 \\ \varepsilon_2 \\ \vdots \\ \varepsilon_n \end{bmatrix}$$

Using this notation, the simple linear regression model can be written:

$$Y = X\beta + \varepsilon$$

A few key results of the matrix formulation of the regression model are:

(1.) Matrix solution for the regression coefficients: $\mathbf{b} = \begin{bmatrix} b_0 \\ b_1 \end{bmatrix} = (\mathbf{X'X})^{-1}\mathbf{X'Y}$

(2.) Matrix solution for the predicted values: $\hat{\mathbf{Y}} = \mathbf{Xb} = \mathbf{X}(\mathbf{X'X})^{-1}\mathbf{X'Y} = \mathbf{HY}$

where $\mathbf{H} = \mathbf{X}(\mathbf{X'X})^{-1}\mathbf{X'}$ is called the *hat matrix.*

(3.) Matrix solution for the fitted model residuals: $\mathbf{e} = \mathbf{Y} - \hat{\mathbf{Y}} = \mathbf{Y} - \mathbf{Xb} = (\mathbf{I} - \mathbf{H})\mathbf{Y}$

From these equations, we begin to see that the computations required for regression analysis can be written as matrix expressions. As stated earlier, *MATRIX* is an SPSS procedure that performs matrix algebra. Most of our needs are met by other SPSS syntax that is more user friendly than that specific to *MATRIX*, but occasionally we need to perform computations that are not directly available, and *MATRIX* provides a convenient way to write programs according to our specifications. Moreover, it is instructional to write matrix language programs that perform regression analysis. We illustrate the use of *MATRIX* by presenting SPSS statements to compute (1.) – (3.) above and other key matrix expressions for the regression analysis of data given in the Toluca Company example. We begin by introducing some basic mathematical operators used in the SPSS *MATRIX* software.

Using *MATRIX* – *END MATRIX* for Matrix Operations: We initiate the *MATRIX* procedure with the keyword *MATRIX*. Next, we use *MATRIX* syntax to write statements for performing one or more matrix operations. We end the procedure with *END MATRIX*. In the following, we illustrate some of the *MATRIX* statements that are useful for programming a regression analysis.

An SPSS program for finding the transpose of a matrix **A** and a vector **C** is given in Insert 5.1. In this program, the numerical values of the matrix **A** and the vector **C** are specified directly in the *MATRIX* procedure. When this is done, the name you assign to the matrix is set equal to a quantity in braces. The SPSS program does not differentiate between upper and lower case names. Therefore, in SPSS code we can write "A = { }" or "a = { }" to specify the matrix **A**. The matrix is specified by rows with row elements separated by commas and complete rows separated by semicolons. Thus, the three lines of code that immediately follow the *MATRIX* statement specify the 3x2 matrix **A** and the next three lines specify the 3x1 vector **C**.

Insert 5.1 then finds the transpose of **A** and **C**. The transpose operator takes the form of *TRANSPOS*(M) or *T*(M)**. Thus, the statement "*COMPUTE AT = TRANSPOS(A).*" and "*COMPUTE AT=T(A).*" exchanges the rows and columns of matrix **A** and stores the results in a matrix named **AT**. Similarly, the statement "*COMPUTE CT = T(C).*" exchanges the rows and columns of matrix **C** and stores the results in a matrix named **CT**. The print statements in the next lines of Insert 5.1 instruct SPSS to print matrices **A**, **AT**, **C**, and **CT** (Output 5.1). These results are given on text page 178 and 179.

```
title "Transpose of a matrix from text page 178".
matrix.
compute A = {2,5;7,10;3, 4}.
compute C = {4; 7;10}.
compute AT = T(A).
compute CT = T(C).
print A.
print AT.
print C.
print CT.
end matirx.
```

SPSS Output 5.1

```
Run MATRIX procedure:

A
     2     5
     7    10
     3     4

AT
     2     7     3
     5    10     4

C
     4
     7
    10

CT
     4     7    10

------ END MATRIX -----
```

Insert 5.2 shows *MATRIX* syntax for text examples of matrix addition, subtraction, and multiplication, and finding the inverse of a matrix. The output is not presented for brevity. The symbol "+" denotes the addition operator and the symbol "−" denotes the subtraction operator. The asterisk symbol "*" denotes the multiplication operator. The *MATRIX* function $inv(A)$ finds the inverse of the matrix **A**. We have named the result *Ainv* but we could have chosen another convenient name. For example, we could have called the result d by writing $d = inv(A)$.

INSERT 5.2

```
title 'Addition and Substraction of a matrix from text page 180'.
matrix.
compute a={1,4;2,5;3,6}.
compute b={1,2;2,3;3,4}.
compute apb = a + b.
compute amb = a - b.
print a.
print b.
print apb.
print amb.
end matirx.

title 'Multiplication of a matrix by a scalar (4) from text page 182'.
matrix.
compute A = {2, 7; 9, 3}.
compute B = 4*A.
print A.
print B.
end matrix.

title 'Multiplication of a matrix by a matrix from text page 182'.
matrix.
compute A = {2, 5; 4, 1}.
compute B = {4, 6; 5, 8}.
compute C = A*B.
print A.
print B.
print C.
end matrix.

title 'Inverse of a matrix from text page 189'.
matrix.
compute a={2,4;3,1}.
compute ainv = inv(a).
print a.
print ainv.
end matirx.
```

The Toluca Company Example

Data File: ch01ta01.txt

We use matrix methods to perform a simple linear regression analysis of the Toluca Company example (text page 19). The required steps are presented in text Chapter 5 and are not detailed here except in Insert 5.3. For the *MARTIX* program in Insert 5.3 to function, the Toluca Company example data file must be the active file in your Data Editor. See Chapter 0, "Working with SPSS," for instructions on opening the Toluca Company data file and renaming the variables *lotsize* and *workhrs*. After opening ch01ta01.txt, the first 5 lines in your SPSS Data Editor should look like SPSS Output 1.1.

The *MATRIX* statement initiates the *MATRIX* procedure. The statement "*GET X / VARIABLES = LOTSIZE.*" stores the column of data named *lotsize* in matrix (vector) **X**. Similarly, the next statement stores the column of data labeled *workhrs* in matrix (vector) **Y**.

The statement *n = nrow(x)* assigns the value of the number of rows in matrix **X** to variable label *n*. The statement *p = ncol(x)* assigns the value of the number of columns in matrix **X** to variable label *p*.

The function *MAKE(a,b,c)* creates an *axb* matrix whose elements all have the value *c*. Thus, *MAKE(n,1,1)* creates an *n*x1 matrix whose elements all have the value '1'. The "*COMPUTE X = {MAKE(N,1,1), X}*" statement creates the *n*x1 matrix and concatenates it with the column vector **X** defined above as the column of *lotsize* values. Thus, **X** is now equal to **X** defined in equation 5.61 of text page 200.

The remaining statements are easy to follow. The statement *xpx* = transpos(*x*)*x computes **X'X** and stores the result as a matrix named *xpx*; the statement *xpy* = transpos(*x*)*y computes **X'Y** and stores the result as a matrix named *xpy*, and so on.

The print statement requests the printing of **Y** (eq 5.61, text page 200), **X** (eq 5.61, text page 200), $\hat{\textbf{Y}}$ (eq 5.71, text page 202), and **E** (**e**, eq 5.76, text page 203) **XPXINV** ($(\textbf{X'X})^{-1}$, eq 5.64, text page 201), **B** (eq 5.60, text 200), **YPY** (**Y'Y**, (eq 5.13, text page 185), **CORRECT** (corrected SSTO), **SSTO** (eq 5.83, text page 204 and 5.89a, text page 206), **SSR** (**b'X'Y**, text page 205), **SSE** (**Y'Y-b'X'Y**, text page 205), **MSE**, **VARB** ($s^2\{b\}$, eq 5.94, text page 207), **XH**, **VARYHATH** ($s^2\{\hat{Y}\}$, eq 5.98, text page 208). Output 5.2 displays the first 5 lines of **X** and **Y** matrices for illustration.

```
                        INSERT 5.3

matrix.
get x / variables = lotsize.
get y /variables = workhrs.
compute n=nrow(x).
compute x={make(n,1,1),x}.
compute p=ncol(x).
compute xpx=transpos(x)*x.
compute xpy=transpos(x)*y.
compute xinv=inv(t(x)*x).
compute b=inv(t(x)*x)*t(x)*y.
compute yhat=x*b.
compute e=y-yhat.
compute ypy=t(y)*y.
compute correct=(t(y)*(make(n,n,1))*y/n).
compute sstot=t(y)*y-t(y)*(make(n,n,1))*y/n.
compute ssr=t(b)*t(x)*y.
compute sse=t(y)*y-t(b)*t(x)*y.
compute mse=sse/(n-p).
compute varb=mse*inv(t(x)*x).
compute xh={1;65}.
compute varyhath=t(xh)*varb*xh.
print y / title "Y = Work Hours" .
print x / title "X = Lot Size" .
print yhat / title "Predicted Values".
print xinv / title "X Inverse".
print e / title "Residual Values".
print xpy / title "(XPY), Y = Work Hours" .
print xpx / title "(XPX), X = Lot Size" .
print xinv / title "X'X inverse".
print b / title "Parameter estimates".
print correct / title "Corrected sum of squares".
print sstot / title "Total sum of squares".
print ssr / title "Sum of squares regression".
print sse / title "Sum of squares error".
print mse / title "Mean square error".
print varb / title "Variance of b".
end matrix.
execute.
```

SPSS Output 5.2

Y Matrix	X Matrix	
399	1	80
121	1	30
221	1	50
376	1	90
361	1	70

Chapter **6** SPSS®

Multiple Regression I

In many applications using regression analysis, the model may involve more than one predictor variable. We call such a model the *multiple regression model* or the *general linear regression model*. We can write the general linear regression model as:

$$Y_i = \beta_0 + \beta_1 X_{i1} + \beta_2 X_{i2} + ... + \beta_{p-1} X_{i,p-1} + \varepsilon_i$$

where:

$\beta_0, \beta_1, ..., \beta_{p-1}$ are unknown model parameters with fixed values

$X_{i1}, ..., X_{i,p-1}$ are known constants

ε_i are independent $N(0, \sigma^2)$ random variables

$i = 1, ..., n$

In matrix notation, the general linear regression model is

$$\mathbf{Y} = \mathbf{X}\boldsymbol{\beta} + \boldsymbol{\varepsilon}$$

where:

$\mathbf{Y}$ is the nx1 vector of response observations on the dependent variable

$\boldsymbol{\beta}$ is the px1 vector of unknown model parameters

$\mathbf{X}$ is the nxp matrix of known constants

$\boldsymbol{\varepsilon}$ is a vector of independent $N(0, \sigma^2)$ random variables

Chapter 6 focuses on the special case where $p = 3$ and therefore only two predictor variables are in the general linear regression model. In the following, we give illustrative SPSS syntax for performing a statistical analysis of the general linear regression model with $p = 3$.

Dwaine Studios, Inc. Example

Data File: ch06fi05.txt

Multiple Regression with Two Predictor Variables: Dwaine Studios, Inc. has locations in 21 medium size cities and is considering expansion into additional cities (text page 236). The company wants to predict sales (Y) in a community using the number of persons greater than 15 years in the community ($X1$) and per capita disposable income in the community ($X2$). See Chapter 0, "Working with SPSS," for instructions on opening the Dwaine Studios data file and renaming the variables $X1$, $X2$, and Y. After opening ch06fi05.txt, the first 5 lines in your SPSS Data Editor should look like SPSS Output 6.1.

SPSS Output 6.1

```
dwaine studio example.sav - SPSS Data Editor
File  Edit  View  Data  Transform  Analyze  Graphs  Utilities  Window  Help
```

```
5: y                                      181.6
```

	x1	x2	y	var	var
1	68.50	16.70	174.40		
2	45.20	16.80	164.40		
3	91.30	18.20	244.20		
4	47.80	16.30	154.60		
5	46.90	17.30	181.60		

The researchers began with the first order regression model. (Insert 6.1) Results are shown in SPSS Output 6.2 and are consistent with Figure 6.5, text page 237. In Insert 6.1 we also present the matrix language syntax for estimating the parameters. Notice that the **X** matrix on text page 237 contains a column of 1's and that this column is not in the Dwaine Studios data file. Thus, in Insert 6.1 we first compute a variable, *X0* that is equal to 1 for each of the 20 cities in the data file. We then enter the SPSS *MATRIX* language (notice the line that specifies *matrix* in Insert 6.1) and set column vectors **CON, A, B,** and **Y** equal to *X0, X1, X2,* and *Y*, respectively. Next, we concatenate the **CON, A,** and **B** column vectors into the **X** matrix such that the **X** matrix is now as shown in equation 6.70, text page 237. (Although the method of constructing the **X** matrix in Insert 6.1 is different from the method presented in Chapter 5, both methods achieve the same outcome.) Next, we find **X'X** (*xtx*) , **X'Y** (*xty*), $(\mathbf{X'X})^{-1}$ (*xtxinv*), and $\mathbf{b} = (\mathbf{X'X})^{-1} \mathbf{X'Y}$. The "*print b.*" statement will print the parameter estimates. Although we do not display the output from the "*print b.*" statement, the estimates obtained through the matrix language program are equal to the estimates shown in SPSS Output 6.2. We did not print **X'X, X'Y,** and $(\mathbf{X'X})^{-1}$, but we could do so and compare the numerical output to the text results shown on text pages 238-239 and the SPSS regression analysis shown in SPSS Output 6.2.

INSERT 6.1	INSERT 6.1 continued	INSERT 6.1 continued
* Regresson analysis. ✔ Analyze ✔ Regression ✔ Linear ✔ *y* ✔ ▶(to the Dependent box) ✔ *x1* ✔ ▶(to the Independent box) ✔ *x2* ✔ ▶(to the Independent box) ✔ OK	* Matrix format. ✔ File ✔ Open ✔ Syntax * Type the following syntax. compute x0=1. execute. matrix. get con / variables =x0. get a / variables = x1. get b / variables = x2. get y /variables=y.	compute x = {con, a, b} . compute xtx=t(x)*x. comute xty=t(x)*y. compute xtxinv=inv(t(x)*x). compute b=inv(t(x)*x)*t(x)*y. print b. end matrix. execute. ✔ Run ✔ All

SPSS Output 6.2

Model Summary

Model	R	R Square	Adjusted R Square	Std. Error of the Estimate
1	.957[a]	.917	.907	11.00739

a. Predictors: (Constant), X2, X1

ANOVA[b]

Model		Sum of Squares	df	Mean Square	F	Sig.
1	Regression	24015.282	2	12007.641	99.103	.000[a]
	Residual	2180.927	18	121.163		
	Total	26196.210	20			

a. Predictors: (Constant), X2, X1

b. Dependent Variable: Y

Coefficients[a]

Model		Unstandardized Coefficients		Standardized Coefficients	t	Sig.
		B	Std. Error	Beta		
1	(Constant)	-68.857	60.017		-1.147	.266
	X1	1.455	.212	.748	6.868	.000
	X2	9.366	4.064	.251	2.305	.033

a. Dependent Variable: Y

To produce the three-dimensional scatter plot shown on text page 238 (text Figure 6.6(a)), we ✔ Graph ✔ Scatter ✔ 3-D ✔ Define and set our dialogue as shown SPSS Output 6.3. (Note that we have entered Variable Labels in the Dwaine Studio data file.) The result is shown in SPSS Output 6.4. Clicking on the three-dimensional scatter plot will open the Chart Editor. You can now rotate the plot by ✔ Format ✔ 3-D rotation to rotate the chart to the desired position.

SPSS Output 6.3

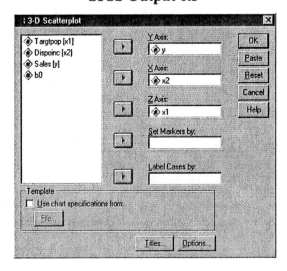

SPSS Output 6.4

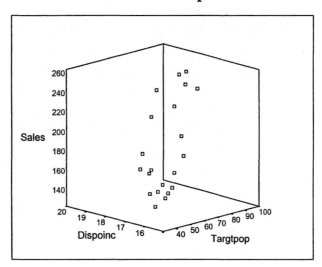

Analysis of Appropriateness of Model: Inspection of Figures 6.8 and 6.9, text pages 242-243, reveals that we need to produce the residuals and predicted values of *Y* based on the first order regression model. Insert 6.2 saves the unstandardized residuals (*res_1*) and predicted values (*pre_1*) in the Data Editor. We now use a Syntax File to produce *absresid*, the absolute value of the residuals, and *x1x2*, the interaction term between *X1* and *X1*. Text Figures 6.8 a-d can now be produced by using ✔ Graph ✔ Scatter ✔ Matrix ✔ Define (not shown in Insert 6.9) and setting the dialogue box as shown in SPSS Output 6.5 (including moving *absresid* into the Matrix Variables box – not visible.) This results in SPSS Output 6.6; note that the first four plots in the first row are equivalent to text Figures 6.8 a-d and that the plot in row 1, column 2 is equivalent to text Figure 6.9a. The last section of Insert 6.2 will produce text Figure 6.9b although the axes will be reversed compared to the text. Reversing the axes will result in SPSS Output 6.7.

INSERT 6.2	INSERT 6.2 continued
✔ Analyze	compute absresid=abs(res_1).
✔ Regression	compute x1x2=x1*x2.
✔ Linear	execute.
✔ *y*	✔ Run
✔ ▶ (to the Dependent box)	✔ All
✔ *x1*	* Produce text Figure 6.9b.
✔ ▶ (to the Independent box)	✔ Analyze
✔ *x2*	✔ Descriptive Statistics
✔ ▶ (to the Independent box)	✔ Explore
✔ Save	✔ *res_1*
✔ Unstandardized (Under residuals)	✔ ▶ (to the Dependent List)
✔ Unstandardized (Under predicted values)	✔ Plots
✔ Continue	✔ Normal probability plot with tests
✔ OK	✔ Continue
	✔ OK
✔ File	* Double click on the chart to open the Chart Editor.
✔ Open	✔ Series
✔ Syntax	✔ Displayed
* Type the following syntax.	* Reverse axis's and close Chart Editor.

SPSS Output 6.5

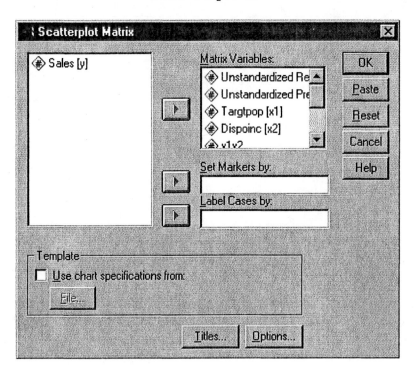

SPSS Output 6.6

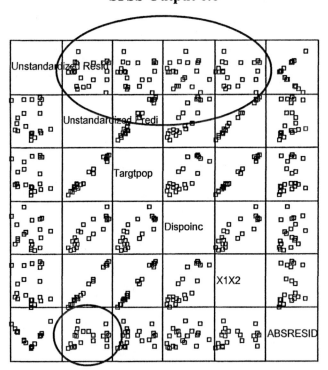

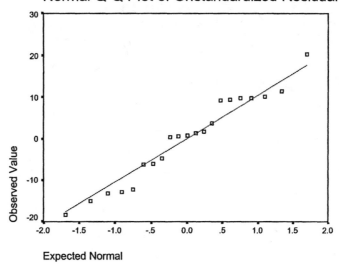

Normal Q-Q Plot of Unstandardized Residual

SPSS Output 6.6 and 6.7 do not indicate any serious problems with the response function including non-constancy of error terms. SPSS Output 6.6 (*x1x2*res_1*) indicates that there is no striking interaction effect in the regression model. The authors report that the normal probability plot of the residuals is moderately linear (SPSS Output 6.7).

On text page 243, the authors calculate the coefficient of correlation between the ordered residuals and their expected value under normality for the Dwaine Studios example. The expected value under normality of the kth smallest observation in a sample of size n is given by:

$$\sqrt{MSE}\left[z\left(\frac{k - 0.375}{n + 0.25} \right) \right]$$

The first step in obtaining the expected value of the ordered residuals under normality is to calculate $(k - 0.375)/(n + 0.25)$. (Insert 6.3) In the Dwaine Studios example, $n = 21$. Since k is the rank of the ordered residuals, we rank the saved residuals (*res_1*, saved in Insert 6.2 above) using the Rank Case function. SPSS creates another new variable, *rres_1,* that is the rank of the value of *res_1*, the unstandardized residual based on a model with *x1* and *x2* as the predictor variables. Thus, $(k - 0.375)/(n + 0.25)$ becomes $(rres\_1 - 0.375)/(21 + 0.25)$. This quantity represents the cumulative probability ($0 \leq$ cumulative probability ≤ 1.0) of the standard normal distribution. We now must obtain a standard score (z) associated with the cumulative probability calculated above. In SPSS, the *Probit* function is an *inverse density function* and takes the form *Probit*(p), where p = cumulative probability. For instance, if the cumulative probability = 0.50 (*Probit*(0.50)) then $z = 0$. If the cumulative probability = 0.8413 (*Probit*(0.8413)) then $z = +1$. Thus, to obtain z we use *Probit* $((rres\_1 - 0.375)/(21 + 0.25))$. Finally, we multiply this quantity by the square root of *MSE* = 121.163 of the first order model (SPSS Output 6.2). From the above, we compute the expected value as:

$$ev = sqrt(121.163) * Probit((rres\_1 - 0.375)/(21 + 0.25)).$$

```
                            INSERT 6.3

    ✔ Transform
    ✔ Rank Cases
    ✔ res_1
    ✔ ▶ (to the Variables box)
    ✔ OK

    ✔ File
    ✔ New
    ✔ Syntax
    compute ev = sqrt(121.163) * Probit ((rres_1 - 0.375)/(21 + 0.25)).
    execute.
    ✔ Run
    ✔ All

    ✔ Analyze
    ✔ Correlate
    ✔ Bivariate
    ✔ ev
    ✔ ▶
    ✔ res_1
    ✔ ▶
    ✔ OK
```

We can now obtain the coefficient of correlation between the residual and the expected value under normality (*ev*). The result of the correlation between *ev* and *res_1* is 0.980 and is consistent with the correlation reported on text page 243. The interpolated critical value from Table B.6, text page 673, for $n = 21$ and $\alpha = 0.05$ is 0.9525. The obtained correlation coefficient (0.980) is greater than the table value (0.9525) and we conclude that the distribution of the error terms appears to be reasonably close to a normal distribution.

On text page 244, the researchers tested the null hypothesis $Ho: \beta_1 = 0$ and $\beta_2 = 0$. The test result is found in the output of the first order regression model above (SPSS Output 6.2). Because the *p*-value (<0.000) associated with $F = 99.103$ is less than the desired significance level ($\alpha = 0.05$), we conclude that sales are related to target population (*X1*) and disposable income (*X2*).

We can use the ANOVA table in SPSS Output 6.2 to find the *Coefficient of Multiple Determination* using text equation 6.40:

$$R^2 = \frac{SSR}{SSTO} = \frac{24,015.28}{26,196.21} = 0.917$$

$R^2 = 0.917$ is also given in the Model Summary table of SPSS Output 6.2. Additionally, $R = 0.957$ and *Adjusted R Square* = 0.907 are presented in the Model Summary table.

Estimation of Regression Parameters: On text page 245, the researchers wish to jointly estimate β_1 and β_2 with a family confidence coefficient of 0.90. The confidence limits for β_k are given by equations 6.52 and 6.52a, text page 228:

$$b_k \pm Bs\{b_k\}$$

$$b_k \pm t(1 - \alpha/2g; n - p)s\{b_k\}$$

where g = the number of intervals to be estimated and p = the number of parameters in the regression model. In the present discussion, we wish to estimate $g = 2$ intervals and there are 3 parameters in the regression model. We use the *inverse distribution function* (*IDF.T*(probability, *df*) where "probability" = $(1 - 0.90/2g)$ and $df = n - p$ or *IDF.T*(0.975, 18). Thus, in Insert 6.4 we find t and place its value in *tvalue*, which is written on each line in the Data Editor. We use the parameter estimates shown in SPSS Output 6.2 ($b1 = 1.455$ and $b2 = 9.366$). We use the standard deviation of each estimate (shown as "*Std. Error*" in SPSS Output 6.2, $s\{b_1\} = 0.212$ and $s\{b_2\} = 4.06$) to find the lower (*b1lo* and *b2lo*) and upper (*b1hi* and *b2hi*) confidence limit for each estimated parameter. These limits are written into the Data Editor.

```
                          INSERT 6.4

  ✔ File
  ✔ New
  ✔ Syntax
  compute tvalue = IDF.T(.975,18) .
  compute b1lo=1.45455-(tvalue*.211178).
  compute b1hi=1.45455+(tvalue*.211178).
  compute b2lo=9.3655-(tvalue*4.06395).
  compute b2hi=9.3655+(tvalue*4.06395).
  execute.
  ✔ Run
  ✔ All
```

Estimation of Mean Response: On text page 245 the authors wish to estimate the expected mean sales when the target population is 65.4 thousand persons and the per capita disposable income is 17.6 thousand dollars. They also estimate a 95 percent confidence interval for $E\{Y_h\}$. The $1 - \alpha$ confidence limits for $E\{Y_h\}$ are given by equations 6.59 and 6.58:

$$\hat{Y}_h \pm t(1 - \alpha/2; n - p)s\{\hat{Y}_h\}$$

where:

$$s^2\{\hat{Y}_h\} = MSE\left(\mathbf{X}_h'(\mathbf{X}'\mathbf{X})^{-1}\mathbf{X}_h\right)$$

In chapter 2 we found the confidence intervals for $E\{Y_h\}$ where there was one predictor variable in the model. However, as the number of predictor variables increases, the task of finding the variance of $\hat{Y}_h$ becomes increasingly laborious. Although text equation 2.30 can be generalized to the case of multiple regression ($p > 2$), the authors present the formulation of the variance of $\hat{Y}_h$ in matrix terms in equation 6.58 (text page 229) presented above. We use this equation and SPSS' *Matirix* program to solve the current problem.

The matrix program shown in Insert 6.5 is a continuation of the matrix program shown in Insert 6.1 above. We find t prior to entering the matrix program and after entering set the matrix *tvalue* to the variable *tvalue*. We then find $s^2\{\hat{Y}_h\}$ using the above notation. $MSE = 121.163$ is given in SPSS Output 6.2 above. X'_h is defined in Insert 6.5 as shown on text page 246. We find $\hat{Y}_h$ (*yhath*) using equation 6.55, text page 229. The lower (*yhathlo*=185.29) and upper (*yhathlo*=196.91) confidence limits are then computed and printed.

INSERT 6.5

```
* Find t(1-alpha/2,n-p) from text equation 6.59.
compute tvalue = IDF.T(.975,18) .
* Write a column of 1's into the Dwaine Studios data file.
compute x0=1.
execute.
matrix.
get con / variables =x0.
get a / variables = x1.
get b / variables = x2.
* Define the Y matrix.
get y /variables=y.
get tvalue /variables=tvalue.
* Define the X matrix.
compute x = {con, a, b} .
compute xtx=t(x)*x.
comute xty=t(x)*y.
* Fine the b matrix using text equation 6.25.
compute b=inv(t(x)*x)*t(x)*y.
compute xtxinv=inv(t(x)*x).
* Define xh matrix as on text page 246.
compute xh={1;65.4;17.6}.
* Find standard deviation of yhat(new) per text equation 6.58.
compute sxh=sqrt(121.163*(t(xh)*(xtxinv)*xh)).
* Find yhat(new) using text equation 6.56..
compute yhath=t(xh)*b.
* Find confidence interval using text equation 6.59.
compute yhathlo=yhath-(tvalue*sxh).
compute yhathhi=yhath+(tvalue*sxh).
print yhathlo(1,1).
print yhathhl(1,1).
end matrix.
execute.
```

Prediction Limits for New Observations: Dwaine Studios now wants to predict sales for two new cities (text page 247). Here, X_{h1} and X_{h2} equal 65.4 and 17.6, respectively, for City A and 53.1 and 17.7, respectively, for City B. To determine which prediction intervals are preferable here, we find S and B using text equations 6.65a and 6.66a (text page 231) where $g = 2$ and $1 - \alpha = 0.90$. Insert 6.6 computes S (s) and B (b) and writes the two variables into the Data Editor. Note that $S = 2.29$ and $B = 2.10$ so that the Bonferroni limits are tighter here and therefore more efficient.

INSERT 6.6

✔ File
✔ New
✔ Syntax
```
compute s=sqrt(2*idf.f(.9,2,18)).
compute b=idf.t(1-(.1/(2*2)),18).
execute.
```
✔ Run
✔ All

To find the simultaneous Bonferroni prediction limits for city A, we use (text equation 6.66):

$$\hat{Y}_h \pm Bs\{pred\}$$

where (text equations 6.66a, 6.63a, and 6.58):

$$B = t(1 - \alpha/2g; n - p)$$

$$s^2\{pred\} = MSE + s^2\{\hat{Y}_h\}$$

$$s^2\{\hat{Y}_h\} = MSE\left(X'_h (X'X)^{-1} X_h\right)$$

From Insert 6.7, we find the lower (*yhathlo*=167.71) and upper (*yhathhi*=214.49) 90% confidence limits for city A. To find the intervals for city B, simply replace "*xh* = {1, 65.4, 17.6}." with "*xh* = {1, 53.1, 17.7}." in Insert 6.7. Insert 6.7 so modified will find the 90 percent confidence interval for city B: *yhathlo* = 149.08 and *yhathi* = 199.21 consistent with text page 246.

INSERT 6.7

```
* Find b(1-alpha/2g,n-p) from text equation 6.59.
compute b = IDF.T(1-(.1/(2*2)),18) .
* Write a column of 1's into the Dwaine Studios data file.
compute x0=1.
execute.
matrix.
get con / variables =x0.
get a / variables = x1.
get b / variables = x2.
* Define the Y matrix.
get y /variables=y.
get bon /variables=b.
* Define the X matrix.
compute x = {con, a, b} .
compute xtx=t(x)*x.
compute xty=t(x)*y.
* Fine the b matrix using text equation 6.25.
compute b=inv(t(x)*x)*t(x)*y.
compute xtxinv=inv(t(x)*x).
* Define xh matrix as on text page 246.
compute xh={1;65.4;17.6}.
* Find standard deviation of yhat(new) per text equation 6.63a.
compute spre=sqrt(121.163+(121.163*(t(xh)*(xtxinv)*xh))).
print xh.
* Find yhat(new) using text equation 6.56.
compute yhath=t(xh)*b.
* Find confidence interval using text equation 6.66.
compute yhathlo=yhath-(bon*spre).
compute yhathhi=yhath+(bon*spre).
print yhathlo(1,1).
print yhathhl(1,1).
end matrix.
execute.
```

Multiple Regression II

We use *sum of squared deviations* or *sum of squares* of various types as a quantitative measurement of variability that is attributable to different sources associated with parameters specified in the model. In previous chapters we defined the sum of squared deviations of the observations on the dependent variable Y_i from the overall sample mean $\bar{Y}$ as the (corrected) *total sum of squares*, denoted *SSTO*. We defined the sum of squared deviations of the fitted (predicted) values $\hat{Y}_i$ from $\bar{Y}$ as the *regression sum of squares* denoted *SSR*. We defined the sum of squared deviations of the observations Y_i from the fitted values $\hat{Y}_i$ as the *error sum of squares* denoted *SSE*. We found that:

$$SSTO = SSR + SSE$$

Suppose we fit a general linear model with $p - 1$ predictor variables and obtain *SST*, *SSR* and *SSE*. Now, suppose we add one or more predictor variables to the model and fit this "new" model. The observations Y_i will remain unchanged as will the mean $\bar{Y}$, but the $\hat{Y}_i$ with this new model may be different from the $\hat{Y}_i$ based on the first model. Thus, the numerical values of *SSR* and *SSE* will likely change. In fact, *SSR* will increase and, correspondingly, *SSE* will decrease by the same magnitude as the increase in *SSR*. We call this magnitude of change the *extra sum of squares* due to the predictor variable(s) that were added to the model. To avoid confusion we need additional notation to identify the variables that are in the model when *SSR* and *SSE* are computed.

It is easier to explain the extra sum of squares notation in terms of a specific example. Suppose predictor variables X_1, X_2 and X_3 are available for fitting a general linear regression model. If we fit a model with only X_1 in the model, we denote the sums of squares for regression and error as $SSR(X_1)$ and $SSE(X_1)$, respectively. If we fit a model that includes the predictor variables X_1 and X_2, we write $SSR(X_1, X_2)$ and $SSE(X_1, X_2)$ to denote the regression and error sums of squares, respectively. If we fit a model that includes X_1 and then add X_2 to the model and refit, the additional or extra regression sum of squares due to adding the additional predictor variable is denoted $SSR(X_2|X_1)$ and the reduction or extra error sum of squares is denoted $SSE(X_2|X_1)$. Similarly, if we fit a model that includes all three predictor variables, we write $SSR(X_1, X_2, X_3)$ and $SSE(X_1, X_2, X_3)$ to denote the regression and error sums of squares, respectively. If we fit a model that includes X_1 and then add X_2 and X_3 to the model and refit, the additional or extra regression sum of squares due to adding both additional predictor variables is denoted $SSR(X_2, X_3 |X_1)$ and the reduction or extra error sum of squares is denoted $SSE(X_2, X_3 |X_1)$. The extra sum of squares is also known as the *partial sum of squares*. It is obvious that we need to specify the predictor variables as being already in the model or as being added to the model when we discuss extra (partial) sums of squares.

In Chapter 7, the authors discuss methods used to make statistical inferences when adding and deleting predictor variables to the general linear regression model. The problem of deciding the merits of adding or deleting predictor variables in choosing an appropriate regression model is of particular interest. The extra sum of squares concept plays a fundamental role in the decision making process.

The Body Fat Example

Data File: ch07ta01.txt

Extra Sum of Squares: The text presents an excellent discussion of extra sum of squares. "Extra sum of squares measure the marginal reduction in error sum of squares when one or several predictor variables are added to the regression model, given that other predictor variables are already in the model." Many statistical packages use slightly different terminology to denote two major types of sum of squares: Type I and Type III. Type I sum of squares examines the sequential incremental improvement in the fit of the model as each independent variable is added to the model. Thus, the order of entry of variables into the model will affect the Type I sum of squares. Type I sum of squares is also called the hierarchical decomposition of sum of squares method. In contrast, Type III sum of squares for a particular predictor variable is the amount of increase in the regression sum of squares (decrease in error sum of squares) due to that predictor variable after all other terms in the model have been included. Thus, the Type III sums of squares for a given predictor is the reduction in error sum of squares achieved when that predictor variable is the last predictor variable added to the model. We will see how Type I and III sum of squares are used in the following examples.

On text page 256, researchers in the Body Fat example wished to develop a regression model to predict body fat from three predictor variables: triceps skin fold thickness, thigh circumference, and midarm circumference. They considered four regression models: (1.) body fat (Y) as a linear function of triceps skin fold thickness alone, (2.) body fat as a linear function of thigh circumference alone, (3.) body fat as a linear function of the two predictor variables triceps skin fold and thigh circumference, and (4.) body fat as a linear function of all three predictor variables.

See Chapter 0, "Working with SPSS," for instructions on opening the Body Fat example (text Table 7.1) and renaming the variables $X1$, $X2$, $X3$, and Y. After opening ch07ta01.txt, the first 5 lines in your SPSS Data Editor should look like SPSS Output 7.1.

SPSS Output 7.1

In Insert 7.1, we regressed Y on $X1$ in the first analysis (SPSS Output 7.2). In the second analysis, we regressed Y on $X1$ and $X2$ (SPSS Output 7.3). In the third regression analysis of Insert 7.1, we use the Block feature (SPSS Output 7.4) available in SPSS. That is, we entered $X1$ at block 1 and $X2$ at block 2 (SPSS Output 7.5). Thus, Model 1 of SPSS Output 7.5 represents a regression model where Y has been regressed on $X1$ and Model 2 represents a regression model where Y has been regressed on $X1$ and $X2$. Note that $SSE(X1) = 143.120$ and $SSE(X1, X2) = 109.951$ in SPSS Output 7.5 and $SSE(X1) = 143.120$ and $SSE(X1, X2) = 109.951$ in SPSS Output 7.2 and 7.3 for the two separate analyses. This is consistent with Table 7.2, text page 257. The difference $SSE(X2|X1) = SSE(X1) - SSE(X1, X2) = 143.120 - 109.951 = 33.169$ is referred to as the *extra reduction in error sum of squares* achieved by adding $X2$ to a model that has $X1$ as the only predictor variable. Correspondingly, $SSR(X2|X1) = SSR(X1, X2) - SSR(X1) = 385.439 - 352.270 = 33.169$ is the *extra increase in the regression sum squares* achieved by adding $X2$ to a model that has $X1$ as the only predictor variable. Note that $SSE(X2|X1) = SSR(X2|X1) = 33.169$ is not given in SPSS Output 7.2, 7.3, or 7.5. This is because SPSS Regression output displays only the *total regression* and *error sum of squares* for a given model and not the incremental change due to adding predictor variables to the model. To directly produce the extra sum of squares, we need Type I sum of squares that is available in the SPSS *General Linear Model* (*GLM*) procedure. The syntax in Insert 7.2 requests Type I sum of squares based on a model with $X1$ and $X2$ as predictor variables (SPSS Output 7.6). The Type I sum of squares $SSR(X2|X1) = 33.169$ here is the same as found earlier.

INSERT 7.1	INSERT 7.1 continued
✔ Analyze ✔ Regression ✔ Linear ✔ y ✔ ▶(to the Dependent box) ✔ x1 ✔ ▶(to the Independent box) ✔ OK ✔ Analyze ✔ Regression ✔ Linear ✔ y ✔ ▶(to the Dependent box) ✔ x1 ✔ ▶(to the Independent box) ✔ x2 ✔ ▶(to the Independent box) ✔ OK	✔ Analyze ✔ Regression ✔ Linear ✔ y ✔ ▶(to the Dependent box) ✔ x1 ✔ ▶(to the Independent box) ✔ Next ✔ x2 ✔ ▶(to the Independent box) ✔ OK

SPSS Output 7.2
Y on *X1*

ANOVA[b]

Model		Sum of Squares	df	Mean Square	F	Sig.
1	Regression	352.270	1	352.270	44.305	.000[a]
	Residual	143.120	18	7.951		
	Total	495.389	19			

a. Predictors: (Constant), X1

b. Dependent Variable: Y

SPSS Output 7.3
Y on *X1* and *X2*

ANOVA[b]

Model		Sum of Squares	df	Mean Square	F	Sig.
1	Regression	385.439	2	192.719	29.797	.000[a]
	Residual	109.951	17	6.468		
	Total	495.389	19			

a. Predictors: (Constant), X2, X1

b. Dependent Variable: Y

SPSS Output 7.4

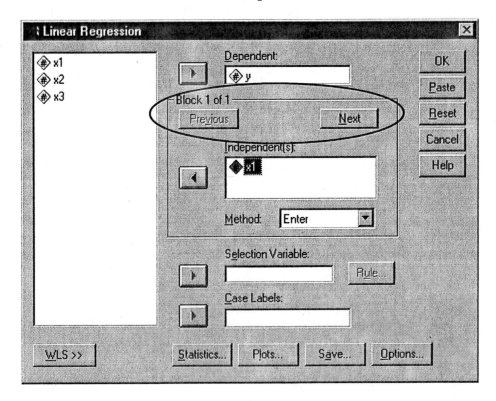

SPSS Output 7.5

Y on *X1* and *X2* using Block feature

ANOVA[c]

Model		Sum of Squares	df	Mean Square	F	Sig.
1	Regression	352.270	1	352.270	44.305	.000[a]
	Residual	143.120	18	7.951		
	Total	495.389	19			
2	Regression	385.439	2	192.719	29.797	.000[b]
	Residual	109.951	17	6.468		
	Total	495.389	19			

a. Predictors: (Constant), X1

b. Predictors: (Constant), X1, X2

c. Dependent Variable: Y

INSERT 7.2

- ✔ Analyze
- ✔ General Linear Model
- ✔ Univariate
- ✔ Y
- ✔ ► (to the Dependent box)
- ✔ x1
- ✔ ► (to the Covariate box)
- ✔ x2
- ✔ ► (to the Covariate box)
- ✔ Model
- ✔ Sum of Squares ▼
- * Select Type I.
- ✔ Continue
- ✔ OK

SPSS Output 7.6

Tests of Between-Subjects Effects

Dependent Variable: Y

Source	Type I Sum of Squares	df	Mean Square	F	Sig.
Corrected Model	385.439[a]	2	192.719	29.797	.000
Intercept	8156.761	1	8156.761	1261.154	.000
X1	352.270	1	352.270	54.466	.000
X2	33.169	1	33.169	5.128	.037
Error	109.951	17	6.468		
Total	8652.150	20			
Corrected Total	495.389	19			

a. R Squared = .778 (Adjusted R Squared = .752)

Test whether a single $\beta_k = 0$: On text page 264, the researchers wished to test whether $X3$ should be dropped from a model that contained $X1$, $X2$, and $X3$. That is, they wanted to test H_0: $\beta_3 = 0$. From Chapter 6, we know that we can use the t^* to test this hypothesis. But, to continue with the general linear test approach, we use the block feature and regress Y on $X1$ at block 1, $X2$ at block 2, and $X3$ at block 3 (Insert 7.3). Results are shown in SPSS Output 7.7. We first find $SSR(X3|X1, X2) = SSR(X1, X2, X3) - SSR(X1, X2)$ or $396.985 - 385.439 = 11.546$ and $SSE(X1, X2, X3) = 98.405$. Thus, from equation 7.15, text page 264, $F^* = \dfrac{11.546}{1} \div \dfrac{98.405}{16} = 1.88$. This is consistent with the regression procedure SPSS Output 7.7 which shows $t = -1.37$ for $X3$ at Model 3. That is, we know that $t^2 = F$ and $(-1.37)^2 = 1.88$. Based on this finding, we do not reject H_0 and drop $X3$ from the model. Requesting Type I sum of squares in GLM with $X1$, $X2$, and $X3$ as predictors results in SPSS Output 7.8 (syntax not shown). The F-value associated with $X3$ is 1.877. Regression analysis yields Type III sum of squares and GLM yields Type I sum of squares (if requested) but we arrive at the same conclusion regarding H_0: $\beta_3 = 0$. Remember the test of H_0: $\beta_3 = 0$ is a conditional test, given that $X1$ and $X2$ are already in the model. The test for $X3$ (Regression Output 7.7 – Coefficients - Model 3) is a test of the significance of adding $X3$ to a model that already includes all other predictors being considered, i.e., $X1$ and $X2$. The test for $X3$ using GLM and requesting Type I sum of squares (GLM Output 7.8) has the same interpretation. Thus, because $X3$ was the last variable entered into the model, inferential tests of the effect of $X3$ based on Type I and Type III sum of squares test the same hypothesis. However, when the statistical test is not based on the last variable in the model, Type I and Type III sum of squares test different hypotheses. Assume that we wish to determine if $X2$ can be dropped from the above model. From the regression procedure using Type III sum of squares (SPSS Output 7.7 – Model 3) we see that $t = -1.106$ for $X2$. This is the effect of $X2$ controlling for all other variables in the model, i.e., $X1$ and $X3$. From the GLM procedure (SPSS Output 7.8) we see that $F = 5.393$ for $X2$. This is the effect of $X2$ controlling for all variables that preceded it in the model, i.e., $X1$. ($t^2 = F$, but $(-1.106)^2 \neq 5.393$). So, using Type III sum of squares we get a test of the statistical significance of $SSR(X2|X1, X3)$, or, equivalently, $SSE(X2|X1, X3)$. On the other hand, using the GLM procedure, the test associated with the Type I sum of squares tests the significance of $SSR(X2|X1)$. The default sum of squares in SPSS GLM is Type III. This also illustrates a very important point: to properly interpret the output of a computer software program, we must know the type of sum of squares used in the analysis.

INSERT 7.3

- ✔ Analyze
- ✔ Regression
- ✔ Linear
- ✔ y
- ✔ ▶ (to the Dependent box)
- ✔ x1
- ✔ ▶ (to the Independent box)
- ✔ Next
- ✔ x2
- ✔ ▶ (to the Independent box)
- ✔ Next
- ✔ x3
- ✔ ▶ (to the Independent box)
- ✔ OK

SPSS Output 7.7

SPSS Regression

ANOVA[d]

Model		Sum of Squares	df	Mean Square	F	Sig.
1	Regression	352.270	1	352.270	44.305	.000[a]
	Residual	143.120	18	7.951		
	Total	495.389	19			
2	Regression	385.439	2	192.719	29.797	.000[b]
	Residual	109.951	17	6.468		
	Total	495.389	19			
3	Regression	396.985	3	132.328	21.516	.000[c]
	Residual	98.405	16	6.150		
	Total	495.389	19			

a. Predictors: (Constant), X1

b. Predictors: (Constant), X1, X2

c. Predictors: (Constant), X1, X2, X3

d. Dependent Variable: Y

Coefficients[a]

Model		Unstandardized Coefficients		Standardized Coefficients	t	Sig.
		B	Std. Error	Beta		
1	(Constant)	-1.496	3.319		-.451	.658
	X1	.857	.129	.843	6.656	.000
2	(Constant)	-19.174	8.361		-2.293	.035
	X1	.222	.303	.219	.733	.474
	X2	.659	.291	.676	2.265	.037
3	(Constant)	117.085	99.782		1.173	.258
	X1	4.334	3.016	4.264	1.437	.170
	X2	-2.857	2.582	-2.929	-1.106	.285
	X3	-2.186	1.595	-1.561	-1.370	.190

a. Dependent Variable: Y

241

SPSS Output 7.8

SPSS GLM

Tests of Between-Subjects Effects

Dependent Variable: Y

Source	Type I Sum of Squares	df	Mean Square	F	Sig.
Corrected Model	396.985[a]	3	132.328	21.516	.000
Intercept	8156.761	1	8156.761	1326.237	.000
X1	352.270	1	352.270	57.277	.000
X2	33.169	1	33.169	5.393	.034
X3	11.546	1	11.546	1.877	.190
Error	98.405	16	6.150		
Total	8652.150	20			
Corrected Total	495.389	19			

a. R Squared = .801 (Adjusted R Squared = .764)

Test whether several $\beta_k = 0$: Continuing with the body fat example (text page 265), the researchers wished to determine if $X2$ and $X3$ could be dropped from the model that contained $X1$, $X2$, and $X3$. This is equivalent to testing the hypothesis Ho: $\beta_2 = \beta_3 = 0$. From SPSS Output 7.8 above, we have the necessary information to find equation 7.18, text page 265:

$$F^* = \frac{SSR(X2 \mid X1) + SSR(X3 \mid X1, X2)}{2} \div MSE(X1, X2, X3)$$

$$F^* = \frac{33.169 + 11.546}{2} \div 6.15 = 3.63$$

However, there is an easier way to test H_0: $\beta_2 = \beta_3 = 0$. We use SPSS regression procedure and enter $X1$ at block 1 and $X2$, $X3$ at block 2, and request a R^2 Change statistic (Insert 7.4). In SPSS Output 7.9, Model 1 contains $X1$ and Model 2 contains $X1$, $X2$, and $X3$. The R^2 Change $F = 3.635$ represents the incremental effect of adding $X2$ and $X3$ to a model that contains $X1$, i.e., $SSR(X2, X3|X1)$.

INSERT 7.4	INSERT 7.4 continued
✔ Analyze ✔ Regression ✔ Linear ✔ y ✔ ▶(to the Dependent box) ✔ x1 ✔ ▶(to the Independent box) ✔ Next	✔ x2 ✔ ▶(to the Independent box) ✔ x3 ✔ ▶(to the Independent box) ✔ Statistics ✔ R square Change ✔ Continue ✔ OK

Model Summary

Model	R	R Square	Adjusted R Square	Std. Error of the Estimate	Change Statistics				
					R Square Change	F Change	df1	df2	Sig. F Change
1	.843[a]	.711	.695	2.81977	.711	44.305	1	18	.000
2	.895[b]	.801	.764	2.47998	.090	3.635	2	16	.050

a. Predictors: (Constant), X1

b. Predictors: (Constant), X1, X3, X2

Coefficients of Partial Determination and Correlation: In the body fat example (text page 270) we obtain first the *Coefficient of Partial Correlation* for *X2* given *X1* is in the model, $r_{Y2|1}$. (Insert 7.5) This results in $r_{Y2|1} = 0.482$ (output not shown). We now can find the *Coefficient of Partial Determination* by squaring the *Coefficient of Partial Correlation*, $0.482^2 = 0.232$.

INSERT 7.5

- ✔ Analyze
- ✔ Correlate
- ✔ Partial
- ✔ *y*
- ✔ ▶(to Variable box)
- ✔ *x2*
- ✔ ▶(to Variable box)
- ✔ *x1*
- ✔ ▶(to Controlling for: box)
- ✔ OK

Standardized Multiple Regression Model: To produce the standardized regression model, we use the regression procedure to regress *Y* on *X1* and *X2* using the Dwaine Studio data (ch06fi05.txt) and request Descriptives under the Statistics option (syntax not shown). Partial output for this procedure is shown in SPSS Output 7.10. Notice that standardized beta coefficients are standard output using SPSS regression procedure. With the descriptive statistics, we can use equation 7.53, text page 276, to shift the standardized regression coefficients back to the unstandardized regression coefficients.

$$b_1 = \left(\frac{s_y}{s_1}\right) b_1^* = \frac{36.1913}{18.620}(.748) = 1.455$$

SPSS Output 7.10

Descriptive Statistics

	Mean	Std. Deviation	N
Y	181.9048	36.19130	21
X1	62.0190	18.62033	21
X2	17.1429	.97035	21

Coefficients[a]

Model		Unstandardized Coefficients		Standardized Coefficients	t	Sig.
		B	Std. Error	Beta		
1	(Constant)	-68.857	60.017		-1.147	.266
	X1	1.455	.212	.748	6.868	.000
	X2	9.366	4.064	.251	2.305	.033

a. Dependent Variable: Y

Multicollinearity and Its Effects: When multiple predictor variables in a general linear regression model are correlated among themselves, we say there is *intercorrelation* or *multicollinearity* among these variables (text page 279). When there is multicollinearity among a set of predictor variables being considered for a general linear regression model, the extra regression (error) sum of squares for a given predictor or set of predictor variables will be different depending on which other of these predictor variables are in the model. On the other hand, if there is no multicollinearity among the predictor variables, the extra regression sum of squares will be the same no matter which other variables are included.

Work Crew Productivity Example: On text page 279, the authors present the results of an experiment on the effect of work crew size ($X1$) and level of pay ($X2$) on productivity (Y). The text presents results from a correlation matrix of the 3 variables and Table 7.7, the output of 3 separate regression models. We present the SPSS syntax for this example as a further illustration. The rationale closely follows that in the body fat example.

Note that we have presented the entire solution in syntax form, as opposed to using the graphical point-and-click interface. Additionally, we present a compact method of requesting the 3 regression models. That is, in one regression procedure, we regress Y on $X1$ and $X2$ (*enter=x1 x2*); this generates the regression model presented in Table 7.7 (a), text page 280. We then remove $X2$ (*remove=x2*) from the model. This generates a model with $X1$ only [text Table 7.7 (b)]. Finally, we remove $X1$ (*remove=x1*) and enter $X2$ (*enter=x2*) to generate a model with $X2$ as the only predictor variable.

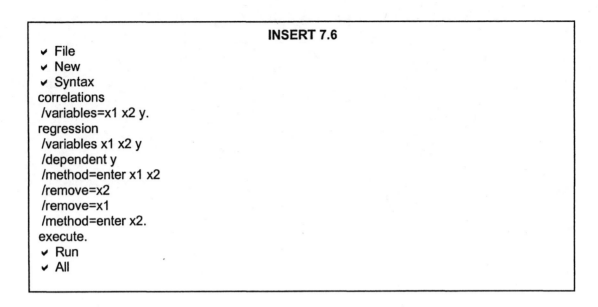

INSERT 7.6

```
✔ File
✔ New
✔ Syntax
correlations
 /variables=x1 x2 y.
regression
 /variables x1 x2 y
 /dependent y
 /method=enter x1 x2
 /remove=x2
 /remove=x1
 /method=enter x2.
execute.
✔ Run
✔ All
```

More on the Body Fat Example: On text page 283, the authors revisit the Body Fat example (text page 256) to illustrate the effect of *multicollinearity*. In Insert 7.7, we explore the relation between the predictor variables by producing a *scatter matrix* (SPSS Output 7.11) and *correlation matrix* (SPSS Output 7.12) of $X1$, $X2$, and $X3$. Note that $r_{12} = 0.924$ and although $X3$ is not strongly correlated with $X1$ and $X2$ individually, the multiple R^2 when $X3$ is regressed on $X1$ and $X2$ is 0.990. Thus, there is a strong correlation between $X3$ and a linear function of $X1$ and $X2$ (Output 7.13). To see the change in parameter estimates as variables are added to the model, we regressed Y on $X2$ and Y on $X1$, $X2$, and $X3$ (SPSS Output 7.14). As noted in the text, the regression coefficient for $X2$ changes in the two analyses from 0.659 to -2.857 due to the intercorrelations of the predictor variables. In Chapter 10, the authors explore formal methods to detect multicollinearity.

INSERT 7.7	**INSERT 7.7 continued**
✔ Graph ✔ Scatter ✔ Matrix ✔ Define ✔ *x1* ✔ ► (to the Matrix variables box) ✔ *x2* ✔ ► (to the Matrix variables box) ✔ *x3* ✔ ► (to the Matrix variables box) ✔ OK	✔ Analyze ✔ Correlate ✔ Bivariate ✔ *x1* ✔ ► (to the Variables box) ✔ *x2* ✔ ► (to the Variables box) ✔ *x3* ✔ ► (to the Variables box) ✔ OK

SPSS Output 7.11

SPSS Output 7.12

Correlations

		X1	X2	X3
X1	Pearson Correlation	1	.924**	.458*
	Sig. (2-tailed)	.	.000	.042
	N	20	20	20
X2	Pearson Correlation	.924**	1	.085
	Sig. (2-tailed)	.000	.	.723
	N	20	20	20
X3	Pearson Correlation	.458*	.085	1
	Sig. (2-tailed)	.042	.723	.
	N	20	20	20

**. Correlation is significant at the 0.01 level (2-tailed).

*. Correlation is significant at the 0.05 level (2-tailed).

SPSS Output 7.13

X3 regressed on *X1* and *X2*

Model Summary

Model	R	R Square	Adjusted R Square	Std. Error of the Estimate
1	.995a	.990	.989	.37699

a. Predictors: (Constant), X2, X1

SPSS Output 7.14

Y regressed on *X1* and *X2*

Coefficients^a

Model		Unstandardized Coefficients		Standardized Coefficients	t	Sig.
		B	Std. Error	Beta		
1	(Constant)	-19.174	8.361		-2.293	.035
	X1	.222	.303	.219	.733	.474
	X2	.659	.291	.676	2.265	.037

a. Dependent Variable: Y

Y regressed on *X1*, *X2*, and *X3*

Coefficients^a

Model		Unstandardized Coefficients		Standardized Coefficients	t	Sig.
		B	Std. Error	Beta		
1	(Constant)	117.085	99.782		1.173	.258
	X1	4.334	3.016	4.264	1.437	.170
	X2	-2.857	2.582	-2.929	-1.106	.285
	X3	-2.186	1.595	-1.561	-1.370	.190

a. Dependent Variable: Y

Chapter 8 SPSS®

Regression Models
for Quantitative
and Qualitative Predictors

In this chapter, the authors discuss some special types of general linear regression models. They discuss *polynomial regression models, interaction regression models, regression models with qualitative predictor variables,* and *more complex regression models.* Recall the simple linear regression model is of the form:

$$Y_i = \beta_0 + \beta_1 X_i + \varepsilon_i$$

This model is said to be a *first-order model* because the predictor variable appears in the model expressed to the first power; i.e., $X_i^1 = X_i$. Similarly, the model:

$$Y_i = \beta_0 + \beta_1 X_{i1} + \beta_2 X_{i2} + \varepsilon_i$$

is a first-order model because the two predictor variables X_{i1} and X_{i2} each appear in the model expressed to the first power.

The model:

$$Y_i = \beta_0 + \beta_1 x_i + \beta_2 x_i^2 + \varepsilon_i$$

where $x_i = X_i - \overline{X}$ is the *centered predictor variable,* is called a *polynomial regression model.* It is referred to as a *second-order model with one predictor* because the single predictor variable appears in the model expressed to the second power; i.e., x_i^2. Similarly, the polynomial regression model:

$$Y_i = \beta_0 + \beta_1 x_i + \beta_2 x_i^2 + \beta_3 x_i^3 + \varepsilon_i$$

where $x_i = X_i - \overline{X}$, is a *third-order model with one predictor variable.* The regression model:

$$Y_i = \beta_0 + \beta_1 x_{i1} + \beta_2 x_{i2} + \beta_{11} x_{i1}^2 + \beta_{22} x_{i2}^2 + \beta_{12} x_{i1} x_{i2} + \varepsilon_i$$

where $x_{i1} = X_{i1} - \overline{X}_1$ and $x_{i2} = X_{i2} - \overline{X}_2$, is a *second-order model with two predictor variables.* The cross-product term represents the interaction effect between x_1 and x_2. This nomenclature extends to models with any number of predictor variables.

The authors use indicator variables that take on values 0 and 1 in a regression model to identify classes or categories of a qualitative variable. They use several examples to illustrate how to fit and interpret a variety of higher-order models including models where some of the predictor variables are quantitative and others are qualitative.

The Power Cell Example

Data File: ch08ta01.txt

Using data from Table 8.1, text page 300, researchers studied the relationship between charge rate and temperature on the life of a power cell. There were three levels (0.6, 1.0, and 1.4 amperes) of charge rate ($X1$) and three levels (10, 20, and 30° C) of ambient temperature ($X2$). The dependent variable (Y) was defined as the number of discharge-charge cycles that a power cell underwent before failure. See Chapter 0, "Working with SPSS," for instructions on opening the Power Cell example and renaming the variables Y, $X1$, and $X2$.

The researchers were uncertain of the expected response function, but decided to fit a second-order polynomial regression model. They first centered $X1$ and $X2$ around their respective means and scaled them using the absolute difference between adjacent levels of the variable. From Insert 8.1, we have computed the coded values for $X1$ and $X2$ (*x1code* and *x2code*, respectively), the square of the coded values (*x1code2* and *x2code2*, respectively), and the cross-product of *X1code* and *X2code* (*x12inter*). After executing the syntax in Insert 8.1, the active data editor should look like Output 8.1 and is consistent with Table 8.1, text page 300.

INSERT 8.1

- ✔ File
- ✔ New
- ✔ Syntax
- * Type the following syntax.
- compute x1code=(x1-1)/.4.
- compute x2code=(x2-20)/10.
- compute x1code2=x1code**2.
- compute x2code2=x2code**2.
- compute x12inter=x1code*x2code.
- execute.
- ✔ Run
- ✔ All

SPSS Output 8.1

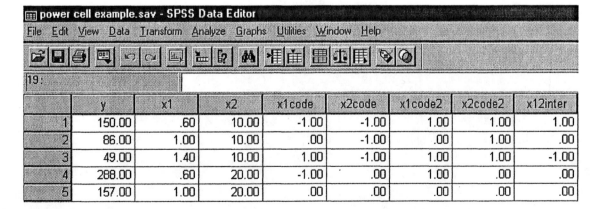

	y	x1	x2	x1code	x2code	x1code2	x2code2	x12inter
1	150.00	.60	10.00	-1.00	-1.00	1.00	1.00	1.00
2	86.00	1.00	10.00	.00	-1.00	.00	1.00	.00
3	49.00	1.40	10.00	1.00	-1.00	1.00	1.00	-1.00
4	288.00	.60	20.00	-1.00	.00	1.00	.00	.00
5	157.00	1.00	20.00	.00	.00	.00	.00	.00

249

Fitting the Model: We begin by obtaining the regression results of the second-order polynomial regression model and saving the unstandardized predicted values and residuals for future use (Insert 8.2). Partial results are shown in Output 8.2. The Unstandardized Coefficients in Output 8.2 are the estimated regression coefficients used in equation 8.16, text page 301.

INSERT 8.2

- ✔ Analyze
- ✔ Regression
- ✔ Linear
- ✔ *y*
- ✔ ▶(to the Dependent box)
- ✔ *x1code*
- ✔ ▶(to the Independent box)
- ✔ *x2code*
- ✔ ▶(to the Independent box)
- ✔ *x1code2*
- ✔ ▶(to the Independent box)
- ✔ *x2code2*
- ✔ ▶(to the Independent box)
- ✔ *x12inter*
- ✔ ▶(to the Independent box)
- ✔ Save
- ✔ Unstandardized (Residual Values)
- ✔ Unstandardized (Predicted Values)
- ✔ Continue
- ✔ OK

SPSS Output 8.2

Model Summary

Model	R	R Square	Adjusted R Square	Std. Error of the Estimate
1	.956[a]	.914	.827	32.37418

a. Predictors: (Constant), X12INTER, X2CODE2, X2CODE, X1CODE, X1CODE2

ANOVA[b]

Model		Sum of Squares	df	Mean Square	F	Sig.
1	Regression	55365.561	5	11073.112	10.565	.011[a]
	Residual	5240.439	5	1048.088		
	Total	60606.000	10			

a. Predictors: (Constant), X12INTER, X2CODE2, X2CODE, X1CODE, X1CODE2

b. Dependent Variable: Y

250

Coefficients[a]

Model		Unstandardized Coefficients		Standardized Coefficients	t	Sig.
		B	Std. Error	Beta		
1	(Constant)	162.842	16.608		9.805	.000
	X1CODE	-55.833	13.217	-.556	-4.224	.008
	X2CODE	75.500	13.217	.751	5.712	.002
	X1CODE2	27.395	20.340	.184	1.347	.236
	X2CODE2	-10.605	20.340	-.071	-.521	.624
	X12INTER	11.500	16.187	.093	.710	.509

a. Dependent Variable: Y

Residual Plots: In this section we produce Figure 8.5, text page 303. The easiest way to produce the various plots of residuals and other values is with the *scatterplot matrix* procedure. That is, we request a *scatterplot matrix* of the unstandardized residuals (*res_1*) saved in Insert 8.2 above with the predicted value of Y ($\hat{Y}$), *x1code*, and *x2code* (Insert 8.3).

INSERT 8.3

✔ Graph
✔ Scatter
✔ Matrix
✔ Define
✔ *res_1*
✔ ► (to the Matrix variables box)
✔ *x1code*
✔ ► (to the Matrix variables box)
✔ *x2code*
✔ ► (to the Matrix variables box)
✔ *pre_1*
✔ ► (to the Matrix variables box)
✔ OK

Power Cell Example

Data File: ch08ta05.txt

Correlation Test for Normality: On text page 301, the researchers in the Power Cell example calculate the coefficient of correlation between the ordered residuals and their expected value under normality. The expected value under normality of the kth smallest observation from a sample of size n is given by:

$$\sqrt{MSE}\left[z\left(\frac{k - 0.375}{n + 0.25} \right) \right]$$

251

We find $MSE = 1048.088$ in Output 8.2 above and in this case $n = 11$. In Chapter 3 we give a detailed description of the steps necessary to obtain z and k. Insert 8.5 assumes that the residuals have been saved (*res_1*) from Insert 8.2 above. In Insert 8.4, we rank the residuals and place the rank in a variable named (by default) *rres_1*. We then compute *ev*, the expected value under normality and correlate this variable with *res_1*. The resulting correlation is 0.975, within rounding error of the value (0.974) reported on the text page 301.

INSERT 8.4

✔ File
✔ New
✔ Syntax
* Type the following syntax.
Rank
 variables=res_1.
Compute *ev* = sqrt(1048.088) * probit((*rres_1* - .375)/(11 + .25)).
Correlations
 /variables=ev res_1.
execute.
✔ Run
✔ All

Test of Fit: To perform a formal test of goodness of fit of the regression model, we use the *GLM* procedure as detailed in Insert 8.5. Output 8.3 contains the partial results of Insert 8.5. The *F*-value = 1.82 is identical to the Lack of Fit *F*-value reported on text page 302. Since the associated probability level of *F* is >0.05, we conclude that the second-order polynomial regression model is a good fit.

INSERT 8.5

✔ Analyze
✔ General Linear Model
✔ Univariate
✔ *y*
✔ ▶ (to the Dependent box)
✔ *x1code*
✔ ▶ (to the Covariate box)
✔ *x2code*
✔ ▶ (to the Covariate box)
✔ *x1code2*
✔ ▶ (to the Covariate box)
✔ *x2code2*
✔ ▶ (to the Covariate box)
✔ *x12code*
✔ ▶ (to the Covariate box)
✔ Options
✔ Lack of Fit
✔ Continue
✔ OK

Output 8.3

Lack of Fit Tests

Dependent Variable: Y

Source	Sum of Squares	df	Mean Square	F	Sig.
Lack of Fit	3835.772	3	1278.591	1.820	.374
Pure Error	1404.667	2	702.333		

Coefficient of Multiple Determination: The coefficient of multiple correlation ($R = 0.956$), the coefficient of multiple determination ($R^2 = 0.914$) and the adjusted coefficient of multiple correlation ($Adj\ R^2 = 0.827$) are given in Output 8.2 above.

Partial F Test: The researchers now want to test whether a first-order model is appropriate. That is, they wish to decide if *x1code2*, *x2code2*, and *x12inter* can be dropped from the model; they test H_0: $\beta_{11} = \beta_{22} = \beta_{12} = 0$. Instead of using the extra sum of squares approach to test this null hypothesis, we use the block option within SPSS *Regression* (Insert 8.6). We enter *x1code* and *x2code* at block 1 and the higher-order variables at block 2, and request the R^2 Change statistic. Partial results are shown in Output 8.4. Consistent with text page 304, the F-value associated with the change in R^2 when the higher-order variables are added to a model that contain only the first-order variables is 0.782 and the associated probability level ("Sig. F Change") is 0.553. Thus, we fail to reject H_0: $\beta_{11} = \beta_{22} = \beta_{12} = 0$ and conclude that the first-order model is adequate.

INSERT 8.6

- ✔ Analyze
- ✔ Regression
- ✔ Linear
- ✔ *y*
- ✔ ► (to the Dependent box)
- ✔ *x1code*
- ✔ ► (to the Independent box)
- ✔ *x2code*
- ✔ ► (to the Independent box)
- ✔ Next
- ✔ *x1code2*
- ✔ ► (to the Independent box)
- ✔ *x2code2*
- ✔ ► (to the Independent box)
- ✔ *x12code*
- ✔ ► (to the Independent box)
- ✔ Statistics
- ✔ R squared change
- ✔ Continue
- ✔ OK

Output 8.4

Model Summary

Model	R	R Square	Adjusted R Square	Std. Error of the Estimate	Change Statistics				
					R Square Change	F Change	df1	df2	Sig. F Change
1	.934[a]	.873	.841	31.02486	.873	27.482	2	8	.000
2	.956[b]	.914	.827	32.37418	.041	.782	3	5	.553

a. Predictors: (Constant), X2CODE, X1CODE

b. Predictors: (Constant), X2CODE, X1CODE, X12INTER, X2CODE2, X1CODE2

First-Order Model: Based on the above analysis, the researchers chose to fit the first-order model:

$$Y_i = \beta_0 + \beta_1 x_{i1} + \beta_2 x_{i2} + \varepsilon_i$$

From Insert 8.7 we produce the first-order model; results are shown in Output 8.5. The regression coefficients are identical to those reported on text page 304, and b_1 and b_2 are identical to the same estimates shown in Output 8.2 above.

INSERT 8.7

- ✔ Analyze
- ✔ Regression
- ✔ Linear
- ✔ *y*
- ✔ ▶ (to the Dependent box)
- ✔ *x1code*
- ✔ ▶ (to the Independent box)
- ✔ *x2code*
- ✔ ▶ (to the Independent box)
- ✔ OK

SPSS Output 8.5

Coefficients[a]

Model		Unstandardized Coefficients		Standardized Coefficients	t	Sig.
		B	Std. Error	Beta		
1	(Constant)	172.000	9.354		18.387	.000
	X1CODE	-55.833	12.666	-.556	-4.408	.002
	X2CODE	75.500	12.666	.751	5.961	.000

a. Dependent Variable: Y

Fitted First-Order Model in Terms of X: The first-order regression model above can be transformed back to the original variables by use of text equation 8.15. We can also obtain a fitted first order regression model by regressing Y on the original (untransformed) predictor variables $X1$ and $X2$. (Insert 8.8) The resulting estimated coefficients (not shown) are consistent with equation 8.19, text page 304; i.e., (unstandardized) $b_1' = -139.58$ and $b_2' = 7.55$ with standard errors $s\{b_1'\} = 31.66$ and $s\{b_2'\} = 1.26$.

INSERT 8.8

- ✔ Analyze
- ✔ Regression
- ✔ Linear
- ✔ y
- ✔ ► (to the Dependent box)
- ✔ x1
- ✔ ► (to the Independent box)
- ✔ x2
- ✔ ► (to the Independent box)
- ✔ OK

Estimation of Regression Coefficients: The researcher now wished to estimate the Bonferroni 90% confidence limits in terms of the original predictor variables. We apply the confidence limits formula for β_1 and β_2 given in equation 6.52, text page 228. In Insert 8.9, we first find $B = t(1 - \alpha/2g; n - p)$. In this case, $\alpha = 0.10$ with two confidence intervals being estimated ($g = 2$), $n = 11$, and there are 3 parameters in the model ($p = 3$). Thus, in Insert 8.9 we find $B = idf.t(1-(0.10/4), 8) = 2.306$. We obtain the numerical values $s\{b_1'\}$ (*sdb1*) and $s\{b_2'\}$ (*sdb2*) as shown on text page 305. Finally, we use equation 6.52, text page 228, to find the two Bonferroni simultaneous 90% confidence intervals: $-212.63 \le \beta_1 \le -66.55$ and $4.63 \le \beta_2 \le 10.47$. The confidence limits are written into the Data Editor.

INSERT 8.9

- ✔ File
- ✔ New
- ✔ Syntax
- * Type the following syntax.
- compute B=idf.t(1-(.1/4),8).
- compute sdb1=(1/.4)*12.67.
- compute sdb2=(1/10)*12.67.
- compute b1lo=-139.59-(b*sdb1).
- compute b1hi=-139.59+(b*sdb1).
- compute b2lo=7.55-(b*sdb2).
- compute b2hi=7.55+(b*sdb2).
- execute.
- ✔ Run
- ✔ All

Insurance Innovation Example

Data File: ch08ta02.txt

An economist studied the relationship between the adoption of new insurance innovations and the type of insurance company and the size of the insurance firm (text page 313, 316). The response variable (time to adoption) is coded Y, and the predictor variables are coded $X1$ for size of the firm and $X2$ for type of firm where $X2 = 0$ for mutual companies and $X2 = 1$ for stock companies. See Chapter 0, "Working with SPSS," for instructions on opening ch08ta02.txt and renaming variables Y, $X1$, and $X2$. After opening the Insurance Innovation example, the first 5 lines in the SPSS Data Editor should look like Output 8.6.

SPSS Output 8.6

	y	x1	x2	var
1	17.0	151.00	0	
2	26.0	92.00	0	
3	21.0	175.00	0	
4	30.0	31.00	0	
5	22.0	104.00	0	

In Insert 8.10, we first produce Output 8.7, which is consistent with Figure 8.12, text page 318. The economist was most interested in the type of insurance firm ($X2$) on the response variable. In Insert 8.11 we request the first-order model including confidence limits for the coefficients in the regression model. Consistent with text page 316, the 95% confidence interval for β_2 is displayed as $4.977 \leq \beta_2 \leq 11.134$ (SPSS Output 8.8). Note that this is not a simultaneous confidence interval like the intervals calculated in the previous example. Because the interval does not include zero, we conclude H_a: $\beta_2 \neq 0$ or that the type of insurance firm has a significant effect on the response variable. This conclusion can also be reached based on the probability level ($Sig. = 0.000$) of the t-value ($t = 5.521$) associated with $X2$.

INSERT 8.10	INSERT 8.10 continued
✔ Graph	
✔ Scatter	✔ OK
✔ Simple	* Open Chart Editor by double clicking on the graph
✔ Define	in the Viewer window.
✔ *y*	✔ Chart
✔ ▶ (to the Y axis box)	✔ Options
✔ *x1*	✔ Subgroups
✔ ▶ (to the X axis box)	✔ OK
✔ *x2*	* Close Chart Editor.
✔ ▶ (to the Set Markers by: box)	

SPSS Output 8.7

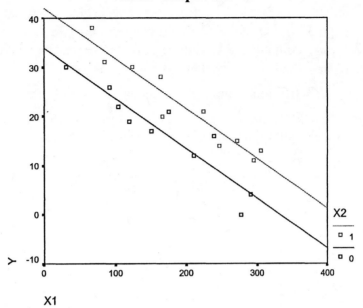

X1

INSERT 8.11

- ✔ Analyze
- ✔ Regression
- ✔ Linear
- ✔ *y*
- ✔ ▶(to the Dependent box)
- ✔ *x1*
- ✔ ▶(to the Independent box)
- ✔ *x2*
- ✔ ▶(to the Independent box)
- ✔ Statistics
- ✔ Confidence Intervals
- ✔ Continue
- ✔ OK

SPSS Output 8.8

Coefficients[a]

Model		Unstandardized Coefficients		Standardized Coefficients			95% Confidence Interval for B	
		B	Std. Error	Beta	t	Sig.	Lower Bound	Upper Bound
1	(Constant)	33.874	1.814		18.675	.000	30.047	37.701
	X1	-.102	.009	-.911	-11.443	.000	-.121	-.083
	X2	8.055	1.459	.439	5.521	.000	4.977	11.134

a. Dependent Variable: Y

On text page 326, the economist wanted to test the hypothesis that there is no interaction between size ($X1$) and type ($X2$) of insurance company. Thus, he fitted a regression model (equation 8.49, text page 324) that included the interaction term $X1X2$, the cross-product of $X1$ and $X2$. In Insert 8.12, we compute $X1X2$ and then fit the regression model with $X1$, $X2$, and $X1X2$ as predictors. We test the hypothesis that there is no interaction in terms of β_3; i.e., we test H_0: $\beta_3 = 0$ against H_a: $\beta_3 \neq 0$. From Output 8.9 and consistent with text page 326, we see the t-value associated with $X1X2$ is -0.023, $p = 0.982$. Thus, the economist concluded that the interaction is not statistically significant and, hence, the interaction term should be dropped from the model.

INSERT 8.12

- File
- New
- Syntax
* Type the following syntax.
compute x1x2=x1*x2.
execute.
- Run
- All

- Analyze
- Regression
- Linear
- y
- ▶ (to the Dependent box)
- x1
- ▶ (to the Independent box)
- x2
- ▶ (to the Independent box)
- x1x2
- ▶ (to the Independent box)
- OK

SPSS Output 8.9

Coefficients[a]

Model		Unstandardized Coefficients		Standardized Coefficients	t	Sig.
		B	Std. Error	Beta		
1	(Constant)	33.838	2.441		13.864	.000
	X1	-.102	.013	-.909	-7.779	.000
	X2	8.131	3.654	.443	2.225	.041
	X1X2	.000	.018	-.005	-.023	.982

a. Dependent Variable: Y

Soap Production Lines Example

Data File: ch08ta05.txt

A company studied the relationship between production line speed (*X1*) and the amount of scrap (*Y*) generated on their 2 production lines (*X2*). The data are presented in Table 8.5, text page 330. After opening ch08ta05.txt and renaming the variables, the first 5 lines in the SPSS Data Editor should look like Output 8.10.

SPSS Output 8.10

Based on text Figure 8.16, the researchers decided to fit a model with an interaction term:

$$Y_i = \beta_0 + \beta_1 X_{i1} + \beta_2 X_{i2} + \beta_3 X_{i1} X_{i2} + \varepsilon_i$$

To produce the Tentative Model (text page 330), we first compute a new variable, *X1X2*, the cross-product of *X1* and *X2*. (Insert 8.13)   We then fit a model using *Y* as the dependent variable and *X1*, *X2*, and *X1X2* as independent variables and save the unstandardized residuals (*res_1*) and predicted values (*pre_1*). This produces Output 8.11, which is consistent with Table 8.6(a), text page 332.

INSERT 8.13	INSERT 8.13 continued
✔ File	✔ ▶(to the Dependent box)
✔ New	✔ x1
✔ Syntax	✔ ▶(to the Independent box)
* Type the following syntax.	✔ x2
compute x1x2=x1*x2.	✔ ▶(to the Independent box)
execute.	✔ x1x2
✔ Run	✔ ▶(to the Independent box)
✔ All	✔ Save
✔ Analyze	✔ Unstandardized (Residual Values)
✔ Regression	✔ Unstandardized (Predicted Values)
✔ Linear	✔ Continue
✔ y	✔ OK

SPSS Output 8.11

ANOVA[b]

Model		Sum of Squares	df	Mean Square	F	Sig.
1	Regression	169164.7	3	56388.228	130.949	.000[a]
	Residual	9904.057	23	430.611		
	Total	179068.7	26			

a. Predictors: (Constant), X1X2, X1, X2

b. Dependent Variable: Y

Coefficients[a]

Model		Unstandardized Coefficients		Standardized Coefficients	t	Sig.
		B	Std. Error	Beta		
1	(Constant)	7.574	20.870		.363	.720
	X1	1.322	.093	1.010	14.273	.000
	X2	90.391	28.346	.552	3.189	.004
	X1X2	-.177	.129	-.242	-1.371	.184

a. Dependent Variable: Y

Diagnostics: Separate production line plots of the line speed (*X1*) against *Y* are requested in Insert 8.14. (Output not shown.) The researchers in the Soap Production Lines Example next calculated the coefficient of correlation between the ordered residuals and their expected values under normality. The expected value under normality of the kth smallest observation from a sample of size *n* is given by:

$$\sqrt{MSE}\left[z\left(\frac{k-.375}{n+.25}\right)\right]$$

We find $MSE = 430.611$ in Output 8.11 above and $n = 27$ in the Soap Production Lines Example. In Chapter 3 we give a detailed description of the steps necessary to obtain z and k. Insert 8.15 assumes that the residuals have been saved (*res_1*) from Insert 8.13 above. In Insert 8.15 we rank the residuals and place the rank in a variable named (by default) *rres_1*. We then compute the expected value under normality, *ev*, and correlate this variable with *res_1*. The resulting correlation is 0.990, as reported on text page 331. From Table B.6, text page 673, the critical value for $\alpha = .05$ and $n = 28$ is 0.962. The calculated value is greater than the critical value found in the table, so we conclude there is sufficient evidence to support the assumption of normality of error terms.

Continuing on text page 331, the authors formally test for constancy of error variance using the Brown-Forsythe test as described in Chapter 3. For simplicity, we again divide the process into steps:

- Step 1: (Insert 8.16) Divide the data file into two subsets. In the Soap Production Lines Example this has been done based on the value of $X2$ where $X2 = 1$ if production line = 1 and $X2 = 0$ if production line = 2. Also, execute Insert 8.13 to save the unstandardized residual values for the second order regression model where $X1$, $X2$, and $X1X2$ are predictor variables.

- Step 2: Use Explore to obtain the median residual values $\tilde{e}_1$ (Production line = 1, therefore, $X2 = 1$) = 5.4959 and $\tilde{e}_2$ (Production line = 2, therefore, $X2 = 0$) = -6.2203.

- Step 3: Find *absdevmd*, the absolute value (*abs*) of the residual minus the median residual value for the respective group, d_{i1} and d_{i2}.

- Step 4: Find the means of *absdevmd* for the two groups, $\bar{d}_1 = 16.132$ and $\bar{d}_2 = 12.648$.

- Step 5: Find *sqrdev*, the square of *absdevmd* minus the respective group mean of *absdevmd*.

- Step 6: Find the sum of *sqrdev* for the two production lines, $\sum(d_{11} - \bar{d}_1)^2 = 2952.19$ and $\sum(d_{12} - \bar{d}_2)^2 = 2045.82$.

We now have the values required to calculate the Brown-Forsythe statistic:

$$s^2 = \frac{2,952.20 + 2,04582}{27 - 2} = 199.921$$

$$s = \sqrt{199.921} = 14.139$$

$$t_{BF}^* = \frac{16.132 - 12.648}{14.139\sqrt{\dfrac{1}{15} + \dfrac{1}{12}}} = .636$$

Using $\alpha = 0.05$ and $df = 25$, the critical value of t can be obtained by the *inverse distribution function*, "compute IDF.T(0.975, 25) = IDF.T(0.975, 25)", which will return 2.060. Since $t_{BF}^* = 0.636 < 2.060$, we conclude that there is constancy of error variance. Further, we can find the two-sided p-value of $t_{BF}^* = 0.636$ using the *cumulative distribution function*. For example, in the Syntax Editor we can use "compute tprob=(1−CDF.T(tvalue, 25))*2 where tvalue = IDF.T(0.975, 25). The *CDF.T* function will return 0.5305, the two-side p-value where $t = 0.636$ and $df = 25$ as reported on text page 333.

INSERT 8.16

* Step 1.
* Use *X2* for Step 1.

* Step 2.
✔ Analyze
✔ Descriptive Statistics
✔ Explore
✔ *res_1*
✔ ▶ (to the Dependent List)
✔ *x2*
✔ ▶ (to the Factor List)
✔ OK

*Step 3.
✔ File
✔ New
✔ Syntax
* Type the following syntax.
if (x2=0) absdevmd=abs(res_1-(-6.2203)).
if (x2=1) absdevmd=abs(res_1-(5.4959)).
execute.
✔ Run
✔ All

* Step 4.
✔ Analyze
✔ Descriptive Statistics
✔ Explore
✔ *absdevmd*
✔ ▶ (to the Dependent List)
✔ *x2*
✔ ▶ (to the Factor List)
✔ OK

* Step 5.
✔ File
✔ New
✔ Syntax
* Type the following syntax.
if (x2=0) *sqrdev*=(absdevmd-12.648)**2.
if (x2=1) *sqrdev*=(absdevmd-16.132)**2.
execute.
✔ Run
✔ All

* Step 6.
✔ Analyze
✔ Compare Means
✔ Means
✔ *sqrdev*
✔ ▶ (to the Dependent List)
✔ *group*
✔ ▶ (to the Independent List)
✔ Options
✔ Sum
✔ ▶ (to the Cell Statistics)
✔ Continue
✔ OK

263

Inferences about Two Regression Lines: The researchers now wish to test the identity of the regression lines for the two soap production lines. The null and alternative hypotheses, respectively, are:

$$H_0 : \beta_2 = \beta_3 = 0$$

and

$$H_a : not\ both\ \beta_2 = 0\ \ and\ \beta_3 = 0$$

To test the null hypothesis using equation 7.27, text page 267, we need to produce text Table 8.6(b), which lists Type I sum of squares. Type I sum of squares is only available in SPSS through the General Linear Model (*GLM*). Output 8.12 is the result using *GLM* to regress *Y* on *X1*, *X2*, and *X1X2* and requesting Type I sum of squares in the Model window. (Syntax not shown.) We can now use equation 8.56a, text page 333, to solve for $F^* = 22.65$ shown on text page 333. Alternatively, we can use the block option in the Regression procedure to test the current hypothesis. In Insert 8.17 we enter *X1* at block 1 and *X2*, *X1X2* at block 2, and request *R Squared Change* statistics. From Output 8.13, we see that the *F*-value for Model 2 is 22.647. This is the effect of adding *X2* and *X1X3* to a model that already contains *X1*; this is, we test the effect of *X2*, *X1X2|X1*. Since the computed *F*-value (22.647) is greater than the table *F*-value (5.67) when $\alpha = 0.01$, we conclude that the regression functions are not the same for the two production lines.

SPSS Output 8.12

Tests of Between-Subjects Effects

Dependent Variable: Y

Source	Type I Sum of Squares	df	Mean Square	F	Sig.
Corrected Model	169164.684[a]	3	56388.228	130.949	.000
Intercept	2687271.3	1	2687271.259	6240.598	.000
X1	149660.983	1	149660.983	347.555	.000
X2	18694.079	1	18694.079	43.413	.000
X1X2	809.623	1	809.623	1.880	.184
Error	9904.057	23	430.611		
Total	2866340.0	27			
Corrected Total	179068.741	26			

a. R Squared = .945 (Adjusted R Squared = .937)

INSERT 8.17	INSERT 8.17 continued
✔ Analyze	✔ *x2*
✔ Regression	✔ ► (to the Independent box)
✔ Linear	✔ *x1x2*
✔ *y*	✔ ► (to the Independent box)
✔ ► (to the Dependent box)	✔ Statistics
✔ *x1*	✔ R squared change
✔ ► (to the Independent box)	✔ Continue
✔ Next	✔ OK

Model Summary

Model	R	R Square	Adjusted R Square	Std. Error of the Estimate	Change Statistics				
					R Square Change	F Change	df1	df2	Sig. F Change
1	.914[a]	.836	.829	34.29738	.836	127.229	1	25	.000
2	.972[b]	.945	.937	20.75117	.109	22.647	2	23	.000

a. Predictors: (Constant), X1

b. Predictors: (Constant), X1, X2, X1X2

Coefficients[a]

Model		Unstandardized Coefficients		Standardized Coefficients	t	Sig.
		B	Std. Error	Beta		
1	(Constant)	64.036	23.249		2.754	.011
	X1	1.196	.106	.914	11.280	.000
2	(Constant)	7.574	20.870		.363	.720
	X1	1.322	.093	1.010	14.273	.000
	X2	90.391	28.346	.552	3.189	.004
	X1X2	-.177	.129	-.242	-1.371	.184

a. Dependent Variable: Y

The researcher also wanted to test whether the slopes of the regression lines for the two production lines were identical. The null and alternative hypotheses are:

$$H_0 : \beta_3 = 0$$

and

$$H_a : \beta_3 \neq 0$$

We can test the null hypothesis by text equation 7.25 or the partial F-test, text equation 7.24. Text Table 8.6(b) and Output 8.12 above provide the necessary to information to find (text equation 8.57a) $F^* = 1.88$. Equivalently, we can use Output 8.13 (Coefficients Table) to obtain $t = -1.371$ for $X1X2$. Remember that $t^2 = F$ and $-1.371^2 = 1.88$. Thus, the t-statistic in Output 8.13 using Type III sum of squares is a partial test of $X1X2|X1, X2$. Since the probability level 0.184 is greater than the required $\alpha = 0.01$ we conclude H_0, that the slopes of the regression functions are not significantly different.

You may notice in Output 8.12 that $F = 1.88$ for $X1X2$ and that this output is based on Type I sum of squares. Remember that the order of entry into the model does not effect Type III sum of squares but does effect Type I sum of squares. Since $X1X2$ is entered into the model after $X1$ and $X2$ for Output 8.12 (Type I sum of squares), it is testing the effect of $X1X2|X1, X2$. Output 8.13 is based on Type III sum of squares so the order of entry does not matter. The two tests are identical here only because $X1X2$ was the last variable entered into the model in Output 8.12. If, however, we used Output 8.12 to test the effect of $X2$, we would be testing the effect of $X2|X1$. If we use Output 8.13 to test the effect of $X2$, we would be testing the effect of $X2|X1, X1X2$. Notice that the effect of $X2$ is Output 8.12 and Output 8.13 are not identical, that is $F = 41.413 \neq t^2 = 3.189^2$). Also note that we requested Type I sum of squares to produce Output 8.12; the default sum of squares in SPSS *General Linear Model* is Type III. Finally, to produce the 95% confidence limits for β_2, use Insert 8.17 and also ✔ Confidence Limits while in the Statistics window.

Chapter 9 SPSS®

Building the Regression Model I: Model Selection and Validation

This chapter presents numerous strategies that are often used to build a regression model. These tactics include making use of theoretical relationships and knowledge gained from previous studies as well as information contained in a set of data available for analysis. We frequently have a large set of potential predictor variables and we wish to choose a subset of these variables to build a regression model that is efficient in its use of predictor variables. This process involves checking for outlying observations as well as investigating the contributions of potential predictor variables. The problem is complicated by the fact that there may be multicollinearity among the predictor variables. We may visually inspect relationships that are sometimes revealed in well-planned graphs. The authors discuss the criteria commonly used to select the predictor variables for inclusion or exclusion from the model. They then discuss three basic ways to validate a regression model once the selection process is in its final stages (see text page 369). These techniques are illustrated with well-chosen numerical examples.

Surgical Unit Example

Data File: ch09ta01.txt

A hospital studied the survival time (Y) of patients who had undergone a liver operation. The predictor or explanatory variables for the predictive regression model were blood clotting score ($X1$), prognostic index ($X2$), enzyme function score ($X3$), and liver function score ($X4$). The researchers also extracted from the preoperative evaluation the patients age ($X5$), sex ($X6$), and history of alcohol use, coded as two indicator variables, $X7$ (moderate use) and $X8$ (severe use). A representation of the Surgical Unit data set is shown in Table 9.1, text page 350. See Chapter 0, "Working with SPSS," for instructions on opening the Surgical Unit data file and renaming the variables $X1$, $X2$, $X3$, $X4$, $X5$, $X6$, $X7$, $X8$, Y, and *ylog*. After opening ch09ta01.txt, the first 5 lines in the SPSS Data Editor should look like SPSS Output 9.1.

SPSS Output 9.1

	x1	x2	x3	x4	x5	x6	x7	x8	y	ylog
1	6.70	62.0	81.00	2.59	50.0	0	1	0	695	6.54
2	5.10	59.0	66.00	1.70	39.0	0	0	0	403	6.00
3	7.40	57.0	83.00	2.16	55.0	0	0	0	710	6.57
4	6.50	73.0	41.00	2.01	48.0	0	0	0	349	5.85
5	7.80	65.0	115.0	4.30	45.0	0	0	1	2343	7.76

The researchers selected a random sample of 108 patients. Because the researchers intended to validate the final model, the sample was split into a model-building set (first 54 cases) and a validation set (second 54 cases). To simplify the illustration, the researchers used only the first four available predictors variables. We begin by regressing Y on $X1$, $X2$, $X3$, and $X4$, and saving the unstandardized residuals and predicted values. (Syntax not shown.) In Insert 9.1 we produce Figures 9.2(a and b), text page 351, as shown in SPSS Output 9.2 (a and b). SPSS Output 9.2(a) suggests both nonconstant error variances and curvature. That is, the variability of the residual values increases as the magnitude of the predicted values increases. Additionally, when we add a Lowess curve (discussed in Chapter 4) to SPSS Output 9.2(a), we obtain SPSS Output 9.2(c). Notice that the Lowess curve supports the text's suspicion of curvature. Because of the above findings, the investigator used a logarithmic transformation of Y to obtain *ylog*. Based on various diagnostics, the investigator then decided to use *ylog* as the response variable; to use $X1$, $X2$, $X3$, and $X4$ as predictors; and to not include any interaction terms in the linear regression model.

INSERT 9.1

* text Figure 9.2a.
- Graph
- Bar
- Simple
- Define
- *res_1*
- ► (to the Y axis)
- *pre_1*
- ► (to the X axis)
- OK

* text Figure 9.2b.
- Analyze
- Descriptive Statistics
- Explore
- *res_1*
- ► (to the Dependent List)
- Plots
- Normality Plots with tests
- Continue
- OK

* Double click the Normal Q-Q plot is the output viewer and then double click
* on any data point that is not on the diagonal line. Switch the axes and then close
* the Chart Editor.

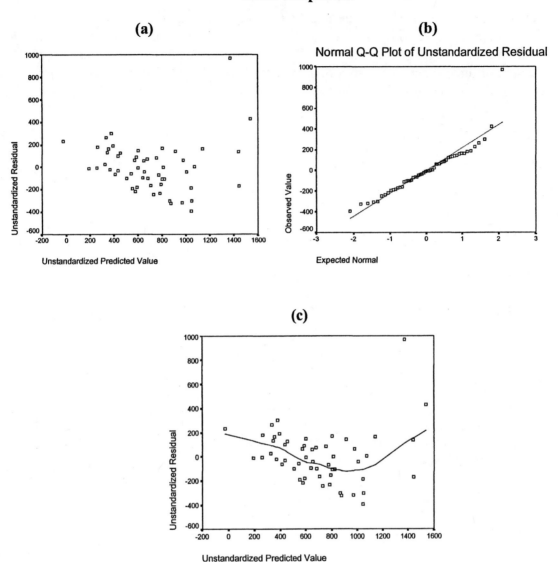

(a)

(b)

(c)

Criteria for Model Selection

The *All-Possible-Regressions Procedure* compares all possible sub-sets of potential predictor variables that could be used to build regression models. Based on various criteria, a small number of regression models are selected for further examination. The number of candidate models, however, increases exponentially as the number of predictor variables increase. A study with only 3 predicator variables has $2^3 = 8$ possible first-order regression models whereas one with 7 predictor variables has $2^7 = 128$. Thus, all-possible-regressions procedures can be used when the number of potential predictor variables is relatively small, but become impractical for use when the number is large.

SPSS does not have available an all-possible-regressions procedure. Using the Syntax Editor, however, we can approximate this procedure. The SSE_p, R_p^2, $R_{a,p}^2$, C_p, AIC_p, SBC_p, and

$PRESS_p$ values for all-possible-regression models are presented in Table 9.2, text page 353, for the Surgical Unit example. To begin the approximation of text Table 9.2, we fit a regression model using a dialogue box with $X1$ as the only predictor variable and *ylog* as the dependent variable. Note from Insert 9.2 that we do not click OK; instead we click Paste to instruct SPSS to paste the syntax of the current Regression dialogue box into a Syntax Editor window. Maximize the SPSS Syntax Editor window; it should look like Insert 9.3.

INSERT 9.2

- ✔ Analyze
- ✔ Regression
- ✔ Linear
- ✔ *ylog*
- ✔ ▶(to the Dependent box)
- ✔ *x1*
- ✔ ▶(to the Independent box)
- ✔ **Paste**

INSERT 9.3

```
REGRESSION
/MISSING LISTWISE
/STATISTICS COEFF OUTS R ANOVA
/CRITERIA=PIN(.05) POUT(.10)
/NOORIGIN
/DEPENDENT ylog
 /METHOD=ENTER x1 .
```

Continuing the approximation, we instruct SPSS to enter and remove each of the 16 possible sub-sets of predictor variables into regression models. (Inserts 9.4 and 9.5) That is, we modify Insert 9.3 to appear as Insert 9.4 by entering $X1$ into a regression model and then removing $X1$, entering $X2$ and then removing $X2$, etc. Note that Insert 9.4 contains 6 *enter* statements and 6 *remove* statements. This is because SPSS allows only 12 such steps in a *regression* command. Thus, to approximate all 16 sub-sets, we have to break the analysis into 3 separate *regression* commands. To request available selection criteria, type "*selection*" at the end of the "*/STATISTICS*" line. This option will request the *Akaike Information Critrion, Amemiya Prediction Criterion, Mallows' Prediction Criterion* (C_p), and *Schwarz Bayesian Criterion*. Copy the syntax in the Syntax Editor (Insert 9.4), paste it twice into the space below, and edit as needed. The Syntax Editor should look like Inserts 9.4 and 9.5 combined. Now ✔ *Run* and ✔ *All*. Partial results of this maneuver are shown in SPSS Output 9.3. Notice in the following table the correspondence between text Table 9.2 and SPSS Output 9.3.

Text Table 9.2 heading	SPSS Output 9.3 heading	Comment
SSE_p	Not present – See comment	R^2_p varies inversely with SSE_p
R^2_p	R Square	
$R^2_{a,p}$	Adjusted R Square	
C_p	Mallows' Prediction Criterion	
AIC_p	Akaike Information Criterion	
SBC_p	Schwarz Bayesian Criterion	
$PRESS_p$	Not present – See comment	Easily calculated in SPSS

INSERT 9.4

```
REGRESSION
/MISSING LISTWISE
/STATISTICS COEFF OUTS R ANOVA (selection)
/CRITERIA=PIN(.05) POUT(.10)
/NOORIGIN
/DEPENDENT ylog
 /METHOD=enter x1
 /METHOD=remove x1
 /METHOD=enter x2
 /METHOD=remove x2
 /METHOD=enter x3
 /METHOD=remove x3
 /METHOD=enter x4
 /METHOD=remove x4
 /METHOD=enter x1 x2
 /METHOD=remove x1 x2
 /METHOD=enter x1 x3
 /METHOD=remove x1 x3.
```

```
                            INSERT 9.5

REGRESSION
 /MISSING LISTWISE
 /STATISTICS COEFF OUTS R ANOVA selection
 /CRITERIA=PIN(.05) POUT(.10)
 /NOORIGIN
 /DEPENDENT ylog
 /METHOD=enter x1 x4
 /METHOD=remove x1 x4
 /METHOD=enter x2 x3
 /METHOD=remove x2 x3
 /METHOD=enter x2 x4
 /METHOD=remove x2 x4
 /METHOD=enter x3 x4
 /METHOD=remove x3 x4
 /METHOD=enter x1 x2 x3
 /METHOD=remove x1 x2 x3
 /METHOD=enter x1 x2 x4
 /METHOD=remove x1 x2 x4.

REGRESSION
 /MISSING LISTWISE
 /STATISTICS COEFF OUTS R ANOVA selection
 /CRITERIA=PIN(.05) POUT(.10)
 /NOORIGIN
 /DEPENDENT ylog
 /METHOD=enter x1 x3 x4
 /METHOD=remove x1 x3 x4
 /METHOD=enter x2 x3 x4
 /METHOD=remove x2 x3 x4
 /METHOD=enter x1 x2 x3 x4
 /METHOD=remove x1 x2 x3 x4.
```

Notice in the footnotes the predictor variables associated with each model in SPSS Output 9.3. Model 1 contains $X1$ and the row values are consistent with text Table 9.2 where $X1$ is the only X variable in the model. For Model 2 of SPSS Output 9.3, $R_p^2=.0$, $AIC_p = -75.703$, $C_p=151.498$, and $SBC_p=-73.714$. Remember that we entered $X1$ at step 1 and removed $X1$ at step 2. Thus, Model 2 contains no variables and the row values are consistent with those reported in text Table 9.2 when the model contains only an intercept. Now, Model 3 contains $X2$; again the row values are consistent with text Table 9.2 where $X2$ is the only X variable in the model. Model 4 (SPSS Output 9.3) contains no variables and Model 5 contains $X3$. The progression of the models continues for each of the three regression procedures.

SPSS Output 9.4 contains the first of the 3 ANOVA summary tables for this maneuver. Notice that the $SSE = 12.031$ for Model 1 ($X1$ only) is consistent with the SSE_p reported in text Table 9.2 for a model with $X1$ only. Continuing the progression, Model 2 contains no variables and is not of interest. Model 3 contains $X2$ and the SSE is consistent with SSE_p reported in text Table 9.2 for a model with $X2$ only.

SPSS Output 9.3

Model Summary

Model	R	R Square	Adjusted R Square	Std. Error of the Estimate	Selection Criteria			
					Akaike Information Criterion	Amemiya Prediction Criterion	Mallows' Prediction Criterion	Schwarz Bayesian Criterion
1	.246[a]	.061	.043	.48101	-77.079	1.012	141.164	-73.101
2	.000[b]	.000	.000	.49158	-75.703	1.038	151.498	-73.714
3	.470[c]	.221	.206	.43807	-87.178	.839	108.556	-83.200
4	.000[b]	.000	.000	.49158	-75.703	1.038	151.498	-73.714
5	.654[d]	.428	.417	.37549	-103.827	.616	66.489	-99.849
6	.000[b]	.000	.000	.49158	-75.703	1.038	151.498	-73.714
7	.649[e]	.422	.410	.37746	-103.262	.623	67.715	-99.284
8	.000[b]	.000	.000	.49158	-75.703	1.038	151.498	-73.714
9	.513[f]	.263	.234	.43029	-88.162	.824	102.031	-82.195
10	.000[b]	.000	.000	.49158	-75.703	1.038	151.498	-73.714
11	.741[g]	.549	.531	.33668	-114.658	.504	43.852	-108.691
12	.246[a]	.061	.043	.48101	-77.079	1.012	141.164	-73.101

a. Predictors: (Constant), X1

b. Predictor: (constant)

c. Predictors: (Constant), X2

d. Predictors: (Constant), X3

e. Predictors: (Constant), X4

f. Predictors: (Constant), X2, X1

g. Predictors: (Constant), X1, X3

SPSS Output 9.4

ANOVA[h]

Model		Sum of Squares	df	Mean Square	F	Sig.
1	Regression	.776	1	.776	3.355	.073[a]
	Residual	12.031	52	.231		
	Total	12.808	53			
2	Regression	.000	0	.000	.	.[b]
	Residual	12.808	53	.242		
	Total	12.808	53			
3	Regression	2.829	1	2.829	14.74	.000[c]
	Residual	9.979	52	.192		
	Total	12.808	53			
4	Regression	.000	0	.000	.	.[b]
	Residual	12.808	53	.242		
	Total	12.808	53			
5	Regression	5.476	1	5.476	38.84	.000[d]
	Residual	7.332	52	.141		
	Total	12.808	53			
6	Regression	.000	0	.000	.	.[b]
	Residual	12.808	53	.242		
	Total	12.808	53			
7	Regression	5.399	1	5.399	37.89	.000[e]
	Residual	7.409	52	.142		
	Total	12.808	53			
8	Regression	.000	0	.000	.	.[b]
	Residual	12.808	53	.242		
	Total	12.808	53			
9	Regression	3.365	2	1.683	9.087	.000[f]
	Residual	9.443	51	.185		
	Total	12.808	53			
10	Regression	.000	0	.000	.	.[b]
	Residual	12.808	53	.242		
	Total	12.808	53			
11	Regression	7.027	2	3.513	31.00	.000[g]
	Residual	5.781	51	.113		
	Total	12.808	53			
12	Regression	.776	1	.776	3.355	.073[a]
	Residual	12.031	52	.231		
	Total	12.808	53			

a. Predictors: (Constant), X1

b. Predictor: (constant)

c. Predictors: (Constant), X2

d. Predictors: (Constant), X3

e. Predictors: (Constant), X4

f. Predictors: (Constant), X2, X1

g. Predictors: (Constant), X1, X3

h. Dependent Variable: YLOG

Following this progression through the three regression procedures in Inserts 9.4 and 9.5, text Table 9.2 will be produced with the exception of the $PRESS_p$ statistic. Although SPSS does not directly produce the $PRESS_p$ statistic, it is easily calculated for individual regression models. From text page 360, the $PRESS_p$ statistic for a regression model is the sum of the squared deleted residuals. For a model with *X1* only, for example, we regress *ylog* on *X1;* save the deleted residuals (*dre_1* by default) in the active Date Editor; and square and sum the deleted residuals. (Insert 9.6) SPSS Output 9.5 contains the sum of the squared deleted residual (*press* = 13.51) which is consistent with the $PRESS_p$ = 13.512 reported in text Table 9.2 for *X1* only.

INSERT 9.6

- ✔ Analyze
- ✔ Regression
- ✔ Linear
- ✔ *ylog*
- ✔ ►(to the Dependent box)
- ✔ *x1*
- ✔ Save
- ✔ Deleted Residuals
- ✔ Continue
- ✔ OK

- ✔ New
- ✔ File
- ✔ Syntax

```
compute press = dre_1**2 .
execute .
descriptives press
 /stat=sum.
```

- ✔ Run
- ✔ All

SPSS Output 9.5

Descriptive Statistics

	N	Sum
PRESS	54	13.51
Valid N (listwise)	54	

Automatic Search Procedures

SPSS does not have a readily available best subset algorithm for model selection. It does, however, have other automatic search procedures for variable reduction. In the Linear Regression dialogue window, there are five methods of variable selection available: *Enter, Stepwise, Remove, Backward,* and *Forward*. (SPSS Output 9.6)

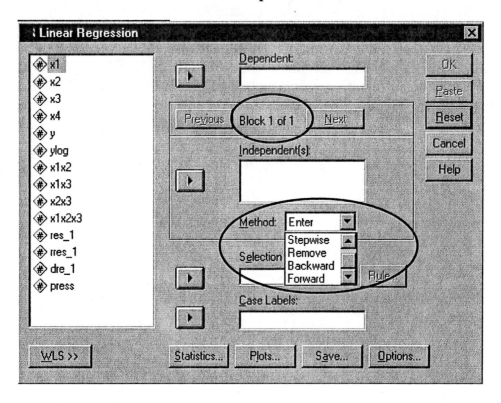

Enter and Remove Methods: The *Enter* and *Remove* methods enter and remove an entire block of variables. The *Enter* method creates a forced entry of all variables in a single step in order of decreasing tolerance. The *Enter* method can be used to control the order of variable entry by using the Block option (SPSS Output 9.6) in the Regression dialogue box. For instance, using the Surgical Unit Example, one could enter *X4* at Block 1 and enter *X1*, *X2*, and *X3* at Block 2. A partial output of the maneuver is shown in SPSS Output 9.7. Notice that Model 1 contains *X4* only and that Model 2 contains *X4* plus *X1*, *X2*, and *X3*. The *Enter* method can also be used to simultaneously enter an entire set of predictor variables in the model. The *Remove* method removes all specified variables in a single step as in Inserts 9.4 and 9.5.

SPSS Output 9.7

Coefficients[a]

Model		Unstandardized Coefficients		Standardized Coefficients		
		B	Std. Error	Beta	t	Sig.
1	(Constant)	5.612	.143		39.381	.000
	X4	.298	.048	.649	6.156	.000
2	(Constant)	3.852	.266		14.467	.000
	X4	.032	.051	.070	.625	.535
	X1	.084	.029	.273	2.902	.006
	X2	.013	.002	.435	5.471	.000
	X3	.016	.002	.676	7.443	.000

a. Dependent Variable: YLOG

Stepwise Selection: The *Stepwise* selection method first selects the best predictor variable (the predictor variable with the largest partial correlation with the response variable). Additional variables are added to the model based on their incremental improvement to the reduction in error sum of squares. Variables are dropped from the model if their test for entry is not significant given the other variables in the model. The default and modifiable α limits for adding and removing a variable are 0.05 and 0.10, respectively.

Using the Stepwise method in SPSS to regress *ylog* on all eight X variables from the Surgical Unit Example results in SPSS Output 9.8. From SPSS Output 9.8 – Coefficients table, note that *X3* is entered at step (Model) 1 because *X3* had the highest partial correlation (not shown) with the response variable. From SPSS Output 9.8 – Excluded Variables, we note that *X2* has the highest partial correlation (.642) with the response function of the variables not in the model at step (Model) 1. Thus, *X2* is added to the model at step 2, SPSS Output 9.8 – Coefficients table. Of the variables not in the model at step 2, *X8* has the highest partial correlation (.584) with the response variable and is added to the model at step 3, SPSS Output 9.8 – Coefficients table. Again, of the variables not in the model at step 3, *X1* has the highest partial correlation (.483) with the response variable and is added to the model at step 4, SPSS Output 9.8 – Coefficients table. Beginning with step 2, the search algorithm compares the probability level of the partial correlation of the variables in the model to the α level needed to remove a variable from the model. A variable is removed from the model if the p-value associated with the partial correlation is greater than the predetermined α level to remove. In the present example, no variables were removed from the model. Notice that the p-value associated with each variable not in the model at step 4 (Model 4 – Excluded Variables, SPSS Output 9.8) is greater than the α level needed to add that variable to the model. Thus, the stepping process terminates with only *X3*, *X2*, *X8*, and *X1* in the final model.

SPSS Output 9.8

Coefficients[a]

Model		Unstandardized Coefficients		Standardized Coefficients	t	Sig.
		B	Std. Error	Beta		
1	(Constant)	5.264	.194		27.138	.000
	X3	.015	.002	.654	6.232	.000
2	(Constant)	4.351	.214		20.296	.000
	X3	.015	.002	.665	8.186	.000
	X2	.014	.002	.486	5.975	.000
3	(Constant)	4.291	.176		24.356	.000
	X3	.014	.002	.626	9.326	.000
	X2	.015	.002	.513	7.677	.000
	X8	.429	.084	.342	5.084	.000
4	(Constant)	3.852	.193		19.992	.000
	X3	.015	.001	.668	11.072	.000
	X2	.014	.002	.488	8.196	.000
	X8	.353	.077	.282	4.573	.000
	X1	.073	.019	.239	3.865	.000

a. Dependent Variable: YLOG

SPSS Output 9.8 continued

Excluded Variables[e]

Model		Beta In	t	Sig.	Partial Correlation	Collinearity Statistics Tolerance
1	X1	.352[a]	3.699	.001	.460	.978
	X2	.486[a]	5.975	.000	.642	.999
	X4	.456[a]	4.679	.000	.548	.827
	X5	-.137[a]	-1.310	.196	-.180	1.000
	X6	.143[a]	1.365	.178	.188	.980
	X7	-.071[a]	-.675	.503	-.094	.993
	X8	.300[a]	3.060	.004	.394	.986
2	X1	.311[b]	4.401	.000	.528	.970
	X4	.283[b]	3.108	.003	.402	.683
	X5	-.114[b]	-1.409	.165	-.195	.998
	X6	.082[b]	.987	.328	.138	.964
	X7	-.134[b]	-1.664	.102	-.229	.977
	X8	.342[b]	5.084	.000	.584	.980
3	X1	.239[c]	3.865	.000	.483	.907
	X4	.233[c]	3.101	.003	.405	.671
	X5	-.075[c]	-1.119	.269	-.158	.984
	X6	.106[c]	1.578	.121	.220	.960
	X7	.051[c]	.655	.516	.093	.729
4	X4	.091[d]	.968	.338	.138	.390
	X5	-.078[d]	-1.316	.194	-.187	.984
	X6	.089[d]	1.494	.142	.211	.954
	X7	.050[d]	.715	.478	.103	.729

a. Predictors in the Model: (Constant), X3

b. Predictors in the Model: (Constant), X3, X2

c. Predictors in the Model: (Constant), X3, X2, X8

d. Predictors in the Model: (Constant), X3, X2, X8, X1

e. Dependent Variable: YLOG

Forward Elimination: The *Forward* method begins with no predictor variables in the model. The predictor variable with the largest partial correlation with the response variable is entered in the model. This method continues to add variables with significant ($p<\alpha$ to add) partial correlations. Notice from SPSS Output 9.9 that the progression of predictor variable entry is the same as the *Stepwise* method shown in SPSS Output 9.8. However, the *Forward* method makes no attempt to remove variables from the model. Since the *Stepwise* procedure above did not remove any predictor variables from the model and the *Forward* method makes no attempt to remove variables, the result of the *Stepwise* method (SPSS Output 9.8) and *Forward* method (SPSS Output 9.9) are the same in the present example, although this will not always be true.

SPSS Output 9.9

Coefficients[a]

Model		Unstandardized Coefficients		Standardized Coefficients	t	Sig.
		B	Std. Error	Beta		
1	(Constant)	5.264	.194		27.138	.000
	X3	.015	.002	.654	6.232	.000
2	(Constant)	4.351	.214		20.296	.000
	X3	.015	.002	.665	8.186	.000
	X2	.014	.002	.486	5.975	.000
3	(Constant)	4.291	.176		24.356	.000
	X3	.014	.002	.626	9.326	.000
	X2	.015	.002	.513	7.677	.000
	X8	.429	.084	.342	5.084	.000
4	(Constant)	3.852	.193		19.992	.000
	X3	.015	.001	.668	11.072	.000
	X2	.014	.002	.488	8.196	.000
	X8	.353	.077	.282	4.573	.000
	X1	.073	.019	.239	3.865	.000

a. Dependent Variable: YLOG

Backward Elimination: The *Backward* method begins with all predictor variables in the model. It then removes each predictor variable that does not have a significant ($p > \alpha$ to remove) partial correlation with the response variable. Again, we are using the default and modifiable α limits for adding and removing a variable, 0.05 and 0.10, respectively. From SPSS Output 9.10 we see that at step (Model) 1 all eight X variables are entered in the model. Because the probability level ("Sig.") of $X4$ is greater than the α needed to remove a variable and is greater than other probability levels of variables in the model, $X4$ is removed at step 2. $X7$, $X5$, and $X6$ are then removed based on their significance levels at steps 2, 3, and 4, respectively. The final model contains $X1$, $X2$, $X3$, and $X8$.

We have used *Stepwise* selection, *Forward* elimination, and *Backward* elimination to reduce the number of predictor variables in the Surgical Unit example. It should be somewhat reassuring to the researcher that each method yielded the same final model.

SPSS Output 9.10

Coefficients[a]

Model		Unstandardized Coefficients		Standardized Coefficients		
		B	Std. Error	Beta	t	Sig.
1	(Constant)	4.051	.252		16.089	.000
	X1	.069	.025	.223	2.695	.010
	X2	.013	.002	.463	6.909	.000
	X3	.015	.002	.647	8.264	.000
	X4	.008	.047	.017	.172	.865
	X5	-.004	.003	-.081	-1.296	.202
	X6	.084	.061	.086	1.386	.173
	X7	.058	.067	.059	.857	.396
	X8	.388	.088	.310	4.394	.000
2	(Constant)	4.037	.236		17.093	.000
	X1	.071	.019	.233	3.824	.000
	X2	.014	.002	.468	7.896	.000
	X3	.015	.001	.655	10.906	.000
	X5	-.004	.003	-.084	-1.429	.160
	X6	.087	.058	.089	1.505	.139
	X7	.059	.067	.060	.880	.383
	X8	.388	.087	.309	4.439	.000
3	(Constant)	4.054	.235		17.266	.000
	X1	.072	.019	.233	3.837	.000
	X2	.014	.002	.473	8.047	.000
	X3	.015	.001	.654	10.912	.000
	X5	-.003	.003	-.078	-1.342	.186
	X6	.087	.058	.089	1.513	.137
	X8	.351	.076	.280	4.594	.000
4	(Constant)	3.867	.191		20.292	.000
	X1	.071	.019	.232	3.791	.000
	X2	.014	.002	.478	8.073	.000
	X3	.015	.001	.654	10.821	.000
	X6	.087	.058	.089	1.494	.142
	X8	.363	.077	.289	4.740	.000
5	(Constant)	3.852	.193		19.992	.000
	X1	.073	.019	.239	3.865	.000
	X2	.014	.002	.488	8.196	.000
	X3	.015	.001	.668	11.072	.000
	X8	.353	.077	.282	4.573	.000

a. Dependent Variable: YLOG

Model Validation: Surgical Unit Example continued

Data File: ch09ta05.txt

Continuing with the Surgical Unit example, the researchers identified three candidate models using the model-selection criteria presented above: SSE_p, R_p^2, $R_{a,p}^2$, C_p, AIC_p, SBC_p, and $PRESS_p$. The researchers now wish to estimate the predictive capability of the three models using the *mean squared prediction error* ($MSPR$) described on text page 370. Model 1, given on text page 373, uses *ylog* as the response variable and *X1*, *X2*, *X3*, and *X8* as predictor variables. In Insert 9.9, we first find the regression model based on Model 1 using the Surgical Unit **training data set** presented in Table 9.1, text page 350. The estimated regression coefficients are given in SPSS Output 9.11. We now open the Surgical Unit **validation data set** presented in Table 9.5, text page 374. After opening ch09ta05.txt and renaming the variables, the first 5 lines of the data file should look like SPSS Output 9.12.

INSERT 9.9

- ✔ Analyze
- ✔ Regression
- ✔ Linear
- ✔ *ylog*
- ✔ ▶(to the Dependent box)
- ✔ *x1*
- ✔ ▶(to the Independent box)
- ✔ *x2*
- ✔ ▶(to the Independent box)
- ✔ *x3*
- ✔ ▶(to the Independent box)
- ✔ *x8*
- ✔ ▶(to the Independent box)
- ✔ OK

SPSS Output 9.11

Coefficients[a]

Model		Unstandardized Coefficients		Standardized Coefficients		
		B	Std. Error	Beta	t	Sig.
1	(Constant)	3.852	.193		19.992	.000
	X1	.073	.019	.239	3.865	.000
	X2	.014	.002	.488	8.196	.000
	X3	.015	.001	.668	11.072	.000
	X8	.353	.077	.282	4.573	.000

a. Dependent Variable: YLOG

281

	x1	x2	x3	x4	x5	x6	x7	x8	y	ylog
1	7.10	23.0	78.00	1.93	45.0	0	1	0	302.0	5.71
2	4.90	66.0	91.00	3.05	34.0	1	0	0	767.0	6.64
3	6.40	90.0	35.00	1.06	39.0	1	0	1	487.0	6.19
4	5.70	35.0	70.00	2.13	68.0	1	0	0	242.0	5.49
5	6.10	42.0	69.00	2.25	70.0	0	0	1	705.0	6.56

From equation 9.20, text page 370, we now compute $\hat{Y}_i(m1)$, the predicted value of the ith observation in the validation data set using the regression coefficients (SPSS Output 9.11) estimated from the training data set. (Insert 9.10) We then find $devsq = \left(Y_i - \hat{Y}_i\right)^2$, where Y_i is the value of the response variable of the ith observation in the validation data. Finally, we find the $\sum_{i=1}^{n^*}\left(Y_i - \hat{Y}_i\right)^2$ (SPSS Output 9.13). We find by hand calculator $\sum_{i=1}^{n^*}\left(Y_i - \hat{Y}_i\right)^2 / n = .07735$.

A note on precision is in order. If in Insert 9.10 we had used the regression coefficients displayed in SPSS Output 9.11 (3 positions to the right of the decimal) to find $\sum_{i=1}^{n^*}\left(Y_i - \hat{Y}_i\right)^2$ and then divided this quantity by 4.18 as displayed in SPSS Output 9.13, we would have found that $MSPR = 0.08259$ for Model 1. This is in disagreement with $MSPR = 0.0773$ reported on text page 373 and is due to the relative magnitude of the numbers involved in the above calculations. That is, because the numbers are relatively small, we require more precise estimates. To gain this precision, double click on the Coefficients table (SPSS Output 9.11) while in the SPSS Viewer. Then double click on the number for which you desire greater precision. SPSS Output 9.14 and 9.15 display the result of this maneuver when attempting to gain more precise regression coefficients and $\sum_{i=1}^{n^*}\left(Y_i - \hat{Y}_i\right)^2$. SPSS Output 9.13 and 9.15 are a result of using 5-digit precision in all calculations and result in $MSPR = .07735$, which is consistent with $MSPR = .0773$ reported on text page 373.

We repeat the above process by estimating regression functions for Models 2 and 3 on text page 373 based on the Surgical Unit training data set and finding $\sum_{i=1}^{n^*}\left(Y_i - \hat{Y}_i\right)^2 / n$ when using the validation data set. This will result in $MSPR = 0.0764$ and 0.0794, respectively. From the relative closeness of the three $MSPR$ values, the researchers concluded that the three candidate models performed equally in terms of predictive accuracy. Model 3 was discarded due to reversal of the sign of a regression coefficient from the training data set to the validation data set. Models 1 and 2 preformed equally well in the validation study. The final selection of Model 1 was based on Occam's razor; Model 1 required one less parameter than Model 2.

INSERT 9.10

✔ File
✔ New
✔ Syntax
compute m1=3.85241+(0.07332*x1)+(0.01418*x2)+(0.01545*x3)+(0.35296*x8).
compute devsq=(ylog-m1)**2.
descriptives
 variables=devsq
 /statistics˗ sum .
execute.
✔ Run
✔ All

SPSS Output 9.13

Descriptive Statistics

	N	Sum
DEV	54	4.18
Valid N (listwise)	54	

SPSS Output 9.14

Coefficients[a]

Model		Unstandardized Coefficients		Standardized Coefficients	t	Sig.
		B	Std. Error	Beta		
1	(Constant)	3.85241856	.193		19.992	.000
	X1	.073	.019	.239	3.865	.000
	X2	.014	.002	.488	8.196	.000
	X3	.015	.001	.668	11.072	.000
	X8	.353	.077	.282	4.573	.000

a. Dependent Variable: YLOG

SPSS Output 9.15

Descriptive Statistics

	N	Sum
DEV	54	4.17732685
Valid N (listwise)	54	

Chapter 10 SPSS®

Building the Regression Model II: Diagnostics

This chapter is concerned with diagnostic techniques for identifying inadequacies of a fitted regression model. Graphical methods were used in previous chapters as an aid in deciding whether to use higher-order terms for a predictor variable in the model or whether to add predictor variables that have not been included in the model. Chapter 10 discusses graphs that can provide useful model building information about a predictor variable given that other predictor variables are already in the model. When we fit a regression model to a set of data, some observations may be outlying or extreme relative to most of the data. As shown in Figure 10.5, text page 391, an observaton may contain a Y value, one or more X values, or both that are outliers. Outlying observations may have too much influence on the estimates of the parameters of the model and, therefore, may require remedial steps. The authors discuss methods for detecting outliers and excessively influential observations. They also discuss remedial steps for dealing with these observations. In Section 10.5, text page 406, the authors continue their discussion of problems created when there is multicollinearity among the predictor variables of a regression model. They present informal methods of detecting multicollinearity problems on text page 407. They then define the variance inflation factor and show how it is used in more formal ways to check for serious multicollinearity among the predictor variables.

The Life Insurance Example

Data File: ch10ta01.txt

Table 10.1, text page 387, presents data from a study of the relationship between amount of insurance carried (Y) and annual income ($X1$) and a measure of risk aversion ($X2$). After opening the Life Insurance data file, the first 5 cases of the Data Editor should look like SPSS Output 10.1.

SPSS Output 10.1

	x1	x2	y	var
1	45.01	6.0	91.00	
2	57.20	.4.0	162.00	
3	26.85	5.0	11.00	
4	66.29	7.0	240.00	
5	40.96	5.0	73.00	

We begin by regressing *Y* on *X1* and *X2* and saving the unstandardized residuals. (Insert 10.1) We then plot *X2* on the x-axis and the residuals (*res_1*) on the y-axis to produce Figure 10.3 (a), text page 387, (SPSS Output 10.2). Since we will not need *res_1* for the remainder of the problem, we delete this variable from the Data Editor. To produce text Figure 10.3 (b), we regress *Y* on *X2* and save the unstandardized residuals by default as *res_1*. If we had not deleted *res_1* above, the second residual would be stored as *res_2*. We then regress *X1* on *X2* and save the unstandardized residual values by default as *res_2*. We request a scatter plot with *res_1* [$e(Y \mid X_2)$] on the y-axis and *res_2* [$e(X_1 \mid X_2)$] on the x-axis. This produces SPSS Output 10.3.

SPSS Output 10.2

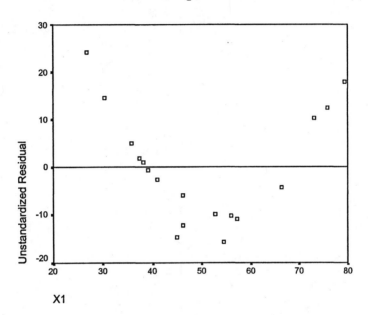

SPSS Output 10.3

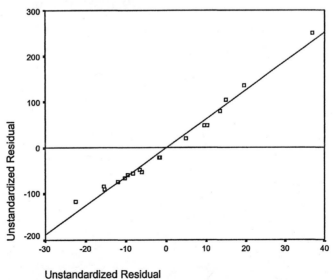

INSERT 10.1	INSERT 10.1 continued
✔ Analyze	✔ Analyze
✔ Regression	✔ Regression
✔ Linear	✔ Linear
✔ *y*	✔ *x1*
✔ ▶(to the Dependent box)	✔ ▶(to the Dependent box)
✔ *x1*	✔ *x2*
✔ ▶(to the Independent box)	✔ ▶(to the Independent box)
✔ *x2*	✔ Save
✔ ▶(to the Independent box)	✔ Unstandardized (Residual Values)
✔ Save	✔ Continue
✔ Unstandardized (Residual Values)	✔ OK
✔ Continue	
✔ OK	✔ Graphs
	✔ Scatter
✔ Graphs	✔ Simple
✔ Scatter	✔ Define
✔ Simple	✔ *res_1*
✔ Define	✔ ▶(to the Y axis box)
✔ *res_1*	✔ *res_2*
✔ ▶(to the Y axis box)	✔ ▶(to the X axis box)
✔ *x1*	✔ OK
✔ ▶(to the X axis box)	* Double click on the scatter plot to open the Chart Editor.
✔ OK	✔ Chart
* Double click on the scatter plot to open the Chart Editor.	✔ Options
✔ Chart	✔ Fit Options
✔ Reference Line	✔ OK
* Add a reference line on the Y-axis at position 0.	✔ Chart
* Close the Chart Editor.	✔ Reference Line
	* Add a reference line on the Y-axis at position 0.
✔ Analyze	* Close the Chart Editor.
✔ Regression	
✔ Linear	
✔ *y*	
✔ ▶(to the Dependent box)	
✔ *x2*	
✔ ▶(to the Independent box)	
✔ Save	
✔ Unstandardized (Residual Values)	
✔ Continue	
✔ OK	

The Body Fat Example

Data File: ch07ta01.txt

We use the Body Fat example (Table 7.1, text page 257) to illustrate all regression diagnostics. See Chapter 0, "Working with SPSS," for instructions on opening the Body Fat example and renaming the variables *X1*, *X2*, *X3*, and *Y*. After opening ch07ta01.txt, the first 5 lines in your SPSS Data Editor should look like SPSS Output 10.4.

SPSS Output 10.4

	x1	x2	x3	y	var
1	19.50	43.10	29.10	11.90	
2	24.70	49.80	28.20	22.80	
3	30.70	51.90	37.00	18.70	
4	29.80	54.30	31.10	20.10	
5	19.10	42.20	30.90	12.90	

Added-Variable Plots: We begin by regressing body fat (*Y*) on triceps skinfold thickness (*X1*) and thigh circumference (*X2*) and saving the unstandardized residuals (Insert 10.2). This results in the regression analysis shown in SPSS Output 10.2. The fitted regression function is:

$$\hat{Y} = -19.174 + 0.222X_1 + 0.659X_2$$

INSERT 10.2

- ✔ Analyze
- ✔ Regression
- ✔ Linear
- ✔ *y*
- ✔ ► (to the Dependent box)
- ✔ *x1*
- ✔ ► (to the Independent box)
- ✔ *x2*
- ✔ ► (to the Independent box)
- ✔ Save
- ✔ Unstandardized (Residual Values)
- ✔ Continue
- ✔ OK

SPSS Output 10.5

ANOVA[b]

Model		Sum of Squares	df	Mean Square	F	Sig.
1	Regression	385.439	2	192.719	29.797	.000[a]
	Residual	109.951	17	6.468		
	Total	495.389	19			

a. Predictors: (Constant), X2, X1

b. Dependent Variable: Y

Coefficients[a]

Model		Unstandardized Coefficients		Standardized Coefficients	t	Sig.
		B	Std. Error	Beta		
1	(Constant)	-19.174	8.361		-2.293	.035
	X1	.222	.303	.219	.733	.474
	X2	.659	.291	.676	2.265	.037

a. Dependent Variable: Y

We now plot the residuals (*res_1*) against *X1* and *X2* (Insert 10.3). This produces Figures 10.4 (a) and (c), text page 389. (SPSS Output 10.6) To produce text Figures 10.4 (b) and (d), we use the regression procedure and request that all partial plots be generated (Insert 10.4). This produces the text Figures 10.4 (b) and (d) without the regression line. To insert the regression line on the charts, we use the Chart Editor to request that a linear regression line be fitted through the data points. SPSS Output 10.7 is equivalent to text Figures 10.4 (b) and (d).

INSERT 10.3

✔ Graphs
✔ Scatter
✔ Simple
✔ Define
✔ *res_1*
✔ ▶ (to the Y axis box)
✔ *x1*
✔ ▶ (to the X axis box)
✔ OK
✔ Graphs
✔ Scatter
✔ Simple
✔ Define
✔ *res_1*
✔ ▶ (to the Y axis box)
✔ *x2*
✔ ▶ (to the X axis box)
✔ OK

288

SPSS Output 10.6

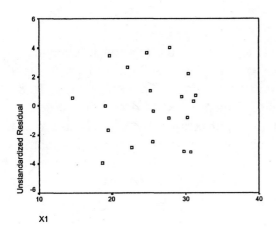

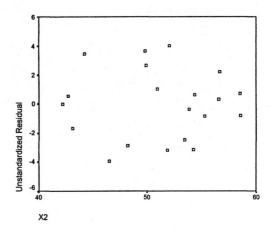

INSERT 10.4

- ✔ Analyze
- ✔ Regression
- ✔ Linear
- ✔ *y*
- ✔ ►(to the Dependent box)
- ✔ *x1*
- ✔ ►(to the Independent box)
- ✔ *x2*
- ✔ ►(to the Independent box)
- ✔ Plots
- ✔ Produce all partial plots
- ✔ Continue
- ✔ OK

* Double click on one of the charts to open the Chart Editor.

- ✔ Chart
- ✔ Options
- ✔ Total (under Fit Line)
- ✔ OK

* Close Chart Editor.

* Repeat the above 4 steps for the second chart.

SPSS Output 10.7

Partial Regression Plot

Dependent Variable: Y

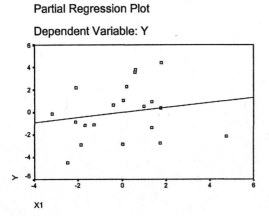

Partial Regression Plot

Dependent Variable: Y

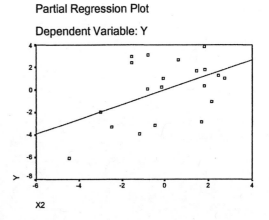

Illustration of Hat Matrix

Data File: ch10ta02.txt

On text page 393, the authors present a small data set to illustrate the relationship between the response variable and the predictor variables. After opening ch10ta02.txt and renaming the variables $X1$, $X2$, and Y, the 4 cases in the Data Editor should look like SPSS Output 10.8. In Insert 10.5, we create a new variable named $X0$ that is equal 1 for each of the 4 cases. This variable will be used to create the **X** matrix. We then set matrixes **CON**, **X1**, **X2**, and **Y** to the column variables $X0$, $X1$, $X2$, and Y, respectively, in the Data Editor. We create the **X** matrix by concatenating **X0**, **X1**, and **X2**. We then use various text equations to find the **B** (parameter estimates), **H** (hat matrix), **YHAT** (predicted values), and **VARCOVAR** (variance-covariance) matrices. Although the syntax only prints **VARCOVAR**, a *print* statement for any matrix can be added at any location in the matrix program.

SPSS Output 10.8

	y	x1	x2	var
1	301.00	14.0	25.0	
2	327.00	19.0	32.0	
3	246.00	12.0	22.0	
4	187.00	11.0	15.0	

INSERT 10.5	INSERT 10.5 CONTINUED
✔ File ✔ New ✔ Syntax * Write a column of 1's into the Hat Matrix Example. * data data file. compute x0=1. execute. matrix. get con / variables =x0. get x1 / variables = x1. get x2 / variables = x2. * Define the Y matrix. get y /variables=y. * Define the X matrix. compute x = {con, x1, x2}. * Fine the b matrix using text equation 6.25. compute b=inv(t(x)*x)*t(x)*y.	* Find hat matrix (hat) by text equation 10.10. compute hat=x*inv(t(x)*x)*t(x). * Find fitted values of y (yhat) by text equation 10.11. compute yhat=hat*y. * Create a 4x4 idenity (I) matrix. compute I=ident(4,4). * Find variance-covariance matrix using text equation. * 10.13. compute varcovar=574.9*(i-hat). print varcovar. end matrix. execute. ✔ Run ✔ All

290

Formal Diagnostics: The remainder of the diagnostics on text pages 390-410 deals with *residuals*, *deleted residuals*, *leverage values*, *DFFITS*, *DFBETAS*, *Cook's distance measure*, and *variance inflation factors*. To produce these statistics we request the appropriate regression model as before. While defining the regression model we also click *Save* and select the needed available options. (SPSS Output 10.9) For instance, if you select *studentized residuals*, *studentized deleted residuals*, *Cook's distance measure*, *leverage values*, and *DIFF*, the resulting values will be saved in the Data Editor as *sre_1*, *sdr_1*, *coo_1*, *lev_1*, and *dff_1*, respectively. SPSS also provides a summary table of residual statistics for use in diagnostics, as in the above Body Fat example. (SPSS Output 10.10)

SPSS Output 10.9

291

SPSS Output 10.10

Residuals Statistics[a]

	Minimum	Maximum	Mean	Std. Deviation	N
Predicted Value	12.2294	26.3838	20.1950	4.50403	20
Std. Predicted Value	-1.769	1.374	.000	1.000	20
Standard Error of Predicted Value	.56915	1.55098	.95209	.25891	20
Adjusted Predicted Value	11.9443	26.3747	20.2928	4.56721	20
Residual	-3.9469	4.0147	.0000	2.40559	20
Std. Residual	-1.552	1.579	.000	.946	20
Stud. Residual	-1.712	1.661	-.017	1.026	20
Deleted Residual	-5.0567	4.4425	-.0978	2.84958	20
Stud. Deleted Residual	-1.826	1.760	-.018	1.062	20
Mahal. Distance	.002	6.117	1.900	1.622	20
Cook's Distance	.000	.490	.065	.113	20
Centered Leverage Value	.000	.322	.100	.085	20

a. Dependent Variable: Y

Identifying Outlying Y Observations – Studentized Deleted Residuals: On text page 396 the researchers wished to examine whether there were outlying Y observations when using a regression model with two predictor variables ($X1$ and $X2$). In Insert 10.6 we obtain a first order regression model with predictors $X1$ and $X2$ and request that the *unstandardized* and *studentized deleted residuals* be saved. (Insert 10.6 – Step 1) The Data Editor will contain the *unstandardized* and *studentized deleted residuals* (*res_1* and *sdr_1*, respectively, SPSS Output 10.11), consistent with columns 1 and 3 of Table 10.3, text page 397. The researchers now wish to determine if case 13 is an outlier based on the *studentized deleted residual* (text page 397). Using SPSS it is actually just as simple to test for outliers for each of the 20 cases. The appropriate Bonferroni critical value is $t(1-\alpha/2n; n-p-1)$ where $\alpha = 0.10$, $n = 20$, and $p =$ number of parameters in the model ($X1$, $X2$, and β_0) or $t(1-0.10/2*20; 20-3-1) = t(0.9975,16)$. In Insert 10.6 – Step 2, we use the *inverse distribution function* to fine *tvalue* = $t(0.9975;16) = 3.2519$. We then compare the absolute value of the *studentized deleted residual* (*abs*(*sdr_1*)) to *tvalue*. If the absolute value of the *studentized deleted residual* is greater than *tvalue*, we set *outlier* = 1, otherwise *outlier* = 0. Consistent with text page 398, all data lines in the Data Editor with *outlier* = 1 should be considered to be outliers and, conversely, all data lines with *outlier* = 0 should be considered not to be outliers. If we were dealing with a large data file, more than 100 data lines, for example, we could search the entire Data Editor using ✔ Data ✔ Select Cases to find all data lines with *outlier* = 1.

```
                        INSERT 10.6

* Step 1.
✔ Analyze
✔ Regression
✔ Linear
✔ y
✔ ▶ (to the Dependent box)
✔ x1
✔ ▶ (to the Independent box)
✔ x2
✔ ▶ (to the Independent box)
✔ Save
✔ Unstandardized (Residual Values)
✔ Studentized deleted (Residual Values)
✔ Continue
✔ OK

* Step 2.
✔ File
✔ New
✔ Syntax
* Type the following syntax.
compute tvalue=idf.t(.9975,16).
if (abs(sdr_1) gt tvalue)outlier=1.
if (abs(sdr_1) le tvalue)outlier=0.
execute.
✔ Run
✔ All
```

SPSS Output 10.11

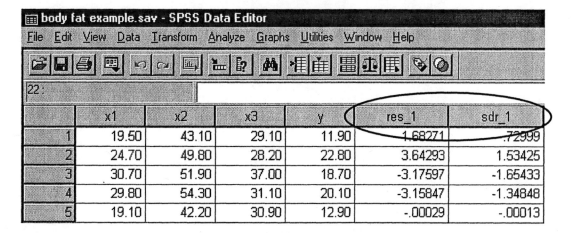

Identifying Outlying X Observations - Hat Matrix Leverage Values: To identify potentially outlying cases, the researchers in the Body Fat example plotted triceps skinfold thickness ($X1$) against thigh circumference ($X2$) and identified each line by a case number (text page 399).

Notice that the Body Fat data file does not have a variable that represents case number. The "*compute case=$casenum*" statement in Insert 10.7 creates a new variable in the Data Editor that is a sequential count of each of the 20 cases. We then request a scatter plot of *X1* against *X2* and that data points be labeled by *case*. (The default value for decimal places in 2. If you produce the scatter plot (SPSS Output 10.12) with this default setting for *case*, the plot will appear cluttered because each case number will be printed with 2 zeros to the right of the decimal. Remedy this by setting the decimal places for *case* to 0 using the *Variable View* tab of the Data Editor.) Note that cases 3 and 15 appear to be outliers with respect to the *X* values.

INSERT 10.7	INSERT 10.7 continued
✔ File ✔ New ✔ Syntax * Type the following syntax. compute case=$casenum. execute. ✔ Run ✔ All	✔ Graphs ✔ Scatter ✔ Simple ✔ Define ✔ *x2* ✔ ▶ (to Y axis box) ✔ *x1* ✔ ▶ (to X axis box) ✔ *case* ✔ ▶ (to Label Cases by box) ✔ Options ✔ Display Chart with Case Labels ✔ Continue ✔ OK

SPSS Output 10.12

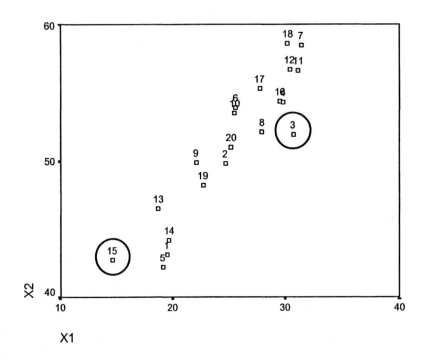

294

The diagonal element (h_{ii}) of the hat matrix is a measure of the ith case from the center of all X cases. The larger the value of h_{ii} the larger the effect the case has on the corresponding fitted value ($\hat{y}_i$). Thus, h_{ii} is referred to as the *leverage* of the ith case. In Insert 10.8, we regress Y on *X1* and *X2* and request that leverage values (*lev_1* by default) be saved in the Data Editor. Note that SPSS will save **centered leverage values**. The leverage values (h_{ii}) reported in column 2 of Table 10.3, text page 397, are non-centered values. Centered leverages can be found by subtracting $1/n$ from the non-centered leverages, where n is the number of cases. Non-centered leverage values are found by adding $1/n$ to the centered values reported by SPSS. (Centered leverages are multiplied by $n-1$ to obtain the *Mahalanobis Distance measure*.) Thus, in Insert 10.8 we add 1/20 to *lev_1* and place the result in *nclev* (non-centered leverage values). The values of *nclev* in the Data Editor will be identical to the leverage values (h_{ii}) reported in column 2 of text Table 10.3.

The text defines a "large" leverage value as one that is more than twice the magnitude of the mean leverage value, defined as:

$$\overline{h} = \frac{p}{n}$$

where p is the number of regression parameters (including the intercept) and n is the number of cases. So, the non-centered leverage (*nclev*) is considered large in this example if $nclev > 2*3/20$. If so, we set *hilev* = 1, otherwise *hilev* = 0. After executing Insert 10.8 note that case numbers 3 and 15 have high leverage values as indicated by *hilev* = 1 in the Data Editor.

INSERT 10.8

- ✔ Analyze
- ✔ Regression
- ✔ Linear
- ✔ *y*
- ✔ ▶(to the Dependent box)
- ✔ *x1*
- ✔ ▶(to the Independent box)
- ✔ *x2*
- ✔ ▶(to the Independent box)
- ✔ Save
- ✔ Leverage Values (under Distance)
- ✔ Continue
- ✔ OK

- ✔ File
- ✔ New
- ✔ Syntax
* Type the following syntax.
compute nclev=lev_1 + 1 / 20.
if (nclev > 2*3/20) hilev = 1.
if (nclev le 2*3/20)hilev = 0.
execute.
- ✔ Run
- ✔ All

Identifying Influential Cases – DFFITS, Cook's Distance, and DFBETAS Measures: The researchers now wish to ascertain whether previously identified outlying cases were also influential cases. In Insert 10.9 we begin by regressing Y against $X1$ and $X2$ and requesting that SPSS save *standardized DFBETA (sdb0_1, sdb1_1,* and *sdb2_1*, by default), *standardized DFFITS (sdf_1*, by default), and *Cook's distance measure (coo_1*, by default) in the Data Editor. Note that SPSS can generate *standardized* or *non-standardized DFBETAs* and *DFFITS*. The *DFBETAs* and *DFFITS* reported in the Table 10.4, text page 402, are standardized.

```
                          INSERT 10.9

    ✔ Analyze
    ✔ Regression
    ✔ Linear
    ✔ y
    ✔ ▶ (to the Dependent box)
    ✔ x1
    ✔ ▶ (to the Independent box)
    ✔ x2
    ✔ ▶ (to the Independent box)
    ✔ Save
    ✔ Standardized DfBeta(s) (under Influence)
    ✔ Standardized DfFit (under Influence)
    ✔ Cook's  (under Distance)
    ✔ Continue
    ✔ OK
```

A *DFFITS* value is judged to indicate an influential case if it exceeds 1 for a small to medium sample and $2\sqrt{p/n}$ for a large sample. In Insert 10.10, we create a new variable, *fitflag*, to indicate if the value of *DFFITS* exceeds $2\sqrt{p/n}$. We set *fitflag* = 1 if *DFFITS* $(sdf_1) > 2\sqrt{p/n}$, otherwise *fitflag* = 0. For *Cook's Distance measure* (D_i), the text authors suggest that D_i be compared to the F distribution with $df1 = p$ and $df2 = n{-}p$ degrees of freedom. A case is not considered influential if the corresponding percentile value is less than 10 to 20 percent. However, a case has a major influence on the fitted regression function if the corresponding percentile value is near 50 percent or greater. In Insert 10.10 we use the *cumulative distribution function* for the F distribution (*CDF.F*) with $df1 = 3$ (parameters) and $df2 = 17$ ($n{-}p$) degrees of freedom to find the percentile value of D_i. We multiply this proportion by 100 to express the value as a percent. The resulting value is saved as a new variable, *percent*. A case that has an absolute *DFBETAS* value greater than 1 for small to medium samples or an absolute *DFBETAS* greater than $2/\sqrt{n}$ for large samples may be considered to be an influential case. In Insert 10.10, we create two new variables, *b1flag* and *b2flag*. We set *b1flag* = 1 or *b2flag* = 1 if the absolute value (*abs*) of *sdb1_1* or *sdb1_2*, respectively, is greater than 0, otherwise *b1flag*=0 and *b2flag*=0.

After executing Insert 10.10, note that all values of *fitflag* = 0, indicating that no *DFFIT* value was greater than $2\sqrt{p/n}$. As reported on text page 404, we also find that the percent value (*percent*) for case 3 is 30.6 percent and that the next largest percent value is for case 13 at 11 percent. Using *DFBETAS*, we find that both *b1flag* = 1 and *b2flag* = 1 for case 3.

If you were working with a large data set, the *flag* variable in Insert 10.10 could be used to mark any case that had a *DFFITS*, *DFBETAS*, or *Cook's Distant measure* that needed further investigation. We use *select cases* to find all cases where *flag*=1 and then use *case summaries* to list the case number of each selected record.

INSERT 10.10	INSERT 10.10 continued

✔ File
✔ New
✔ Syntax
* Type the following syntax.
compute case=$casenum.
compute fitflag=0.
If (sdf_1 > 2*sqrt(3/20))fitflag=1.
compute percent = 100*CDF.F(coo_1,3,17) .
if abs((sdb1_1) > 1) b1flag=1.
if abs((sdb2_1) > 1) b2flag=1.
if (fitflag=1 OR percent > 20 OR b1flag=1 OR b2flag=1)flag=1.
execute.
✔ Run
✔ All
	✔ Data
	✔ Select Cases
	✔ Syntax
	✔ If condition is satisfied
	✔ If (below if condition is satisfied)
	✔ Flag
	✔ ▶
	✔ =
	✔ 1
	✔ Continue
	✔ OK
	✔ Analyze
	✔ Reports
	✔ Case Summaries
	✔ Flag
	✔ ▶ (to Variables box)
	✔ OK

Influence on Inferences: Case 3 was identified as an outlying X observation and all three influence measures identified case 3 as influential. Thus, the researcher fitted regression functions with and without case 3. The researcher then calculated the percent difference between the predicted value of Y ($\hat{Y}_i$) based on the regression function with all 20 cases and the predicted value of Y ($\hat{Y}_{i(3)}$) based on the regression function when case 3 was omitted. From Insert 10.11, we begin this analysis by regressing Y on $X1$ and $X2$ and saving the *unstandardized predicted values* (*pre_1* by default). We then use the *Select If* function to select cases where case number (*case*) is not equal to 3, and again regress Y on $X1$ and $X2$. (See Insert 10.5 if *case* is not in your Data Editor.) We do not save the *unstandardized predicted values* for this regression analysis because case 3 will not have a fitted value and we need the fitted values for all 20 cases; this analysis is only to find the parameter estimates for the model without case 3 in the data set. To find $\hat{Y}_{i(3)}$ for all 20 cases, we use the parameter estimates (shown on text page 406) from the regression function with case 3 omitted to compute *pre_2*. At this point, we must turn off (reset) the *Select If* function to use all 20 cases for subsequent analysis. We then find *pcinf*, the absolute value (*abs*) of (($\hat{Y}_{i(3)}$ - $\hat{Y}_i$)/$\hat{Y}_i$)*100. Finally, we use the *Descriptives* procedure to find the mean of *pcinf* = 3.1039. We can also generate a frequency distribution of *pcinf* by using ✔ Analyze ✔ Descriptives ✔ Frequencies to see that the mean difference for 17 of the 20 cases is less than 5 percent. Based on this direct evidence, the researcher concluded that case 3 did not exert undue influence on the regression function.

INSERT 10.11	INSERT 10.11 continued
✔ Analyze ✔ Regression ✔ Linear ✔ *y* ✔ ►(to the Dependent box) ✔ *x1* ✔ ►(to the Independent box) ✔ *x2* ✔ ►(to the Independent box) ✔ Save ✔ Unstandardized (under Predicted values) ✔ Continue ✔ OK ✔ Data ✔ Select Cases ✔ If condition is satisfied ✔ If (under If condition is satisfied) ✔ case ✔ ► ✔ ▱ ✔ 3 ✔ Continue ✔ OK ✔ Analyze ✔ Regression ✔ Linear ✔ *y* ✔ ►(to the Dependent box) ✔ *x1* ✔ ►(to the Independent box) ✔ *x2* ✔ ►(to the Independent box) ✔ OK	✔ Data ✔ Select Cases ✔ Reset ✔ OK ✔ File ✔ New ✔ Syntax * Type the following syntax. compute pre_2=-12.428+.5641*(x1)+.3635*(x2). compute pcinf=(abs((pre_2-pre_1)/pre_1))*100. execute. ✔ Run ✔ All ✔ Analyze ✔ Descriptive Statistics ✔ Descriptives ✔ pcinf ✔ ►(to the Variable box) ✔ OK

Multicollinearity Diagnostics Variance Inflation Factor: The researchers now fitted a regression model with all three predictor variables. From text chapter 7, we noted that triceps skinfold thickness and thigh circumference were highly correlated. We also noted large changes in parameter estimates when a variable was added to the model and that one estimated coefficient was negative when a positive value was expected. Accordingly, the researchers were concerned that there may be *multicollinearity* among the predictor variables. They generated *Variance Inflation Factors (VIF)* as a formal test for the presence of *multicollinearity*. Insert 10.12 requests *collinearity diagnostics* and SPSS Output 10.13 contains partial results. The *VIF"s* for *X1*, *X2*, and *X3* are identical to those reported on text page 409. SPSS does not report the mean VIF value.

✔ Analyze
✔ Regression
✔ Linear
✔ *y*
✔ ►(to the Dependent box)
✔ *x1*
✔ ►(to the Independent box)
✔ *x2*
✔ ►(to the Independent box)
✔ *x3*
✔ ►(to the Independent box)
✔ Statistics
✔ Collinearity diagnostics
✔ Continue
✔ OK

SPSS Output 10.13

Coefficients[a]

Model		Unstandardized Coefficients B	Unstandardized Coefficients Std. Error	Standardized Coefficients Beta	t	Sig.	Collinearity Statistics Tolerance	Collinearity Statistics VIF
1	(Constant)	117.085	99.782		1.173	.258		
	X1	4.334	3.016	4.264	1.437	.170	.001	708.843
	X2	-2.857	2.582	-2.929	-1.106	.285	.002	564.343
	X3	-2.186	1.595	-1.561	-1.370	.190	.010	104.606

a. Dependent Variable: Y

Building the Regression
Model III: Remedial Measures

When we detect inadequacies in a regression model that we are building, we take steps to improve the model so that it is more appropriate for the problem at hand. In earlier chapters, the authors used *remedial measures* such as transformations of Y "to linearize the regression relation, to make the error distributions more nearly normal, or to make the variances of the error terms more nearly equal." Further remedial measures are presented in Chapter 11 to improve model adequacy in the presence of unequal error variances, multicollinearity, and overly influential observations. The use of *weighted least squares* in regression analysis is discussed as an alternative to transformations in building models where the variances of the error terms are unequal. *Ridge regression* is presented as a method for dealing with serious multicollinearity among the predictor variables. *Robust regression* techniques are used to dampen the influence of outlying observations. Two methods of *nonparametric regression* are described: *lowess* and *regression trees*. The method of *bootstrapping* is presented to deal with the problem of estimating the precision of some of the complex estimators discussed in this chapter.

Unequal Error Variances Remedial Measures – Weighted Least Squares: Section 11.1 defines the *generalized multiple regression model*:

$$Y_i = \beta_0 + \beta_1 X_{i1} + \ ... \ + \beta_{p-1} X_{i,p-1} + \varepsilon_i$$

where

$\beta_0, \beta_1, \ ... \ , \beta_{p-1}$ are unknown parameters

$X_{i1}, X_{i2}, \ ... \ , X_{i,p-1}$ are known constants

ε_i are independent $N(0, \sigma_i^2)$

$i = 1, 2, \ ... \ , n$

The method of *weighted least squares* is frequently used to find estimators of the parameters in a generalized multiple regression model because the resulting estimators are *minimum variance unbiased esimators*. Let

$$w_i = \frac{1}{\sigma_i^2}$$

and

$$\mathbf{W} = \begin{bmatrix} w_1 & 0 & \cdots & 0 \\ 0 & w_2 & \cdots & 0 \\ \vdots & \vdots & \ddots & \vdots \\ 0 & 0 & \cdots & w_n \end{bmatrix}$$

Assume the σ_i^2 and hence the w_i are known. Then the weighted least squares (and maximum likelihood) estimators of the coefficients in the generalized multiple regression model are:

$$\mathbf{b_W} = (\mathbf{X'WX})^{-1}\mathbf{X'WY}$$

and the variance-covariance matrix of these coefficients is:

$$\sigma^2\{\mathbf{b_W}\} = (\mathbf{X'WX})^{-1}$$

The σ_i^2 are unknown in most applications and the authors present methods for dealing with this circumstance.

The Blood Pressure Example

Data File: ch11ta01.txt

A researcher studied the relationship between diastolic blood pressure (*dbp*) and age (*age*). See Chapter 0, "Working with SPSS," for instructions on opening the Blood Pressure example data file and renaming the variables. After opening ch11ta01.txt, the first 5 lines in the SPSS Data Editor should look like SPSS Output 11.1.

SPSS Output 11.1

	age	dbp	var	var	var	var
1	27	73.0				
2	21	66.0				
3	22	63.0				
4	24	75.0				
5	25	71.0				

301

We begin the preliminary analysis of the residuals by fitting an *unweighted least squares regression* model ($Y_i = \beta_0 + \beta_1 X_{i1} + \varepsilon_i$) and saving the *unstandardized residuals (res_1)*. (Insert 11.1) To produce the plots in Figure 11.1 (text page 428), we use the SPSS absolute value function to generate a new variable, *absresid=abs(res_1)*, which is the *absolute value of the unstandardized residual (res_1)*. We can now generate Figure 11.1 using the *Scatter Matrix* procedure. Notice that the first column in SPSS Output 11.2 replicates the three scatter plots shown on text page 428.

INSERT 11.1

- ✔ Analyze
- ✔ Regression
- ✔ Linear
- ✔ *dbp*
- ✔ ▶ (to the Dependent box)
- ✔ *age*
- ✔ ▶ (to the Independent box)
- ✔ Save
- ✔ Unstandardized Residuals
- ✔ Continue
- ✔ OK

- ✔ Transform
- ✔ Compute
- * Type "absresid" in Target Variable box.
- * Scroll down the list of functions.
- ✔ ABS(numexpr)
- ✔ ▲
- ✔ res_1
- ✔ ▶
- ✔ OK

- ✔ Graphs
- ✔ Scatter
- ✔ Matrix
- ✔ Define
- ✔ *age*
- * Hold down the shift key.
- ✔ *absresid*
- ✔ ▶ (to the Matrix Variables)
- ✔ OK

The plot of *age* against the *unstandardized residual (res_1)* suggests that the error variance is not constant. From Figure 11.1c, text page 428, the researcher decided that a linear relation between the error standard deviation and *age* was reasonable. He then estimated the standard deviation function by regressing *absresid* against *age* and saved the *fitted (unstandardized predicted)* values. (Insert 11.2) Notice that SPSS has placed the predicted values (*pre_1*) in the active Data Editor. From equation 11.16a, text page 425, we now can compute the weights:

302

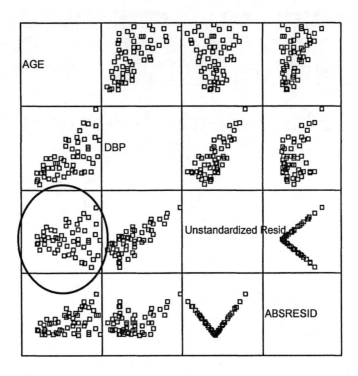

$$w_i = \frac{1}{\left(\hat{s}_i\right)^2} \ \text{ or } \ w_i = \frac{1}{\left(pre\_1\right)^2} \, .$$

After executing Insert 11.2, the active Data Editor should look like SPSS Output 11.3, which is identical to Table 11.1, text page 427.

INSERT 11.2

- ✔ Analyze
- ✔ Regression
- ✔ Linear
- ✔ *absresid*
- ✔ ► (to the Dependent box)
- ✔ *age*
- ✔ Save
- ✔ Unstandardized Predicted values
- ✔ Continue
- ✔ OK

- ✔ Transform
- ✔ Compute
- * Type "wt" in Target Variable box.
- * in the "Numeric Express" box type.
- 1 / (pre_1 ** 2)
- ✔ OK

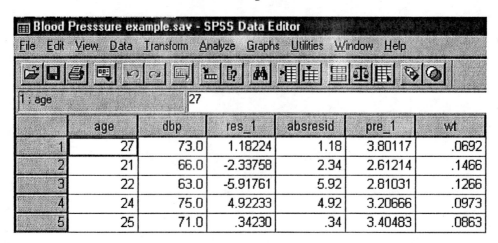

We can now perform a weighted least squares regression by regressing *dbp* on *age* using *wt* as a weighting variable. (Insert 11.3) Partial output is shown in SPSS Output 11.4. Notice that

$$\hat{Y} = 55.566 + .596\,X$$

and the 95% confidence interval is:

$$0.437 \le \beta_1 \le 0.755$$

consistent with text pages 428-429.

INSERT 11.3

- ✔ Analyze
- ✔ Regression
- ✔ Linear
- ✔ *dbp*
- ✔ ▶ (to the Dependent box)
- ✔ *age*
- ✔ ▶ (to the Independent box)
- ✔ WLS>>
- ✔ *wt*
- ✔ ▶ (to the WLS weight box)
- ✔ Statistics
- ✔ Confidence Intervals
- ✔ Continue
- ✔ OK

Coefficients[a,b]

Model		Unstandardized Coefficients		Standardized Coefficients			95% Confidence Interval for B	
		B	Std. Error	Beta	t	Sig.	Lower Bound	Upper Bound
1	(Constant)	55.566	2.521		22.042	.000	50.507	60.624
	AGE	.596	.079	.722	7.526	.000	.437	.755

a. Dependent Variable: DBP

b. Weighted Least Squares Regression - Weighted by WT

Multicollinearity Remedial Measures – Ridge Regression: One approach to estimating regression coefficients in the presence of multicollinearity problems among the predictor variables is to use a method known as *ridge regression*. The idea is to use estimators that have small bias but greater precision than their unbiased competitors. Let $\mathbf{r}_{XX}$ denote the correlation matrix of the X variables in the model of interest and let $\mathbf{r}_{XY}$ denote the vector of coefficients of simple correlation between Y and each X variable. The ridge standardized regression coefficients are:

$$\mathbf{b}^R = \left(\mathbf{r}_{XX} + c\mathbf{I}\right)^{-1}\mathbf{r}_{XY}$$

The constant c is the amount of bias in the estimators and $\mathbf{I}$ is the $(p-1)$ x $(p-1)$ *identity matrix*. When $c = 0$, $\mathbf{b}^R$ reduces to the standardized least squares estimators. When $c > 0$, the estimators are biased but more precise than ordinary least squares estimators. The strategy is to first obtain estimators based on $c = 0$. Then a small increment is added to c and another set of estimators is obtained. This process is continued for a large number of c values, usually between 0 and 1. Then, we examine the *ridge trace* and *variance inflation factors* $(VIF)_k$ and try to choose the smallest value of c where the coefficients $\mathbf{b}^R$ first become stable and the $(VIF)_k$ are small. This requires judgement but we take the value of $\mathbf{b}^R$ corresponding to the chosen c as the ridge estimate for the problem at hand. (See text page 434).

The Body Fat Example

Data File: ch07ta01.txt

On text page 434, a researcher studied the relationship between the amount of body fat (Y) and triceps skinfold thickness ($X1$), thigh circumference ($X2$), and midarm circumference ($X3$). See Chapter 0, "Working with SPSS," for instructions on opening the Body Fat example data file and renaming the variables. After opening the Body Fat example, the first 5 lines in the SPSS Data Editor should look like SPSS Output 11.5.

	x1	x2	x3	y	var
1	19.50	43.10	29.10	11.90	
2	24.70	49.80	28.20	22.80	
3	30.70	51.90	37.00	18.70	
4	29.80	54.30	31.10	20.10	
5	19.10	42.20	30.90	12.90	

In text Chapter 7, the researchers noted severe multicollinearity and decided that an appropriate alternative to ordinary least square regression was ridge regression. SPSS does not have a procedure available through a dialogue box (point-and-click) for ridge regression. A macro, however, is available and can be invoked using command syntax in the Syntax Editor. The syntax to invoke the macro takes the general form as given in SPSS Output 11.6.

SPSS Output 11.6

```
INCLUDE 'Ridge regression.sps'.
RIDGEREG DEP=varname /ENTER = varlist
[/START={0**}] [/STOP={1**}] [/INC={0.05**}]
 {value}       {value}       {value }
[ /K=value] .
```

If you have not moved SPSS program files since the installation of the software, the first command line [*INCLUDE 'Ridge regression.sps'.*] will automatically locate the macro. If the SPSS software is installed on a network drive or has been moved into a different folder, you will have to type the correct file path to connect to the *'Ridge regression.sps'* file. If we get a "file not found" error when invoking the macro, use Windows Explorer© (or equivalent) to find 'Ridge regression.sps'. Insert 11.4 shows an example of a modified '*Include*' statement that indicates the path necessary to connect to the ridge regression macro. (This path will not work on all computers.) We also replace "*varname*" in the *Dep = varname* statement with *y* and "*varlist*" in the *Enter = varlist* statement with *x1 x2 x3* as shown in Insert 11.4. This will generate *R-Square* vs. *k* (here, *k* rather than *c* is used to represent the biasing constant) with the default *START*, *STOP* and *INC* values and plots of parameter estimates vs. the ridge parameter (biasing constant). Partial results (truncated) are shown in SPSS Output 11.7 and are consistent with Tables 11.2 and 11.3, text page 434. The *K = value* statement can be used to specify a single constant value (biasing constant) with which to produce estimated regression coefficients. Insert 11.5, for instance, requests a *ridge regression* using a *biasing constant* = 0.02. Partial (truncated) results are shown in SPSS Output 11.8 and are consistent with the fitted regression model on text page 435. The *START*, *STOP*, and *INC* commands in SPSS Output 11.6 allow the user to specify the starting and ending constant value (biasing constant) and the value by which the constant will increment.

```
                              INSERT 11.4
  ✔ File
  ✔ New
  ✔ Syntax
  include 'c:\program files\spss\Ridge regression.sps'.
  ridgereg dep=y /ENTER = x1 x2 x3.
  ✔ Run
  ✔ OK
```

SPSS Output 11.7

```
R-SQUARE AND BETA COEFFICIENTS FOR ESTIMATED VALUES OF K

     K          RSQ          X1           X2           X3


   .00000      .80136      4.263705     -2.92870     -1.56142
   .05000      .78042       .460460      .439245      -.100508
   .10000      .77839       .423354      .448960      -.081246
   .15000      .77567       .404876      .443453      -.069952
   .20000      .77226       .391425      .434720      -.061290
   .25000      .76828       .380223      .425122      -.054086
   .30000      .76382       .370323      .415403      -.047887
```

```
                              INSERT 11.5
  ✔ File
  ✔ New
  ✔ Syntax
  include 'c:\program files\spss\ridge regression.sps'.
  ridgereg dep=y /enter = x1 x2 x3
  /K=.02 .
  ✔ Run
  ✔ OK
```

```
****** Ridge Regression with k = .02 ******


Mult R       .884191813
RSquare      .781795162
Adj RSqu     .740881755
SE           2.599235292

          ANOVA table
              df        SS        MS
Regress     3.000    387.293    129.098
Residual   16.000    108.096      6.756


        F value          Sig F
     19.10853230       .00001540


--------------Variables in the Equation---------------
                   B          SE(B)        Beta        B/SE(B)
X1            .555353057   .124647201    .546833894    4.455399350
X2            .368144438   .118414502    .377403656    3.108947226
X3           -.191626885   .164356363   -.136871542   -1.165923127
Constant    -7.403425423  6.407173706    .000000000   -1.155490043
```

Table 11.3, text page 434, displays *variance inflation factors* and R^2 for different biasing constants. SPSS Output 11.7 above also displays R^2 for each biasing constant; however, there is no readily available routine in SPSS that will produce *variance inflation factors* for a given biasing constant.

Remedial Measures for Influential Cases - Iteratively Reweighted Least Squares (IRLS) Robust Regression: "For robust regression, weighted least squares is used to reduce the influence of outlying cases by employing weights that vary inversely with the size of the residual. Outlying cases that have large residuals are thereby given smaller weights. The weights are revised as each iteration yields new residuals until the estimation process stabilizes." Two widely used weight functions, the *Huber* and *bisquare weight functions*, are described. These functions are used to dampen the effect of outlying observations.

Mathematics Proficiency Example

Data File: ch11ta04.txt

On text page 441, a researcher studied the relationship between the home environment and aggregate school performance (mathematics proficiency) for 37 states, the District of Columbia, Guam, and the Virgin Islands. The data are given by state and states are the study unit in this example. The average mathematics proficiency score, denoted *Y*, is the dependent variable of interest. The predictor variables are parents (*X1*), home library (*X2*), reading (*X3*), watch TV (*X4*), and absences (*X5*). This example considers only the predictor variable *X2* (home library) which represents the percentage of eighth-grade students who had three or more types of reading materials at home. See Chapter 0, "Working with SPSS," for instructions on opening the Mathematics Proficiency example data file and renaming the variables. After opening the Mathematics Proficiency data file, the first 5 lines in the SPSS Data Editor should look like SPSS Output 11.9.

SPSS Output 11.9

	state	y	x1	x2	x3	x4	x5
1	Alabama	252	75	78	34	18	18
2	Arizona	259	75	73	41	12	26
3	Arkansas	256	77	77	28	20	23
4	California	256	78	68	42	11	28
5	Colorado	267	78	85	38	9	25

To produce the scatter plots on text page 442 we first regress *Y* on *X2* and save the residuals. (Insert 11.6) The *unstandardized residuals* are saved in the active Data Editor as *res_1*. We then request a *scatter plot matrix* using *res_1*, *X2*, and *Y*. The top and bottom of the middle column in SPSS Output 11.10 correspond to text Figures 11.5 (a) and 11.5 (b) (text page 442), respectively. From the plots, we notice that there are 3 states with outlying *Y* values (mathematics proficiency scores). Also, there are 6 states with low *X2* scores (resources) whose math proficiency scores are above the fitted regression line. Thus, the researchers decided to fit a linear regression function robustly, using *IRLS regression* and the *Huber weight function*.

INSERT 11.6	INSERT 11.6 continued
✔ Analyze	✔ Graphs
✔ Regression	✔ Scatter
✔ Linear	✔ Matrix
✔ *y*	✔ Define
✔ ▶(to the Dependent box)	✔ *y*
✔ *x2*	* Hold down the **ctl** key.
✔ ▶(to the Independent box)	✔ *x2*
✔ Save	✔ *res_1*
✔ Unstandardized residuals	✔ ▶(to the Matrix Variables)
✔ Continue	✔ OK
✔ OK	

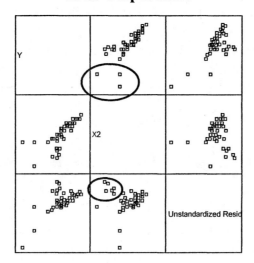

The researchers considered a second order model using $x2$ (*homelib*) centered, and its square as the single predictor variable (equation 11.48, text page 443). To diminish the effect of outliers the researchers used *iteratively reweighted least squares* and the *Huber weight function*. We detail each step in the following and in Insert 11.7, although the output for each step is not shown. We begin in Insert 11.7 by centering $x2$ (*x2cnt*) and finding the square of the centered variable (*x2cnt2*). We then:

1) Save the unstandardized residuals (*res_1*). This was done for the first order model in Insert 11.6 above . Since we are fitting a second order model, we delete *res_1* from the active Data Editor before running the syntax in Insert 11.7. Running Step 1 in Insert 11.7 will then generate *unstandardized residuals* based on the second order model and save them in *res_1*.

2) Find the median value ($median\{e_i\}$) of *res_1* = 0.70629.

3) Compute (name arbitrary) $v1$, the absolute value of res_1 minus the median of the residuals ($|e_i - median\{e_i\}|$).

4) Find the median of ($median\{|e_i - median\{e_i\}|\}$) $v1$ = 3.1488.

5) Find the median absolute deviation estimator (*MAD*) = *median of v1 − constant* = 0.6745 (3.1488 / 0.6745) = 4.6683.

6) Compute the scaled residual, $u1$ = *res_1* ÷ *mad* = *res_1* ÷ 4.6683.

7) Use the scaled residual in the computation of the Huber weight (w), where w = 1 if the absolute value of $u1 \leq 1.345$ and w = (1.345 / *absolute value of u1*) if the *absolute value of u1* > 1.345.

8) Regress y on *x2cnt* and *x2cnt2* using w as a weighting variable and save the unstandardized residuals. This will produce the regression model: $\hat{Y} = 259.39 + 1.670\,x_2 + 0.064\,x_2^2$. If the saved residuals from Step 1 above are still in the active Data Editor, the residual values for this step will be saved as *res_2*. The saved residual values for Step 8 will be those shown in column 4 of text Table 11.5.

9) Return to Step 2 for the second iteration and use the residuals saved in Step 8 for the next iteration.

INSERT 11.7	INSERT 11.7 continued

* Begin.
* Center x2 and square centered value.
✔ File
✔ New
✔ Syntax
* Type the following.
compute x2cnt = x2-80.4 .
compute x2cnt2=x2cnt**2.
execute .
✔ Run
✔ All
 	
* Step 1.
* Run second order model text equation 11.48.
* and save residuals.
✔ Analyze
✔ Regression
✔ Linear
✔ y
✔ ►(to the Dependent box)
✔ x2cnt
✔ ►(to the Independent box)
✔ x2cnt2
✔ ►(to the Independent box)
✔ Save
✔ Unstandardized residuals
✔ Continue
✔ OK
 	
* Step 2.
✔ Analyze
✔ Descriptive Statistics
✔ Explore
✔ res_1
✔ ►(to the Dependent box)
✔ OK
 	
* Step 3.
* In the syntax file opened in the Begin step.
* type the following.
compute v1=abs(res_1-.70629).
execute.
* Highlight the above 2 statements only.
✔ ▶ on the tool bar.
	* Step 4.
	✔ Analyze
	✔ Descriptive Statistics
	✔ Explore
	✔ v1
	✔ ►(to the Dependent box)
	✔ OK
	
	* Steps 5, 6, and 7.
	* In the syntax file opened in the Begin step.
	* type the following.
	compute mad=3.1488/.6745.
	compute u1 = res_1 / mad.
	if (abs(u1) <=1.345) w=1.
	if (abs(u1) > 1.345) w=1.345/abs(u1).
	execute.
	* Highlight the above 5 statements only.
	✔ ▶ on the tool bar.
	
	* Step 8.
	* Run regression using w as a weighting variable.
	* and save residuals as res_2, by default.
	✔ Analyze
	✔ Regression
	✔ Linear
	✔ y
	✔ ►(to the Dependent box)
	✔ x2cnt
	✔ ►(to the Independent box)
	✔ x2cnt2
	✔ ►(to the Independent box)
	✔ Save
	✔ Unstandardized residuals
	✔ Continue
	✔ WLS >>
	✔ w
	✔ ►(to WLS Weight)
	✔ OK

Nonparametric Regression: There is no readily available procedure in SPSS Base®, SPSS Advanced Models®, or SPSS Regression Models® for performing nonparametric regression. Decision trees are available in SPSS' AnswerTree and Clementine products. These products, however, are not widely available to students through college computing centers.

Remedial Measures for Evaluating precision in Nonstandard Situations – Bootstrapping:
The authors give an overview of the *bootstrap method* on text page 459. The basic idea is as follows. We estimate regression coefficients using some method such as an iterative procedure (that may not provide an estimate of precision). We temporarily view the sample data being analyzed (the original sample) as "the population." We retake a large number of samples (resamples) from the original sample (temporary population). For each resample, we compute regression coefficients using an iterative procedure (that may not provide an estimate of precision). From the estimates gleaned from the resampling, we (using the estimates as data) apply first principles to calculate their standard errors. We consider the calculated standard errors to be estimates of standard errors of the regression coefficients in the original estimation.

The Toluca Company Example

Data File: ch01ta01.dat

Using the Toluca Company example (Table 1.1, text page 19), we found that the parameter estimates for lot size (*lotsize*) using ordinary least squares (linear) regression. (SPSS Output 11.11) Now the authors wish to evaluate the precision of the estimates by the *bootstrap method*. See Chapter 0, "Working with SPSS," for instructions on opening the Toluca Company data file and renaming the variables *lotsize* and *workhrs*. After opening ch01ta01.dat, the first 5 lines in the SPSS Data Editor should look like SPSS Output 1.1. To evaluate the precision of the parameter estimates, we use *constrained nonlinear regression (CNLR)*. *CNLR* allows for linear and non-linear constraints on parameters and can compute bootstrap estimates of parameter standard errors and correlations. In Insert 11.9, we begin the procedure by assigning the initial values to the parameters in the *Model Program* statement. The *Model Program (Compute* statement) specifies the nonlinear or linear model for the data. We request that predicted values be modeled as: *b0 + b1*lotsize*. We then specify *workhrs* as the dependent variable in the *CNLR* statement and request bootstrap estimates based on 1000 samples. Optionally, you can request (*/OUTFILE =*) that the parameter estimates, *SSE*, and samples sizes for each sample be written to an output file.

SPSS Output 11.11

Coefficients^a

Model		Unstandardized Coefficients		Standardized Coefficients			95% Confidence Interval for B	
		B	Std. Error	Beta	t	Sig.	Lower Bound	Upper Bound
1	(Constant)	62.366	26.177		2.382	.026	8.214	116.518
	LOTSIZE	3.570	.347	.906	10.29	.000	2.852	4.288

a. Dependent Variable: WORKHRS

INSERT 11.9

✔ File
✔ New
✔ Syntax
MODEL PROGRAM b0=0 b1=0 .
COMPUTE pred = b0 + b1*lotsize.
CNLR hours
/BOOTSTRAP 1000
/OUTFILE='c:\bootstrap.sav' .
execute.
✔ Run
✔ All

Partial output of the *CNLR* procedure is shown in SPSS Output 11.12. The parameter estimates using linear regression (SPSS Output 11.11) are the same as those generated by specifying a linear equation in the nonlinear regression procedure (*CNLR* – SPSS Output 11.12).

The standard deviation of the 1000 b^*_1 estimates is 0.3733 (listed as "*Std. E*" for *B1* – SPSS Output 11.12) and is relatively close to the estimate ($s\{b_1\}$= 0.347) (listed as "*Std. Error for LOT_SIZE*" – SPSS Output 11.11 and "Asymptotic Std. Error" for B1 – SPSS Output 11.12) based on the original sample. Additionally, the 95% confidence limits (2.8364 < β_1 < 4.2818) (SPSS Output 11.12) are very close to the confidence limits obtained using the original sample (2.852 < β_1 < 4.288) (SPSS Output 11.11). To produce the histogram of the 1000 b_1 estimates, simply open the output file specified in Insert 10.9 and request a histogram of *b1*. This result is shown in Output 11.13. The histogram is relatively symmetrical and close to a normal distribution. Based on the above analysis, the parameter estimates for the original sample appear to be relatively precise.

313

SPSS Output 11.12

```
                                        Asymptotic 95 %
                         Asymptotic    Confidence Interval
Parameter    Estimate    Std. Error    Lower         Upper

B0      62.365857908 26.177434151  8.213709530 116.51800629
B1       3.570202034   .346972168  2.852435419   4.287968649
```

Asymptotic Correlation Matrix of the Parameter Estimates

```
                 B0          B1

B0            1.0000      -.9278
B1            -.9278      1.0000
```

Bootstrap statistics based on 1000 samples

```
                          95% Conf. Bounds   95% Trimmed Range
Parameter Estimate Std. E  Lower Upper       Lower     Upper

B0      62.3658 29.3921   4.6884 120.0433   3.5152 120.8500
B1       3.5702   .3733   2.8375   4.30289  2.8364    4.2818
```

Bootstrap Correlation Matrix of the Parameter Estimates

```
                 B0          B1

B0            1.0000      -.9498
B1            -.9498      1.0000
```

SPSS Output 11.13

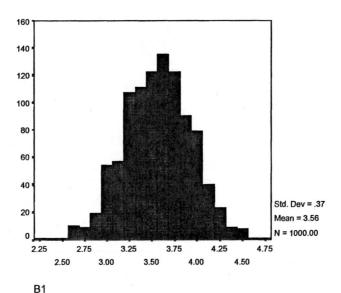

B1

314

Chapter 12 SPSS®

Autocorrelation in Time Series Data

In the context of the generalized multiple regression model, a time series is a set of data collected by observing a random variable and one or more independent variables at each time in a sequence of times. For example, company sales (the dependent variable) and industry sales (the independent variable) could be observed quarterly over a five-year interval. Or, net income (the dependent variable), gross revenues, and labor costs (the independent variables) for a company could be observed monthly over a three-year period. The error terms for regression models for times series data are often correlated and are, therefore, referred to as being *autocorrelated* or *serially correlated*. A generalized multiple regression model in which the error terms are autocorrelated is called an *autoregressive error model*. In this chapter, the authors of the text introduce the concept of *autocorrelation* and associated *autoregressive error models*. They discuss the *Durbin-Watson test for autocorrelation* as a means for detecting *autocorrelation* among error terms of a model. They then discuss remedial measures for dealing with data where *autocorrelation* has been detected.

The Blaisdell Company Example

Data File: ch12ta02.txt

The Blaisdell Company (text page 488) used trade association record of industry sales to predict their own sales. See Chapter 0, "Working with SPSS," for instructions on opening the Blaisdell Company data file and renaming the variables *company* and *industry*. After opening ch12ta02.txt, the first 5 lines in the SPSS Data Editor should look like SPSS Output 12.1.

SPSS Output 12.1

	company	industry	var	var
1	20.96	127.30		
2	21.40	130.00		
3	21.96	132.70		
4	21.52	129.40		
5	22.39	135.00		

In Insert 12.1, we begin the analysis by producing a scatter plot of the company and industry sales data. (SPSS Output 12.2) Since the scatter plot suggested that a linear regression model might be appropriate the analyst working with Blaisdell Company fitted an ordinary least squares regression model and saved the unstandardized residuals (*res_1*). To plot the residuals against time, we used the *compute time=$casenum* statement in Insert 12.1 to create a sequential record number for each line in the Blaisdell Company data file. A plot of the residual values against *time* is shown in SPSS Output 12.3.

INSERT 12.1	INSERT 12.1 Continued
* Produce Scatter Plot. ✔ Graphs ✔ Scatter ✔ Simple ✔ Define ✔ *industry* ✔ ►(to the X axis) ✔ *company* ✔ ►(to the Y axis) ✔ OK ✔ Analyze ✔ Regression ✔ Linear ✔ *company* ✔ ►(to the Dependent box) ✔ *industry* ✔ ►(to the Independent box) ✔ Save ✔ Unstandardized Residuals ✔ Continue ✔ OK	✔ File ✔ New ✔ Syntax * Type the following syntax. compute time=$casenum. execute. ✔ Run ✔ All * Produce Scatter Plot. ✔ Graphs ✔ Scatter ✔ Simple ✔ Define ✔ *time* ✔ ►(to the X axis) ✔ *res_1* ✔ ►(to the Y axis) ✔ OK

SPSS Output 12.2   **SPSS Output 12.3**

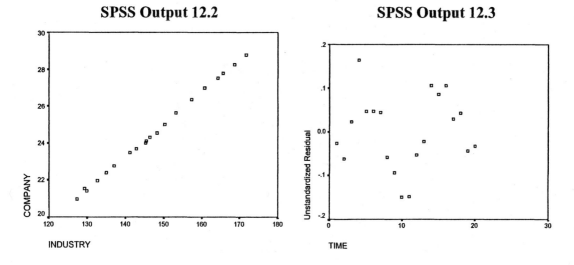

316

Durbin-Watson Test for Autocorrelation: Since SPSS Output 12.3 indicates a positive autocorrelation in the error terms, the analyst produced the *Durbin-Watson Test for Autocorrelation*. To request the Durbin-Watson test, use the first regression procedure in Insert 12.1 and ✔ Statistics ✔ Durbin-Watson ✔ Continue. Partial output is shown in SPSS Output 12.4. Note that the Durbin-Watson value = 0.735 as reported on text page 488. Using Table B.7, text page 675, we conclude that the error terms are positively autocorrelated.

SPSS Output 12.4

Model Summary^b

Model	R	R Square	Adjusted R Square	Std. Error of the Estimate	Durbin-W atson
1	.999^a	.999	.999	.08606	.735

a. Predictors: (Constant), INDUSTRY

b. Dependent Variable: COMPANY

Cochrane-Orcutt Procedure: On text page 492, the analyst first used the *Cochrane-Orcutt procedure* to estimate the autocorrelation parameter ρ. Column 1 (e_t) of Table 12.3, text page 493, contains the residual values for the regression model fitted in Insert 12.1. Each line in text Table 12.3, column 2 is the residual value of column 1 for the preceding time period. In Insert 12.2 we use SPSS' *Cross-Case operation* that takes the form, *compute new=lag(old)* where *new* is the value of *old* for the previous case. That is, we use [*compute res_m1=lag(res_1)*] to compute a new variable, *res_m1*, that is equal to the residual (*res_1*) of the preceding time period (line) in the data file. This produces text Table 12.3, column 2, e_{t-1}. We then *compute etm1et=res_1*res_m1* which is the residual times the lagged residual. This produces text Table 12.3, column 3, $e_{t-1}e_t$. We then square the lagged residual value [*compute etm1et2=res_m1**2*] to produce text Table 12.3, column 4, e_{t-1}^2. Finally, in SPSS Output 12.5, we find the sum of *etm1et* = 0.0834478 and *etm1et2* = 0.1322127. As before, double click on the table displayed in SPSS Output 12.5 and then double click on the sum for *etm1et2*, for example, to display greater precision (0.13 vs. 0.1322127279868). We now obtain *r*, an estimate of ρ, by equation 12.22, text page 492, where

$$r = 0.0834478/\ 0.1322127 = 0.631166$$

317

INSERT 12.2	INSERT 12.2 Continued
✔ File ✔ New ✔ Syntax * Type the following syntax. compute res_m1=lag(res_1). compute etm1et=res_1*res_m1. compute etm1et2=res_m1**2. execute. ✔ Run ✔ All * Find sum of *etm1et* and *etm1et2*. ✔ Analyze ✔ Descriptive Statistics ✔ Descriptives ✔ *etm1et*	✔ ▶ ✔ *etm1et2* ✔ ▶ ✔ Options ✔ Sum ✔ Continue ✔ OK * Produce columns 3 and 4 of text Table 12.4. ✔ File ✔ New ✔ Syntax compute ypt=company-.631166*lag(company). compute xpt=industry-.631166*lag(industry). execute. ✔ Run ✔ All

SPSS Output 12.5

Descriptive Statistics

	N	Minimum	Maximum	Sum	Mean	Std. Deviation
ETM1ET	19	.00	.02	.08	.0044	.00619
ETM1ET2	19	.00	.03	0.13221272	.0070	.00821
Valid N (listwise)	19					

In Insert 12.2, we use equation 12.18, text page 491, to find Y_t' (*ypt*) and X_t' (*xpt*). This produces the 3rd and 4th columns of Table 12.4, text page 493. To complete the *Cochrane-Orcutt* procedure, we then regress *ytp* on *xpt* and request the Durbin-Watson statistic (syntax not show). Partial results are shown in SPSS Output 12.6. The Durbin-Watson statistic = 1.65 consistent with text page 494. This leads to the conclusion that the autocorrelation of error terms is not significantly different from zero. From SPSS Output 12.6 we note that $b_0' = -0.394$, $b_1' = 0.174$, $s\{b_0'\} = 0.167$, and $s\{b_1'\} = 0.003$. We can use equation 12.20, text pages 491-492, to transform back to a fitted regression model in terms of the original variables and equation 12.21 to obtain estimated standard deviations of the back-transformed fitted model coefficients.

SPSS Output 12.6

Model Summary[b]

Model	R	R Square	Adjusted R Square	Std. Error of the Estimate	Durbin-Watson
1	.998[a]	.995	.995	.06715	1.650

a. Predictors: (Constant), XPT

b. Dependent Variable: YPT

Coefficients[a]

Model		Unstandardized Coefficients		Standardized Coefficients	t	Sig.
		B	Std. Error	Beta		
1	(Constant)	-.394	.167		-2.357	.031
	XPT	.174	.003	.998	58.767	.000

a. Dependent Variable: YPT

Hildreth-Lu Procedure: The *Hildreth-Lu procedure* uses the value of ρ that minimizes the error sum of squares (*SSE*) for the transformed regression model. As noted in the text, you can repeatedly fit regression models while varying the value of ρ to find the value that minimizes *SSE*. The most efficient method of producing Table 12.5, text page 495, is to use SPSS' *Matrix – End Matrix* procedure to automatically perform the numerical search across a large range of ρ values.

To use Insert 12.3, we must have the Blaisdell Company data file open as shown in SPSS Output 12.1. The other option is to open the Blaisdell Company example in SPSS and then save the file as *blaisdell.sav*, for example, in a user-specified directory. Assume that we saved the Blaisdell Company data file from the SPSS Data Editor to a directory on our C: drive named *alrm*. We could remove the asterisk from the first and second *get* statements (un-comment the line) so that Insert 12.3 will read the file into the Data Editor for use by the program, and after the program has run and modified the Data Editor, reread the Blaisdell Company data file into the Data Editor. In the present example, this is preferable to opening the data file in the Data Editor as shown in Insert 12.1. This is because the syntax in Insert 12.3 modifies the Data Editor such that if we execute the syntax twice without rereading the Blaisdell Company data file back into the Data Editor, we will produce incorrect results. Using the un-commented first and second *get* statements in Insert 12.3 assures that the data file used in the analysis will not have been modified from a previous analysis. Note that this requires saving the ch12ta02.txt file from the Data Editor as a *.sav format file in a directory indicated on the path of the *get* statements.

In Insert 12.3, we set the maximum number of loops to 100 (the default value in SPSS is 40). We then use the *cross-case operation* to find Y_{t-1} (*ylag*) and X_{t-1} (*xlag*). Since there are 20 lines of data in the file with values for *company* and *industry*, the first line will have a missing value for the lagged variables *ylag* and *xlag*. If we read *company*, *industry*, *ylag*, and *xlag* into matrices and subsequently attempt to use perform arithmetic [e.g., *compute xstep1=industry-(r100*xlag)*] we will get an error due to **industry** being a 20x1 matrix and **xlag** being a 19x1 matrix. Accordingly, we set *company*, *industry*, and *X0* to a system missing value if *ylag* and *xlag* are missing, as they will be on the first line of the modified data file. We then read the data variables into column vector matrices and instruct SPSS to omit cases with missing values. Thus, all initial column vector matrices will be 19x1.

To reproduce the *Hildreth-Lu* results shown in Table12.5, text page 495, we wish to calculate the *SSE* for a range of ρ values from 0.01 to 1.0 in increments of 0.01. This is possible using the SPSS *loop – end loop* procedure when we are not in *Matrix – End Matrix*. However, when in the matrix program, integers only can be used as an increment value. We loop from *r=1 to 100* (and by default increment by 1) and then compute *r100=r/100* to give us the range of ρ values from 0.01 to 1.0 for calculating 100 regression models using equation 12.17, text page 491. We find X_t' (**xstep1**) and concatenate *X0* (a column vectors of 1's) and **xstep1** to form matrix **xmat**. We find Y_t' (**ystep**) from equation 12.18 (a), text page 491, and use this as our "*Y*" matrix. Finally, we use equations 5.65 and 5.84a, text pages 201 and 204, to find the **b** matrix and **SSE**. The numerical output from this insert is not shown but is consistent with the selected values of ρ and corresponding *SSE* values shown on Table 12.5, text page 495. SPSS Output 12.7 displays the scatter plot of *r* and *SSE* values. Note that the magnitude of *SSE* decreases as the magnitude of *r* increases until *r* = 0.96, at which time the magnitude of *SSE* begins to increase. This is consistent with text Table 12.5.

SPSS Output 12.7

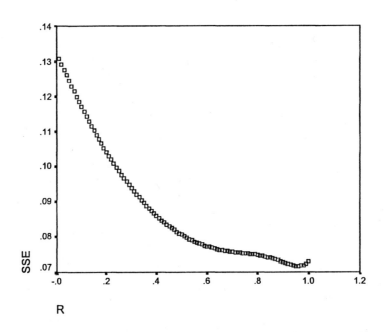

First Difference Procedure: On text page 496, the analyst used the *First Difference procedure* where $\rho = 1.0$ for the transformed model (equation 12.17, text page 491), where $Y_t^{'} = Y_t - Y_{t-1}$ and $X_t^{'} = X_t - X_{t-1}$, and the regression model is based on regression through the origin. In Insert 12.4, we compute $Y_t^{'}$ (*fdyp*) and $X_t^{'}$ (*fdxp*) using the *cross-case operation*. We then regress *fdyp* on *fdxp* and ✔ Option ✔ (Uncheck) Include constant in equation and ✔ Continue (syntax not shown). This results in SPSS Output 12.8 and a regression function where $\hat{Y}' = 0.168X'$. To produce the Durbin-Watson statistic, we run the same regression model as above except we include the intercept term in the model [✔ Option ✔ Include constant in equation, and ✔ Continue (syntax not shown)]. SPSS Output 12.9 shows the Durbin-Watson statistic, which is consistent with text page 497.

INSERT 12.4

✔ File
✔ New
✔ Syntax
* Type the following syntax.
compute fdyp=company-lag(company).
compute fdxp=industry-lag(industry).
execute.
✔ Run
✔ All

SPSS Output 12.8

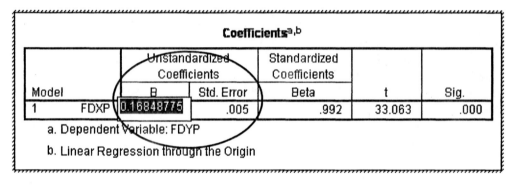

Coefficients[a,b]

Model		Unstandardized Coefficients		Standardized Coefficients	t	Sig.
		B	Std. Error	Beta		
1	FDXP	0.16848775	.005	.992	33.063	.000

a. Dependent Variable: FDYP
b. Linear Regression through the Origin

SPSS Output 12.9

Model Summary[b]

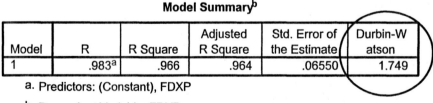

Model	R	R Square	Adjusted R Square	Std. Error of the Estimate	Durbin-Watson
1	.983[a]	.966	.964	.06550	1.749

a. Predictors: (Constant), FDXP
b. Dependent Variable: FDYP

We use equation 12.31, text pages 496-497, to transform the coefficients of the fitted model based on the transformed variables back to a fitted model in terms of the original variables. If in the first regression procedure in Insert 12.1 we had also ✔ Statistics ✔ Descriptives, and ✔

322

Continue we would have obtained SPSS Output 12.10. We use equation 12.31a and 12.31b, respectively, and SPSS Output 12.8 and 12.10 to find:

$$b_0 = \overline{Y} - b_1'\overline{X}$$

$$b_0 = 24.5690 - 0.168(147.6250) = -0.30349$$

and

$$b_1 = b_1'$$

$$b_1 = 0.16849$$

Thus, the fitted model in terms of the original variables is:

$$\hat{Y} = -.30349 + 0.16849X$$

SPSS Output 12.10

Descriptive Statistics

	Mean	Std. Deviation	N
COMPANY	24.5690	2.41040	20
INDUSTRY	147.6250	13.66520	20

Forecasting with Autocorrelated Error Terms

Forecasting Based on Cochrane-Orcutt Estimates: On text page 500, the analyst forecasts the Blaisdell Company sales for the 21st quarter. The trade association's estimate for this quarter is $175.3 million. We use the *Cochrane-Orcutt* results presented in Insert 12.2 and SPSS Output 12.6 where:

$$\hat{Y} = -0.3941 + 0.17376X'$$

We transform the model back to the original variables as shown on text page 497, which leads to:

$$\hat{Y} = -1.0685 + 0.17376X$$

Before using Insert 12.5, reread the Blaisdell Example data file to appear as SPSS Output 12.1. In Insert 12.5, we find X_t' (*xpt*) and Y_t' (*ypt*) for subsequent use and set our estimate of ρ (*r*) = 0.631166. We find (*e*) $e_{20} = Y_{20} - \hat{Y}_{20}$ where $Y_{20} = 28.78$ and $\hat{Y}_{20} = -1.0685 + 0.17376*171.7$. Using the above regression function, the fitted value (*yhat*) for $\hat{Y}_{21}$ when $X_{21} = 175.3$ is

$-1.0685 + 0.17376 * 175.3 = 29.392$. We then compute F_{21} (*forcast*) $= \hat{Y}_{21} + re_{20} = 29.40$. We find X'_{n+1} where $X_{n+1} = 175.3$ (X_{21} above), $r = 0.631166$, and $X_{20} = 171.7$. Thus:

$$X'_{n+1} = 175.3 - 0.631166 * 171.7 = 66.629$$

Using the above numerical results, we find $s\{pred\}$ using equation 2.38, text page 59. Note that in this equation, $MSE = 0.0045097$ (regress *ypt* on *xpt*), $(X_h - \overline{X})^2 = [(66.929 - 56.3186)**2]$, and $\sum(X_i - \overline{X})^2 = 515.8551$. Thus, text equation 2.38a:

$$s^2\{pred\} = MSE\left[1 + \frac{1}{n} + \frac{(X_h - \overline{X})^2}{\sum(X_i - \overline{X})^2}\right]$$

where $n = 19$, $X_h = X'_{n+1}$, $\overline{X}$ = mean of *XPT*, and $X_i = XPT_i$, becomes:

$$s\{pred\} = \sqrt{0.004509\left[1 + \frac{1}{19} + \frac{(66.929 - 56.3186)^2}{515.8551}\right]}$$

The value of *spred* = 0.07570 is consistent with text page 500. Finally, we find *tvalue* = *idf.t*(.975,17) = 2.11 and then compute the lower (*cilo*) and upper (*cihi*) prediction intervals as F_{21} (*forcast*) $= 29.40 \pm 2.11 * 0.0757$ or $29.24 \leq Y_{21(new)} \leq 29.56$. These values will be contained in your Data Editor.

INSERT 12.5

✔ File
✔ New
✔ Syntax
compute ypt=company-.631166*lag(company).
compute xpt=industry-.631166*lag(industry).
execute.
compute r = 0.631166.
compute e = company-(-1.0685+(.17376*industry)).
compute yhat = -1.0685+(.17376*175.3).
compute forecast = yhat+(r*e).

* Use text equation 2.38 to find s{pred} (*spred*).
 * Find mean of xpt=56.3186 using Analyze > Descriptives.
compute xptdif2=(xpt-56.3186)**2.
 * Find sum of xptdif2 = 515.8551 using Analyze > Descriptives and Sum Option.
compute spred=sqrt(0.0045097+ (0.0045097*((1/19)+(((66.929- 56.3186)**2)/515.8551)))).
compute tvalue=idf.t(.975,17).
compute cilo=29.4-(tvalue*spred).
compute cihi=29.4+(tvalue*spred).
exe.
✔ Run
✔ All

Forecasting Based on First Difference Estimates: Before using Insert 12.6, reread the Blaisdell Example data file to appear as SPSS Output 12.1. On text page 501 the authors use the first differences regression function to forecast sales for quarter 21. To find F_{21} we recall that the first part of the expression [(-0.30349 + 0.16849*(175.3)] comes from equation 12.33, text page 497, and from $X_{21} = 175.3$ above. The latter part of the expression {1.0[28.78 + 0.30349–0.16849*171.70)]} comes from text equation 12.33, the fact that $\rho = 1.0$ in the *First Difference* model, and $X_{20} = 171.70$. Our forecasted value for the 21st quarter is $F_{21} = 29.39$ as shown on text page 501. We find the prediction intervals using equation 4.20, text page 162. We first find X_h^2 (*xh2*), which is the forecasted value (F_{21}) squared. We then find X_i^2 (*fdxp2*) and, subsequently, use *Descriptives* to find the sum of *fdxp2* = 185.42. We find X'_{n+1} (*xnp1*) = $X_{21} -$ (r * X_{20}) (note that $X_{20} = industry_{20}$). The MSE = .00482 is from regressing *fdyp* on *fdxp* and requesting that no intercept term be included in the model. The statistic *tvalue* [idf.t(.975,18)] is based on 18 *df*. First, 19 of the original 20 records were used in the analysis because we "lost" a record when we lagged to get *xpt* and *ypt*. (This was also the case for the *Cochrane-Orcutt* procedure above where we used 17 *df* (*n*–2) to find *tvalue*.) We have 18 *df* (*n*–1) here because we estimated only 1 parameter (β_1), i.e., we did not include an intercept term in the model. We then find the 95% prediction intervals for $F_{21} = 29.39$ as shown on text page 501: $29.24 \leq Y_{21(new)} \leq 29.52$. These values will be contained in your Data Editor.

INSERT 12.6

```
✔ File
✔ New
✔ Syntax
* Based on first difference.
compute fdyp=company-lag(company).
compute fdxp=industry-lag(industry).
exe.
compute xh2=29.39**2.
compute fdxp2=fdxp**2.
compute xnp1=175.3-(1.0*171.7).
* compute xnp12=xnp1**2.
exe.
DESCRIPTIVES
 VARIABLES=fdxp2
 /STATISTICS=sum .
compute spred=sqrt(.00482*(1+(12.96/185.42))).
compute tvalue=idf.t(.975,18).
compute cilo=29.39-(tvalue*spred).
compute cihi=29.39+(tvalue*spred).
execute.
✔ Run
✔ All
```

Forecasting Based on Hildreth-Lu Estimates: On text page 501, the authors find the 95 percent prediction interval based on the *Hildreth-Lu* procedure. Before using Insert 12.7, reread the Blaisdell Example data file to appear as SPSS Output 12.1. Similar to the *Cochrane-Orcutt* procedure above, we first find Y_t' (*ytp*) and X_t' (*xpt*) using equation 12.18 (a) and (b), text page 491, and set $r = 0.96$ as shown in Table 12.5, text page 495. We transform the fitted regression model (equation 12.28, text page 496) back to a fitted model in terms of the original variables to obtain: $\hat{Y} = 1.7793 + .16045X$. We use this model to calculate *e* and *yhat*. Since X_{n+1}' is not given we compute $xnp1 = X_{21} - (r* X_{20}) = X_{n+1}'$ (note that $X_{20} = industry_{20}$). We use equation 2.38, text page 59, to find $s^2\{pred\}$ (*spred*) where *MSE* is the mean square error from regressing *ypt* on *xpt*. The lower (*cilo*) and upper (*cihi*) prediction intervals are printed on the 20th line of the Data Editor.

INSERT 12.7

```
✔ File
✔ New
✔ Syntax
compute r = 0.96.
compute ypt=company-r*lag(company).
compute xpt=industry-r*lag(industry).
execute.

compute e = company-(1.7793+(.16045*industry)).
compute yhat = 1.7793+(.16045*175.3).
compute forecast = yhat+(r*e).
compute xnp1=175.3-(.96*171.7).
* Use text equation 2.38 to find s{pred} (spred).
 * Find mean of xpt=8.1912 using Analyze > Descriptives.
compute xptdif2=(xpt-8.1912)**2.
 * Find sum of xptdif2 = 90.1088 using Analyze > Descriptives and Sum Option.
compute spred=sqrt(0.00422+ (0.00422*((1/19)+(((10.468- 8.1911)**2)/90.1088)))).
compute tvalue=idf.t(.975,17).
compute cilo=forecast-(tvalue*spred).
compute cihi=forecast+(tvalue*spred).
exe.
✔ Run
✔ All
```

Chapter 13 SPSS®

Non Linear Regression

The general linear regression model is:

$$Y_i = \beta_0 + \beta_1 X_{i1} + \beta_2 X_{i2} + \ldots + \beta_{p-1} X_{i,p-1} + \varepsilon_i$$

In this model, Y_i is a linear function of the $\beta's$ (the parameters). The general linear regression model can also be expressed as:

$$Y_i = f(\mathbf{X}_i, \boldsymbol{\beta}) + \varepsilon_i$$

where $\mathbf{X}_i$ is the vector of predictor variables for the ith subject, β is the vector of model parameters, and

$$f(\mathbf{X}_i, \boldsymbol{\beta}) = \beta_0 + \beta_1 X_{i1} + \beta_2 X_{i2} + \ldots + \beta_{p-1} X_{i,p-1}$$

The nonlinear regression model can be expressed as (text page 511):

$$Y_i = f(\mathbf{X}_i, \boldsymbol{\gamma}) + \varepsilon_i$$

In this representation of the nonlinear regression model, the authors use γ to denote the vector of parameters rather than β as a reminder that in this model $f(\mathbf{X}_i, \boldsymbol{\gamma})$ is a nonlinear function of the parameters.

Estimation of the parameters of a nonlinear regression model is usually based on the method of least squares or the method of maximum likelihood. However, it is usually not possible to find closed form "formulas" for the estimators of the parameters of nonlinear regression models. Instead, iterative search procedures are required. Thus, although the approach to finding estimators is the same as that used for linear regression models, the computations are much more tedious.

The problem reduces to one of finding values of the parameters that maximize (minimize) a nonlinear function of the parameters. Many search procedures have been developed for finding the maxima (minima) of nonlinear functions. SPSS offers two methods for estimating regression parameters and statistics for models that are not linear in their parameters. Constrained nonlinear regression (CNLR) may be used for constrained and unconstrained nonlinear regression. CNLR uses a sequential quadratic programming algorithm. The NLR (nonlinear regression) can only be used for unconstrained problems and uses the Levenberg-Marquardt algorithm. Both CNLR and NLR estimate model parameter values and can compute and save predicted and residual values and derivatives. The CNLR method is the more general and can generate bootstrap estimates of parameter standard errors and correlations.

Severely Injured Patients Example

Data File: ch13ta01.txt

A hospital administrator wanted to predict long-term recovery among a group of severely injured patients (text page 514). The predictor variable used in the regression model was number of days of hospitalized and the response variable was a long-term recovery index measure. See Chapter 0, "Working with SPSS," for instructions on opening the Severely Injured Patient data file and renaming the variables *index* and *days*. After opening ch13ta01.txt, the first 5 lines in the SPSS Data Editor should look like SPSS Output 13.1.

SPSS Output 13.1

severely injured patient example.sav - SPSS Data Editor				
File Edit View Data Transform Analyze Graphs Utilities Window Help				

15 : days · 65

	index	days	var	var	var
1	54.0	2.0			
2	50.0	5.0			
3	45.0	7.0			
4	37.0	10.0			
5	35.0	14.0			

The administrator began by producing a scatter plot (Insert 13.1 – SPSS Output 13.2) of the data, which suggested an exponential relationship between number of days hospitalized and long-term recovery. A review of related literature also suggested an exponential relationship. Thus, the administrator explored the appropriateness of a two-parameter nonlinear exponential regression model.

INSERT 13.1

✔ Graphs
✔ Scatter
✔ Simple
✔ Define
✔ *index*
✔ ►(to the Y axis box)
✔ *days*
✔ ►(to the X axis box)
✔ OK

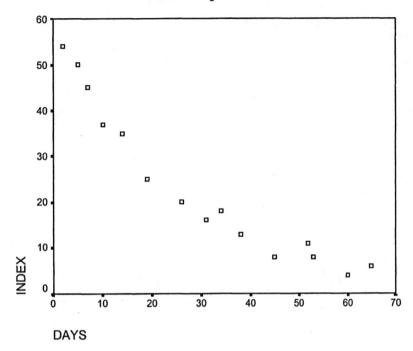

To find the initial starting values $g_0^{(0)}$ and $g_1^{(0)}$ (text page 521), we create a new variable, *logy*, the logarithmic transformation of *index*, and perform an ordinary least squares regression using *logy* as the response variable and *days* as the predictor variable. (Insert 13.2) From SPSS Output 13.3, we find the parameter estimates for this model to be $b_0 = 4.037$ and $b_1 = -0.038$. Using a hand-held calculator, we find $g_0^{(0)} =$ exponential of $b_0 = \exp(4.037) = 56.6646$ and $g_1^{(0)} = b_1 = -0.03797$ (text page 521). We now use the SPSS *Nonlinear Regression* procedure as shown in Insert 13.3 to produce the estimates of the parameters of the nonlinear model and related results. The final least square estimates, including the *MSE*, shown in SPSS Output 13.4 are consistent with Table 13.3, text page 524. The 95% confidence intervals for β_0 and β_1 are also shown in SPSS Output 13.4.

SPSS Output 13.3

Coefficients[a]

Model		Unstandardized Coefficients		Standardized Coefficients	t	Sig.
		B	Std. Error	Beta		
1	(Constant)	4.037	.084		48.002	.000
	DAYS	-.038	.002	-.977	-16.625	.000

a. Dependent Variable: LOGY

330

SPSS Output 13.4

```
All the derivatives will be calculated numerically.

Iteration  Residual SS         G0            G1

1       58.75436856  56.6640000   -.03700000
1.1     49.48365580  58.5123263   -.03944850
2       49.48365580  58.5123263   -.03944850
2.1     49.45932723  58.6037115   -.03958179
3       49.45932723  58.6037115   -.03958179
3.1     49.45929989  58.6064737   -.03958630
4       49.45929989  58.6064737   -.03958630
4.1     49.45929986  58.6065633   -.03958645


Run stopped after 8 model evaluations and 4 derivative evaluations.
Iterations have been stopped because the relative reduction between
successive residual sums of squares is at most SSCON = 1.000E-08


Nonlinear Regression Summary Statistics    Dependent Variable INDEX

Source              DF   Sum of Squares   Mean Square

Regression           2     12060.54070    6030.27035
Residual            13        49.45930       3.80456
Uncorrected Total   15     12110.00000

(Corrected Total)   14      3943.33333

R squared = 1 - Residual SS / Corrected SS =      .98746

                                       Asymptotic 95 %
                         Asymptotic   Confidence Interval
  Parameter   Estimate   Std. Error    Lower         Upper

  G0      58.606563279  1.472157894 55.426159508 61.786967050
  G1       -.039586448   .001711288  -.043283461   -.035889434

Asymptotic Correlation Matrix of the Parameter Estimates

             G0         G1

G0        1.0000     -.7071
G1         -.7071     1.0000
```

331

On text page 530, the authors generated bootstrap samples to check the appropriateness of large-sample inferences. In Insert 13.4, we generate bootstrap samples for the current data file. By default, the bootstrap option generates $10*p(p+1)/2$ samples where p is the number of parameters. Thus, there are 10 samples generated for each statistic (standard error or correlation) to be calculated. Suppose, however, that you wish to generate 1000 bootstrap samples, as did the authors. We use Insert 13.4 except that we ✔ Paste at the last step as opposed to ✔ OK. In the Syntax Editor, we now modify the pasted syntax. Edit the */BOOTSTRAP* statement to read */BOOTSTRAP=1000*. To save the bootstrap estimates for each sample, modify the */OUTFILE* statement to read */OUTFILE='C:\bootstrap.sav'* or indicate a path and file name appropriate for your situation. This results in Insert 13.5. Execution of this edited syntax resulted in SPSS Output 13.5.

INSERT 13.4

✔ Analyze
✔ Regression
✔ Nonlinear
✔ *index*
✔ ▶(to the Dependent box)
* In the Model Expression box type.
g0 * EXP(g1*days)
✔ Parameters
* Type "g0" in the Name box.
* Type "56.6646" in the Starting Value box.
✔ Add
* Type "g1" in the Name box.
* Type "-0.03797" in the Starting Value box.
✔ Add
✔ Continue
✔ Options
✔ Bootstrap estimates of standard error
✔ Continue
✔ OK

INSERT 13.5

```
* NonLinear Regression.
MODEL PROGRAM G0=56.664 G1=-.0379 .
COMPUTE PRED_ = g0*exp(g1*days).
CNLR index
 /OUTFILE='C:\bootstrap.sav'
 /PRED PRED_
 /BOOTSTRAP=1000
 /CRITERIA STEPLIMIT 2 ISTEP 1E+20 .
```

```
                                            Asymptotic 95 %
                             Asymptotic      Confidence Interval
   Parameter     Estimate    Std. Error    Lower         Upper

   G0          58.606566138  1.472160332  55.426157099  61.786975176
   G1           -.039586453   .001711294   -.043283479   -.035889427

   Asymptotic Correlation Matrix of the Parameter Estimates

                    G0          G1

   G0             1.0000      -.7071
   G1             -.7071      1.0000

   Bootstrap statistics based on 1000 samples

                                    95% Conf. Bounds      95% Trimmed Range
   Parameter   Estimate  Std. Error   Lower     Upper      Lower      Upper

   G0         58.6065661  1.5067126  55.6498816  61.5632507  55.0486054  60.7722780
   G1          -.0395865   .0017714   -.0430625   -.0361104   -.0429752   -.0361746
```

Notice that $g_0 = 58.6065$ from SPSS Output 13.4; from text page 524, Table 13.3(b); and from SPSS Output 13.5, the mean of the bootstrap sampling distributions. Also, note that $g_1 = -0.3959$ from SPSS Output 13.4; from text page 524, Table 13.3(b); and from SPSS Output 13.5. Additionally, the bootstrap estimates of $s^*\{g_0^*\} = 1.5067$ and $s^*\{g_1^*\} = .00177$ (SPSS Output 13.5) are very close to the large-sample standard deviations given in SPSS Output 13.4 and on text page 524 where $s\{g_0\} = 1.472$ and $s\{g_1\} = .00171$.

On text page 531, Figure 13.4, the authors present the bootstrap sampling distributions of g_0^* and g_1^*. To produce theses distributions, use the Data Editor to open the file specified on the */outfile* statement in Insert 13.5. The first line in this file will be the estimates for the original sample. The next 1000 lines will be the estimates for each of the 1000 bootstrap samples. To produce the sampling distributions of the bootstrap samples, delete the first line of the Data Editor. You can now ✔ Graph ✔ Histogram and request histograms for g_0^* and g_1^*, separately. Histograms for the current set of bootstrap estimates are shown in SPSS Output 13.6. Of course, the distribution of estimates will change for each bootstrap. The distribution of g_0^* estimates is somewhat skewed to the left. Overall, however, the distribution of both g_0^* and g_1^* are reasonably close to normal. Based on the closeness of the least square estimates to the bootstrap estimates and given the distribution of bootstrap estimates, the authors concluded that the use of large-sample inferences was appropriate.

SPSS Output 13.6

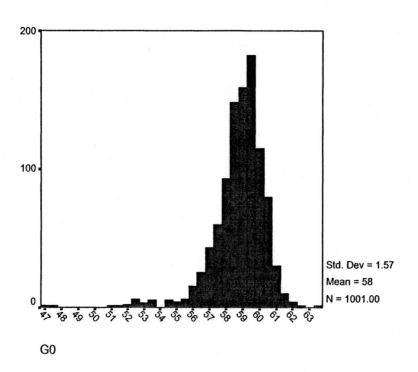

G0

SPSS Output 13.5 continued

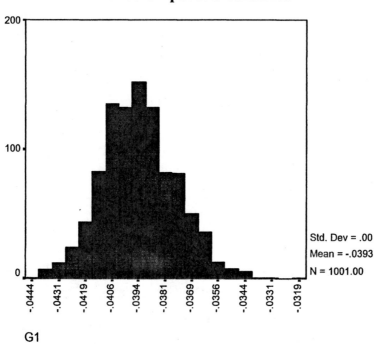

G1

On text page 531 the authors find the 95% confidence interval for γ_1 to be $-0.0433 \le \gamma_1 \le -0.0359$. This confidence interval is found in SPSS Output 13.4 above. The authors now find the 95% confidence interval based on the reflection method (equation 11.59, text page 460). Using our present findings, we have $g_1 = -0.03958$ from SPSS Output 13.4, and $g_1^*(.025) = -0.04306$ and $g_1^*(.975) = -0.03611$ from the bootstrap estimates shown in SPSS Output 13.5. Thus,

$$d_1 = g_1 - g_1^*(.025) = -0.03958 - (-0.04306) = 0.003480$$

$$d_2 = g_1^*(.975) - g_1 = -0.03611 - (-0.03958) = 0.003470$$

$$g_1 - d_2 = -0.03958 - 0.003470 = -0.043050$$

$$g_1 + d_1 = -0.03958 + 0.003480 = -0.036100$$

Previously, we found $-0.0433 \le \gamma_1 \le -0.0359$. Based on the reflection method we find $-0.043050 \le \gamma_1 \le -0.036100$. The confidence intervals are very close, which supports the use of large sample inferences.

Learning Curve Example

Data File: ch13ta04.txt

An electronic products manufacture produces a new product at two different locations: location A ($X1=1$) and location B ($X1=0$). (Text page 533.) The response variable was relative efficiency, a measure related to production costs at each location. See Chapter 0, "Working with SPSS," for instructions on opening the Learning Curve Example data file (Table 13.4, text page 534) and renaming the variables $X1$, $X2$, and Y. After opening ch13ta04.txt, the first 5 lines in the SPSS Data Editor should look like SPSS Output 13.6.

SPSS Output 13.6

It was well known that production efficiency of a new product initially increases with time and then levels off. We use the syntax as detailed in SPSS Output 13.7 to produce the scatter plot of relative efficiency (Y) and time (X2) with separate symbols for Location A and B (SPSS Output 13.8). The engineers decided to fit an exponential model that used location (X1) and the relationship between relative efficiency (Y) and time (X2). The selected nonlinear model (equation 13.38, text page 534) was:

$$Y_i = \gamma_0 + \gamma_1 X_{i1} + \gamma_3 \exp(\gamma_2 X_{i2}) + \varepsilon_i$$

On text page 535, the authors detailed the rationale for determining the starting values for parameters: $g_0^{(0)} = 1.025$, $g_1^{(0)} = -0.0459$, $g_2^{(0)} - 0.122$, and $g_3^{(0)} = -0.5$.

SPSS Output 13.7

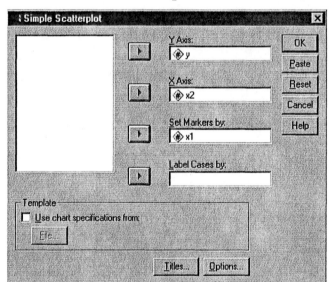

SPSS Output 13.8

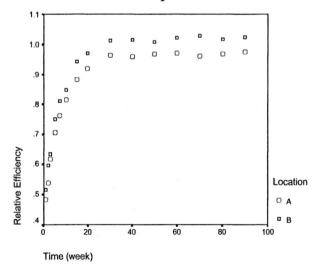

336

To find the fitted regression function for the above model we use ✔ Analyze ✔ Regression ✔ Nonlinear and setup the dialogue box as shown in SPSS Output 13.9. To produce the bootstrap results shown in Table 13.5, text page 535, we also ✔ Options ✔ Bootstrap estimates of standard error ✔ Continue. Finally, ✔ Paste instead of OK, modify the /bootstrap statement to read /bootstrap=1000, and ✔ Run ✔ All. Results of the procedure are shown in SPSS Output 13.10. The nonlinear least squares estimates of the g_k in the first set of estimates are almost identical to the mean $\bar{g}_k^*$ of the respective bootstrap sampling distribution shown in the second set of estimates. Both the nonlinear least squares and the bootstrap estimates of the g_k obtained via SPSS agree with the corresponding estimates reported in the text. Both sets of estimates of the standard deviations of g_0^* and g_1^* are consistent with each other and with the text. The bootstrap estimates of the standard deviations of g_2^* and g_3^*, however, are not in close agreement with the corresponding statistic based on the nonlinear least squares estimates.

SPSS Output 13.9

To produce the histograms on text page 536, we Paste the syntax shown in SPSS Output 13.9 above including a request for bootstrap estimates into the Syntax Editor. We modify the /bootstrap statement to indicate /bootstrap=1000 and modify the /outfile statement to indicate the appropriate path as we did in Insert 13.5 above. We then open the file specified in the /outfile statement and ✔ Graph ✔ Histrogram and request histograms for each of the four bootstrap estimates shown in Figure 13.6, text page 536. (SPSS output of these histograms not shown)

```
All the derivatives will be calculated numerically.

 Iteration Residual SS         G0            G1           G2            G3

    0.1     .0160356176   1.02500000   -.04590000   -.12200000   -.50000000
    1.1     .0121164281   1.02251428   -.04719754   -.12033423   -.50125049
    2.1     .0061021546   1.01758349   -.03690723   -.13606064   -.55161092
    3.1     .0051054017   1.01772462   -.04638415   -.14369939   -.55737130
    4.1     .0038575425   1.02258114   -.05012288   -.13366241   -.55782249
    5.1     .0032939313   1.01553570   -.04690319   -.13474979   -.55266624
    6.1     .0032926913   1.01558861   -.04726263   -.13482922   -.55249904
    7.1     .0032926829   1.01560015   -.04726789   -.13479238   -.55244318
    8.1     .0032926828   1.01559834   -.04726666   -.13479547   -.55244225

Run stopped after 8 major iterations.
Optimal solution found.
```

Nonlinear Regression Summary Statistics   Dependent Variable Y

Source	DF	Sum of Squares	Mean Square
Regression	4	22.85407	5.71352
Residual	26	3.292683E-03	1.266416E-04
Uncorrected Total	30	22.85737	
(Corrected Total)	29	.86673	

R squared = 1 - Residual SS / Corrected SS =   .99620

| | | | Asymptotic 95 % | |
| | | Asymptotic | Confidence Interval | |
Parameter	Estimate	Std. Error	Lower	Upper
G0	1.015598343	.003671721	1.008051011	1.023145675
G1	-.047266663	.004109204	-.055713252	-.038820074
G2	-.134795470	.004359523	-.143756597	-.125834342
G3	-.552442252	.008157281	-.569209784	-.535674719

Asymptotic Correlation Matrix of the Parameter Estimates

	G0	G1	G2	G3
G0	1.0000	-.5596	.4113	-.1442
G1	-.5596	1.0000	.0000	.0000
G2	.4113	.0000	1.0000	.5603
G3	-.1442	.0000	.5603	1.0000

Bootstrap statistics based on 1000 samples

Parameter	Estimate	Std. Error	95% Conf. Bounds Lower	Upper	95% Trimmed Range Lower	Upper
G0	1.0155983	.0030804	1.0095535	1.0216432	1.0091062	1.0212429
G1	-.0472667	.0040116	-.0551387	-.0393946	-.0553678	-.0395930
G2	-.1347955	.0059994	-.1465683	-.1230226	-.1466684	-.1228470
G3	-.5524423	.0112183	-.5744564	-.5304281	-.5692589	-.5252973

Neural Networks: SPSS Base, Advanced Models, or Regression Models does not offer Neural Network Modeling although SPSS Clementine does.

Chapter 14 SPSS®

Logistic Regression, Poisson Regression, And Generalized Linear Models

In this chapter, the authors consider nonlinear regression models for dependent variables that are categorical. In the simplest case, the dependent variable is binary and has two possible outcomes. Examples of binary outcome categories are success versus failure, disease absent versus disease present, and good performance versus bad performance. At the next level, the dependent variable is multinomial and has more than two possible outcomes. Examples of multinomial outcome categories are complete success, partial success, failure; no disease, minimum disease, moderate disease, severe disease; and performance on an increasing scale of 1 to 5. Lastly, the dependent variable may be a count of the number of events where large counts are rare and thought to be distributed as a Poisson random variable. Examples include counts of the number of accidents at a particular intersection, the number of people with a rare disease, and the number of sales.

One type of nonlinear regression model for categorical random variables is known as the logistic regression model. If the dependent variable is binary, we refer to the model as a logistic regression model whereas, if it is multinomial, we refer to the model as a polytomous or polynomial logistic regression model. If the dependent variable has a Poisson distribution, we refer to the nonlinear regression model as the Poisson regression model. This chapter of the text focuses on the statistical analysis of data based on the logistic model but also gives the essentials for the polytomous logistic regression model and the Poisson regression model.

Linear and nonlinear models considered in the text, including the three models mentioned above, belong to a family of models called *generalized linear models*. The salient feature of generalized linear models is that the initial model formulation is the same for each application with the exception of the specification of a link function. Thus, a unified approach to fitting regression models is provided. A concise description of generalized linear models is given on text pages 623-624.

Simple Logistic Regression

The Programming Task Example

Data File: ch14ta01.txt

An analyst studied the relationship between computer programming experience and success in completing a complex programming task within a specified time period. That is, the response variable was coded $Y=1$ if the task was successfully completed and $Y=0$ if the task was not successfully completed. The predictor variable was the number of months of previous programming experience. See Chapter 0, "Working with SPSS," for instructions on opening the Programming Task example data file (Table 14.1, text page 566) and renaming the variables X, Y, and *FITTED*. After opening ch14ta01.txt, the first 5 lines in the SPSS Data Editor should look like SPSS Output 14.1.

SPSS Output 14.1

	x	y	fitted	var
1	14.0	0	.31	
2	29.0	0	.84	
3	6.0	0	.11	
4	25.0	1	.73	
5	18.0	1	.46	

We begin the preliminary analysis by plotting *fitted* against the X values [Insert 14.1 - SPSS Output 14.2 (a)]. In the second graph, we plot Y against the X values and added a Lowess curve [SPSS Output 14.2 (b)]. The chart exhibits a sigmoidal S-shaped function. Accordingly, the analyst decided to use Logistic Regression.

Continuing in Insert 14.1, we perform a logistic regression using Y as the response variable and X as the single predictor variable. We also request that predicted values based on the regression function be saved to the Data Editor, by default as *pre_1*. Actually, the authors included the fitted values in the Programming Task data file. We duplicate the values here to demonstrate that the newly saved *pre_1* variable is equal to the column labeled *fitted* in the Data Editor. Finally, we request that the confidence intervals for the odds ratio be generated. Partial regression results are also shown in SPSS Output 14.3. The parameter estimates are consistent with the estimates reported on text page 565. Additionally, the odds ratio (*Exp(B)*=1.175) and 95% confidence interval are displayed and consistent with text page 567.

INSERT 14.1	INSERT 14.1 continued
✔ Graph	✔ Chart
✔ Scatter	✔ Options
✔ Simple	✔ Total (under Fit Line)
✔ Define	✔ Fit Options
✔ fitted	✔ Lowess
▶ (to the Y axis box)	✔ Continue
✔ X	✔ OK
▶ (to the X axis box)	* Close the Chart Editor.
✔ OK	✔ Analyze
	✔ Regression
✔ Graph	✔ Binary Logistic
✔ Scatter	✔ y
✔ Simple	✔ ▶ (to the Dependent box)
✔ Define	✔ x
✔ Y	✔ ▶ (to the Covariate box)
▶ (to the Y axis box)	✔ Save
✔ X	✔ Probabilites (under Predicted Values)
▶ (to the X axis box)	✔ Continue
✔ OK	✔ Options
* Double click the resulting graph to open the Chart Editor.	✔ CI for exp(B)
	✔ Continue
	✔ OK

SPSS Output 14.2

(a)

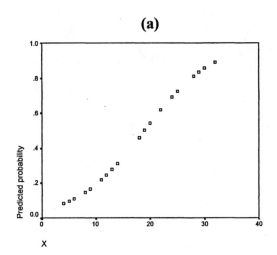

(b)

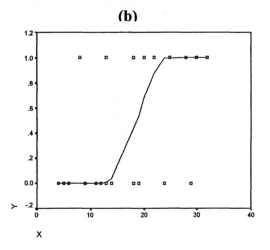

SPSS Output 14.3

Variables in the Equation

		B	S.E.	Wald	df	Sig.	Exp(B)	95.0% C.I.for EXP(B)	
								Lower	Upper
Step 1ᵃ	X	.161	.065	6.176	1	.013	1.175	1.035	1.335
	Constant	-3.060	1.259	5.903	1	.015	.047		

a. Variable(s) entered on step 1: X.

The Coupon Effectiveness Example

Data File: ch14ta02.txt

Marketing researchers mailed packets containing promotional material and related coupons to 1000 selected homes. Two hundred homes each received one of 5 different coupons that offered price reductions of 5, 10, 15, 20, or 25 dollars. Thus, the predictor variable was the coupon available to the home and the response variable (Y) was whether or not the home redeemed the coupon within a six-month period. See Chapter 0, "Working with SPSS," for instructions on opening the Coupon Effectiveness example data file (Table 14.2, text page 569) and renaming the variables X, N, Y, and *pro*. After opening ch14ta02.txt, the 5 lines in the SPSS Data Editor should look like SPSS Output 14.4.

SPSS Output 14.4

	x	n	y	pro
1	5.0	200.00	30.00	.15
2	10.0	200.00	55.00	.28
3	15.0	200.00	70.00	.35
4	20.0	200.00	100.00	.50
5	30.0	200.00	137.00	.69

The data file as shown in Table 14.2, text page 569, is a convenient way to summarize the study outcome. However, SPSS cannot be used to analyze the data in its present form. In Insert 14.2, we present two methods for restructuring the data file so that it can be used by SPSS. In Method 1 the program reads the first line of SPSS Output 14.4 and then writes 2 records (*loop #i=1 to 2*). When *#i*=1 we set *newn=y* and *response*=1. Thus, for the first line of SPSS Output 14.4 we set *newn* = 30 and *y*=1. When *#i*=2 we set *newn* = 200–*y* = 170 and *response* = 0. After each increment of the *loop* statement, we write a record to an output file. We *keep* the value of *x*, *newn*, and *response*, and we rename *newn* to *n* and *response* to *y*. After execution of Insert 14.2, the restructured data file will look like SPSS Output 14.5. The first line indicates that there were 30 (*N*) homes that redeemed (*Y*=1) a $5 coupon (*X*). The second line indicates that there were 170 (*N*) homes that did not redeem (*Y*=0) a $5 coupon (*X*). The progression continues such that we have cell counts for each combination of *X* and *Y* values.

In Insert 14.2 (Method 2), we read the input file as shown in SPSS Output 14.4 and write *X* and *Y* values for each of the 1000 observations in the study. Both Method 1 and Method 2 require that the Data Editor be as shown in SPSS Output 14.4 before execution of the syntax. Method 2 produces a file structure with which we are most familiar, i.e., a line for each observation. We will use Method 1, however, to introduce the SPSS *weight* statement.

To use the data as shown in SPSS Output 14.5, we first declare *N* to be a weight variable prior to running any statistical procedure. This declaration will be in effect until we open a new data file or turn off the *weight* function. Note that if we save the data file with the *weight* function on, the weight function will remain on when the data file is next used. If *weight* is on, "Weight Cases" will appear in the lower right-hand corner of the Data Editor. To turn *weight* off, execute *weight off* in the Syntax Editor or ✔ Data ✔ Weight Cases ✔ Reset ✔ OK.

In Insert 14.3, we declare *N* to be a *weight* variable and then request logistic regression using *Y* as the dependent variable and *X* as the single predictor variable. By declaring *N* to be a *weight* variable, SPSS interprets the first line of SPSS Output 14.5 to represent 30 (*N*) cases where *X*=5 and *Y*=1, for example. Insert 14.3 produces the odds ratio = 1.102 as shown on text page 570 (SPSS Output not shown).

INSERT 14.2

```
* METHOD 1 -Table 14.2 as input, write summary data file.
weight off.
LOOP #i = 1 TO 2.
if (#i=1) newn=y.
if (#i=1) response=1.
if (#i=2) newn=200-y.
if (#i=2) response=0.
XSAVE OUTFILE = 'temp.sav' / keep x  newn response / rename newn=n response=y.
END LOOP.
EXE.
GET FILE = 'temp.sav'.
exe.

* METHOD 2 - Table 14.2 as input, write 1000 records.
weight off.
LOOP #i = 1 TO n.
if (#i le y) response=1.
if (#i gt y) response=0.
XSAVE OUTFILE = 'temp.sav' / keep x response / rename response=y.
END LOOP.
EXE.
GET FILE = 'temp.sav'.
weight off.
exe.
```

SPSS Output 14.5

temp.sav - SPSS Data Editor

File Edit View Data Transform Analyze Graphs Utilities W

1 : x   5

	x	n	y	var
1	5.0	30.00	1.00	
2	5.0	170.00	.00	
3	10.0	55.00	1.00	
4	10.0	145.00	.00	
5	15.0	70.00	1.00	
6	15.0	130.00	.00	
7	20.0	100.00	1.00	
8	20.0	100.00	.00	
9	30.0	137.00	1.00	
10	30.0	63.00	.00	

INSERT 14.3	INSERT 14.3 continued
✔ Data	✔ Analyze
✔ Weight Cases	✔ Regression
✔ Weight cases by:	✔ Binary Logistic
✔ N	✔ y
✔ ▶	✔ ▶ (to the Dependent box)
✔ OK	✔ x
	✔ ▶ (to the Covariate box)
	✔ OK

Multiple Logistic Regression

The Disease Outbreak Example – Model Building

Data File: ch14ta03.txt and appenc10.txt

Note: Data for the Disease Outbreak example are available in two files: ch14ta03.txt and appenc10.txt. There are differences in the files that we will identify. For all analyses involving the Disease Outbreak example, we must pay close attention to the data file we use, i.e., ch14ta03.txt versus appenc10.txt. We will be unable to match the chapter results if we use the wrong file.

Using ch14ta03.txt, a researcher studied the relationship between disease outbreak ($Y=1$ if disease was present and $Y=0$ if disease was not present) and the individual's age ($X1$), socioeconomic status (represented by indicator variables $X2$ and $X3$), and the sector in which the individual lived ($X4$ coded 0 or 1). See Chapter 0, "Working with SPSS," for instructions on opening the Disease Outbreak example data file (Table 14.3, text page 574) and renaming the variables *case, X1, X2, X3, X4, and Y*. After opening ch14ta03.txt, the first 5 lines in the SPSS Data Editor should look like SPSS Output 14.6a.

SPSS Output 14.6 (a)

disease outbreak model building.sav - SPSS Data Editor

File Edit View Data Transform Analyze Graphs Utilities Window Help

1 : case   1

	case	x1	x2	x3	x4	y
1	1.0	33.0	0	0	0	0
2	2.0	35.0	0	0	0	0
3	3.0	6.0	0	0	0	0
4	4.0	60.0	0	0	0	0
5	5.0	18.0	0	1	0	1

Similar to the steps for requesting the logistic regression analysis in the previous example (Insert 14.3), for the Disease Outbreak example we enter Y as the dependent variable and $X1$, $X2$, $X3$, and $X4$ as covariates. Partial results are shown in SPSS Output 14.7(a). The regression function is consistent with equation 14.46, text page 574. The odds ratio [listed under $Exp(B)$] for $X1$=1.03, indicating that, since age is coded in units of years, the risk of disease increases by 1.03 or approximately 3% for each additional year of age. The odds ratio for $X4$=4.83, indicating that individuals living in sector 2 are 4.83 times more likely to have contracted the disease compared to individuals living in sector 1.

As noted above, data file ch14ta03.txt comprises the first 98 cases of a sample of 196 cases used in the Disease Outbreak example. The entire 196 cases are available in the data file appenc10.txt. After opening appenc10.txt and renaming the variables *case*, *age*, *status*, *sector*, *y*, and *other*, the Data Editor should look like SPSS Output 14.6(b). In this file, socioeconomic status is represented by one variable (*status*) where *status* = 1 if status is "upper," *status* = 2 if status is "middle," and *status* = 3 if status is "lower." The sector is coded *sector* = 1 if the individual lives in sector 1 and *sector* = 2 if the individual lives in sector 2.

SPSS Output 14.6 (b)

	case	age	status	sector	y	other
1	1.0	33.0	1	1	0	1
2	2.0	35.0	1	1	0	1
3	3.0	6.0	1	1	0	0
4	4.0	60.0	1	1	0	1
5	5.0	18.0	3	1	1	0

appenc10.sav - SPSS Data Editor
File Edit View Data Transform Analyze Graphs Utilities Window Help
1 : case   1

Note that in ch14ta03.txt, two indicator variables ($X2$, $X3$) were used to represent the 3 levels of socioeconomic status. However, SPSS's *Binary Logistic* procedure allows the user to use one variable to represent socioeconomic status (*status*) as done in appenc10.txt. Using this coding scheme we can enter *age*, *status*, and *sector*, as predictor variables and specify *status* as a categorical variable in the *Binary Logistic* dialogue box. For each variable specified as a categorical variable, SPSS will automatically code a set of indicator variables as defined by the user in *Binary Logistic's* Define Categorical Variables dialogue box. We also define the first category (upper) as the reference category. Thus, we replaced $X2$ and $X3$ in the analysis that produced SPSS Output 14.7(a) with *status* and specified the first category (upper) as the reference category. We also replaced $X1$ with *age*, although these two variables are equal, and replaced $X4$ with *sector*. This results in SPSS Output 14.7(b). So, SPSS Output 14.7(a) is a result of using $X1, X2, X3$, and $X4$ as predictors from the ch14ta03.txt data set. SPSS Output 14.7(b) is a result of using *age*, *status*, and *sector* as predictors from the appenc10.txt data set (and using ✔

Data ✓ Select Cases where *case* < 99). Although the variables used in the two models were coded differently, they contain the same "information." Thus we note the correspondence between the parameter estimates and odds ratios in SPSS Output 14.7 (a) and (b). The only discrepancy in table values is the intercept (*Constant*). This is due to city sector (*X4*) being coded 0 versus 1 in ch14ta03.txt and 1 versus 2 (*sector*) in appenc10.txt. The coding scheme used to produce SPSS Output 14.7(b) also yields an overall test statistic that is indicative of lack of significance for the *status* variable, $p = 0.547$.

SPSS Output 14.7

(a)

Variables in the Equation

		B	S.E.	Wald	df	Sig.	Exp(B)
Step 1a	X1	.030	.014	4.854	1	.028	1.030
	X2	.409	.599	.466	1	.495	1.505
	X3	-.305	.604	.255	1	.613	.737
	X4	1.575	.502	9.855	1	.002	4.830
	Constant	-2.313	.643	12.956	1	.000	.099

a. Variable(s) entered on step 1: X1, X2, X3, X4.

(b)

Variables in the Equation

		B	S.E.	Wald	df	Sig.	Exp(B)
Step 1a	AGE	.030	.014	4.854	1	.028	1.030
	STATUS			1.206	2	.547	
	STATUS(1)	.305	.604	.255	1	.613	1.357
	STATUS(2)	.714	.654	1.193	1	.275	2.042
	SECTOR	1.575	.502	9.855	1	.002	4.830
	Constant	-4.193	.911	21.194	1	.000	.015

a. Variable(s) entered on step 1: AGE, STATUS, SECTOR.

Polynomial Logistic Regression

The Initial Public Offerings Example

Data File: appenc11.txt

On text page 575, a researcher studies the relationship between whether or not a company was financed by venture capital (*VC*) and the face value (*facval*) of the company. See Chapter 0, "Working with SPSS," for instructions on opening the IPO example data file and renaming the variables *case, VC, FACEVAL, SHARES,* and *X3.* After opening appenc11.txt, the first 5 lines in the SPSS Data Editor should look like SPSS Output 14.8.

SPSS Output 14.8

	case	vc	faceval	shares	x3
1	1.0	0	1200000	3000000	0
2	2.0	0	1454000	1454000	1
3	3.0	0	1500000	300000.0	0
4	4.0	0	1530000	510000.0	0
5	5.0	0	2000000	800000.0	0

In Insert 14.4, we find *LNFACE,* the natural logarithm of *FACEVAL.* We then plot (syntax not shown) *LNFACE* and *VC* and request that a Lowess curve be drawn. As noted in SPSS Output 14.9 (a), there appears to be a mound shaped relationship between the two variables. To produce the first order regression line displayed in Figure 14.9 (a), text page 575, we regress *VC* on *LNFACE* and save the predicted probabilities. We plot the saved predicted probabilities (*pre_1*) against *LNFACE.* It is obvious that the first order regression line displayed in Output 14.9 (b) is not consistent with the mound shaped relationship shown in Output 14.9 (a). Thus, the authors decided to explore a second-order model using *LNFACE* (centered) and *LNFACE* (centered) squared as predictors. Continuing in Insert 14.4, we center LNFACE (*xcnt*), find its square (*xcnt2*), regress *VC* on *xcnt* and *xcnt2*, and save the predicted values. Partial results of this model are shown in 14.10 and are consistent with Table 14.5, text page 576. Additionally, we plotted (syntax not shown) the predicted values (from regressing *VC* on *xcnt* and *xcnt2*) against *xcnt* to obtain the second order logistic curve [Output 14.9 (c)]. Notice that the relationship in Output 14.9 (c) is more consistent with the relationship shown in Output 14.9 (a).

SPSS Output 14.9

(a)   **(b)**   **(c)**

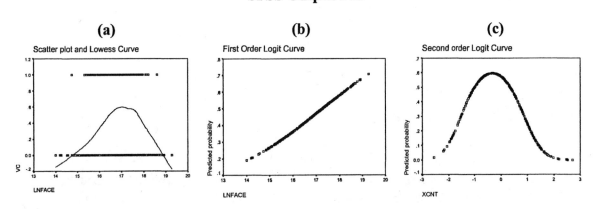

SPSS Output 14.10

Variables in the Equation

		B	S.E.	Wald	df	Sig.	Exp(B)
Step 1[a]	XCNT	-.552	.138	15.872	1	.000	.576
	XCNT2	-.862	.140	37.651	1	.000	.423
	Constant	.300	.124	5.874	1	.015	1.350

a. Variable(s) entered on step 1: XCNT, XCNT2.

Inferences about Regression Parameters

The Programming Task Example

On text page 578 the authors wish to test $H_0 : \beta_1 \leq 0$ versus $H_a : \beta_1 > 0$. The test for β_1 on SPSS Output 14.3 is a two-sided test. For a one-sided test, we simply divide the probability level (0.013 listed under *sig* and across from X) by 2. This yields the one-sided probability level of

0.0065 as noted on text page 579. Also, note that the authors report $z^* = 2.485$ and the square of z is equal to the Wald statistic which is distributed approximately as chi-square distribution with 1 degree of freedom. Thus, the Wald statistic is $X^2 = 2.485^2 = 6.176$ as shown in SPSS Output 14.3.

SPSS does not report the confidence interval for β_1 shown on text page 579. From SPSS Output 14.3, however, we find that $b_1 = 0.161$ and $s\{b_1\} = 0.065$. For a 95% confidence interval, we require $z = 1.96$. The 95% confidence interval is:

$$0.161 \pm 1.96(0.065)$$

or

$$0.034 \le \beta_1 \le 0.288.$$

Finally, a logistic model can be specified in SPSS' *Nonlinear Regression*. Bootstrap estimates and confidence intervals for the logistic model can be generated.

The Disease Outbreak Example

Data File: ch14ta03.txt

On text page 581, the researchers now wish to determine if $X1$ can be dropped from the model in the Disease Outbreak Example. In Insert 14.5, we use the *block* function that is available in the *Binary Logistic* program. When we use the *block* option in *Binary Logistic*, the results are presented for each block starting with Block 0, which represents an intercept only model. The Omnibus Tests of Model Coefficients table, as shown in SPSS Output 14.11, displays the chi-square test, *df*, and probability level of the test. This table is organized by *step*, *block*, and *model*. If we use the stepwise method, the *step* line will be a test of improvement in the model due to adding the current variable to the model. If we use the *block* option and do not use a stepwise method (*/method=entry*) then the *step* line and *block* line are identical because there is only 1 step per block. The test for *block* in the Omnibus Tests of Model Coefficients table is a test of model improvement for the current block of predictors beyond the predictors previously entered in the model. The test for *block* is analogous to the R^2 Change test that we previously used in the ordinary least squares regression procedure. For any Omnibus Tests of Model Coefficients table, the *model* line is a test of whether the overall model (at the end of the current block) is significantly better then the intercept only model. Thus, the model test is analogous to the test of whether $R^2=0$ at the end of each block in ordinary least squares regression.

We enter $X2$, $X3$, and $X4$ at *block* 1 and then enter $X1$ at *block* 2. Partial output is shown in SPSS Output 14.11. The test statistic and *sig* (0.023) for *block* 2 are consistent with text page 581. Similarly, we can enter $X1$ and $X4$ at block 1 and $X2$ and $X3$ at block 2 to determine if $X2$ and $X3$ (socioeconomic status) can be dropped from the model (syntax not shown). Partial results are shown in SPSS Output 14.12 and are consistent with text page 581. This is the effect of socioeconomic status and the p-value associated with the test is 0.547. If we refer back to SPSS Output 14.7(b) we see that the p-value of the partial test for *status* was also 0.547. Finally, we can compute new variables to represent the five possible two-way interactions, e.g., compute *intx1x2 = X1 * X2*. Because *X2* and *X3* are both indicator variables for socioeconomic status, we do not create a term for their interaction. Using the *block* function, we enter all main effects at *block* 1 and all interaction terms at *block* 2 (syntax not shown). This yields the results shown on text page 582.

INSERT 14.5

- ✔ Analyze
- ✔ Regression
- ✔ Binary Logistic
- ✔ Y
- ✔ ▶(to the Dependent box)
- ✔ X1
- ✔ ▶(to the Covariate box)
- ✔ X2
- ✔ ▶(to the Covariate box)
- ✔ X3
- ✔ ▶(to the Covariate box)
- ✔ Next
- ✔ X1
- ✔ ▶(to the Covariate box)
- ✔ OK

SPSS Output 14.11

Block 2: Method = Enter

Omnibus Tests of Model Coefficients

		Chi-square	df	Sig.
Step 1	Step	5.150	1	.023
	Block	5.150	1	.023
	Model	21.263	4	.000

SPSS Output 14.12

Block 2: Method = Enter

Omnibus Tests of Model Coefficients

		Chi-square	df	Sig.
Step 1	Step	1.205	2	.547
	Block	1.205	2	.547
	Model	21.263	4	.000

Automatic Model Selection Methods

The Disease Outbreak Example

Data File: appenc10.txt

Best Subsets Procedure. SPSS' *Binary Logistic* does not offer a best subset procedure. Additionally, the AIC_p and SBC_p statistics are not available. The $-2\log_e L(\mathbf{b})$ statistic, Cox and Snell R square, and Nagelkerke r square are standard output for the *Binary Logistic* procedure. Both R squares are pseudo R square measures and tend to be much lower than non-pseudo R squares seen in other regression models.

Stepwise Model Selection. SPSS offers both forward and backward stepwise variable selection. Additionally, for both forward and backward selection the user can specify the statistic used to test for removal of a variable from the model. The Conditional statistic (default method), Wald statistic, and Likelihood ratio statistic are available. The Conditional statistic uses the same formula as the Likelihood ratio statistic. However, the Conditional statistic uses conditional parameter estimates rather than maximum likelihood estimates. The methods generally follow the description of selection procedures described for multiple linear regression in Chapter 9.

From the appenc10.txt data file, we select the first 98 cases and create four new variables to match the coding scheme of ch14ta03.txt (Insert 14.6). That is, socioeconomic status will be represented by two indicator variables: $X2 = 1$ if class is middle, otherwise $X2 = 0$, and $X3=1$ if class is lower, otherwise $X3 = 0$. Thus, upper is the reference class. Also, $X4 = 0$ if the individual lives in sector 1 and $X4=1$ if the individual lives in sector 2. Finally, to use the same variable names as the text, we set $X1=age$.)

We then present the syntax to perform forward stepwise selection using the Wald statistic to test for removal of variables from the model. Partial results are shown in SPSS Output 14.13. At Step 0, before any variables are entered into the model, $X4$ has the lowest probability level of the 4 variables under consideration and, thus, is entered into the model at Step 1 (Variables in the Equation – Step 1). Of the variables not in the model at Step 1, $X1$ has the lowest probability level of the three candidate variables and is entered into the model at Step 2. Of the variables not in the model at Step 2 ($X2$, $X3$) the probability levels are both greater than the probability level required for entering a variable into the model and the stepping process stops.

Although we arrive at the same conclusion as the text, the *constant* values in SPSS Output 14.13 do not match those displayed in Table 14.11, text page 585. In the model specified in Insert 14.7, we used $X1$, $X2$, $X3$, and $X4$ as predictor variables. The text example used *age, status*, and *sector* as predictors. Both sets of predictors contain the same "information" but socioeconomic status and city sector are parameterized differently in the two models. Thus, although we arrive at the same conclusion regarding the regression coefficients, the intercepts differ due to the different parameterization.

INSERT 14.6

- ✔ Data
- ✔ Select Cases
- ✔ If Condition is Satisfied
- ✔ IF (under If Condition is Satisfied
- ✔ "case >= 99" (in the upper right hand corner box).
- ✔ Continue
- ✔ OK

- ✔ File
- ✔ New
- ✔ Syntax
- * Type the following syntax.
- if (status=2) x2=1.
- if (status ne 2)x2=0.
- if (status=3) x3=1.
- if (status ne 3)x3=0.
- if (sector=2) x4=1.
- if (sector ne 2)x4=0.
- compute x1=age.
- exe.
- ✔ Run
- ✔ All

- ✔ Analyze
- ✔ Regression
- ✔ Binary Logistic
- ✔ Y
- ✔ ▶(to the Dependent box)
- ✔ X1
- ✔ ▶(to the Covariate box)
- ✔ X2
- ✔ ▶(to the Covariate box)
- ✔ X3
- ✔ ▶(to the Covariate box)
- ✔ X4
- ✔ ▼ (by Method)
- ✔ Forward: Wald
- ✔ ▶(to the Covariate box)
- ✔ OK

SPSS Output 14.13

Variables not in the Equation

			Score	df	Sig.
Step 0	Variables	X1	7.580	1	.006
		X2	1.480	1	.224
		X3	3.909	1	.048
		X4	14.780	1	.000
	Overall Statistics		20.407	4	.000

Variables in the Equation

		B	S.E.	Wald	df	Sig.	Exp(B)
Step 1[a]	X4	1.743	.473	13.594	1	.000	5.717
	Constant	-1.589	.347	20.976	1	.000	.204
Step 2[b]	X1	.029	.013	4.946	1	.026	1.030
	X4	1.673	.487	11.791	1	.001	5.331
	Constant	-2.335	.511	20.872	1	.000	.097

a. Variable(s) entered on step 1: X4.

b. Variable(s) entered on step 2: X1.

Variables not in the Equation

			Score	df	Sig.
Step 1	Variables	X1	5.239	1	.022
		X2	.664	1	.415
		X3	1.155	1	.283
	Overall Statistics		6.296	3	.098
Step 2	Variables	X2	.968	1	.325
		X3	.733	1	.392
	Overall Statistics		1.221	2	.543

Pearson Chi-Square Goodness of Fit Test: SPSS does not compute the Pearson Chi-Square Goodness of Fit statistic.

Deviance Goodness of Fit Test: As noted on text page 588, the *deviance goodness of fit test* for logistic regression is analogous to the *F lack of fit test* we discussed in previous chapters. We present the steps without the specific syntax. Find −2 log likelihood (1191.973) for the reduced model by regressing Y on X using the Coupon Effectiveness example described above. To find −2 log likelihood for the full model we first compute $j=1$ if $X=5$, $j=2$ if $X=10$, . . . , $j=5$ if $X=30$. Now use *Binary Regression* to regress Y on X and j, and specify j as a categorical covariate. This will result in −2 log likelihood for the full model of 1189.806. The difference in the 2 values is 2.167, consistent with text page 589 where $DEV(X_0,X_1) = 2.16$.

Hosmer-Lemeshow Goodness of Fit Test. On text page 590, the author calculated the Hosmer-Lemeshow goodness of fit test for the Disease Outbreak example (ch14ta03.txt). In this case they based the calculation on 5 classes. SPSS, however, uses approximately 10 classes and the classes are often referred to as deciles of risk. To generate the Hosmer-Lemeshow goodness of fit statistic, request a logistic regression model and ✔ Options ✔ Hosmer-Lemshow goodness-of-fit ✔ Continue. Partial results are found in SPSS Output 14.14. Note that $X^2 = 9.180$ ($p = 0.327$) in SPSS Output 14.14 and the text reports $X^2 = 1.98$ ($p = 0.58$). The discrepancy is due to the test statistics being based on different contingency tables, i.e., 2x5 versus 2x10. The use of 20 cells (2x10) decreases the expected frequencies per cell relative to the use of 10 cells (2x5). However, Hosmer and Lemeshow take a liberal approach to expected values and argue that their method is an accurate assessment of goodness of fit.

SPSS Output 14.14

Hosmer and Lemeshow Test

Step	Chi-square	df	Sig.
1	9.188	8	.327

Contingency Table for Hosmer and Lemeshow Test

		Y = 0		Y = 1		
		Observed	Expected	Observed	Expected	Total
Step	1	10	9.214	0	.786	10
1	2	9	8.981	1	1.019	10
	3	9	9.489	2	1.511	11
	4	9	8.224	1	1.776	10
	5	7	7.663	3	2.337	10
	6	6	6.909	4	3.091	10
	7	3	6.092	7	3.908	10
	8	8	5.494	3	5.506	11
	9	5	3.684	5	6.316	10
	10	1	1.249	5	4.751	6

Logistic Regression Diagnostics

Logistic Regression Residuals. On text page 591, the authors demonstrate the calculation of various residual values. To request that residual and other values be saved from the logistic procedure, ✔ Options and then select the desired option as shown in SPSS Output 14.15.

SPSS Output 14.15

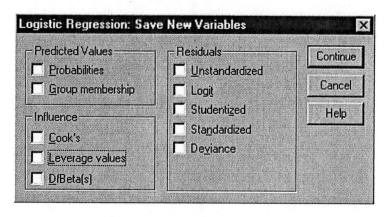

To produce Figure 14.12(d), text page 594, for example, we request that deviance residuals (saved as *dev_1* by default) (text dev_i residuals) and probabilities (*pre_1* by default) (text $\hat{\pi}_i$) be saved. We then (scatter) plot *dev_1* on the Y-axis and *pre_1* on the X-axis. Results are shown in SPSS Output 14.16. We use the Chart Editor to add the Lowess curve line to simulate Figure 14.13(c), text page 595.

SPSS Output 14.16

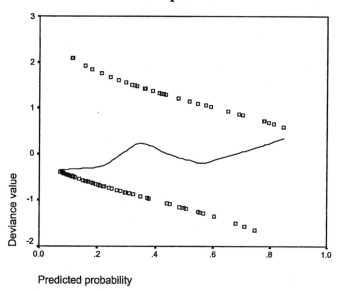

Detection of Influential Observations

Influence on Pearson Chi-Square and the Deviance Statistics. On text page 599, the authors calculate the *delta chi-square statistic* (ΔX_i^2) and the *deviance statistic* (Δdev_i). Here we use appenc10.txt; invoke *Select Cases* as shown in Insert 14.6; and use *age*, *status*, and *sector* as predictors. To calculate these statistics, we first instruct SPSS (syntax not shown) to save the standardized residuals (*zre_1*, text r_{Pi}, Table 14.11, column 1), deviance residuals (*dev_1*, text *dev$_i$*, Table 14.11, column 4), and leverage values (*lev_1*, h_{ii}, Table 14.11, column 3). SPSS does not produce the studentized Pearson residual. However, this is given in equation 14.81, text page 592 as:

$$r_{SPi} = \frac{r_{Pi}}{\sqrt{1-h_{ii}}}$$

or

$$r_{SPi} = \frac{zre\_1}{\sqrt{1-lev\_1}}$$

In Insert 14.7 we compute the studentized Pearson residual (*rspi*), ΔX_i^2 (*dx2*), and Δdev_i (*ddev*). We also use text equation 14.87, text page 600, to compute D_i (*cook*).

Figures 14.15 (a) and (b), text page 600, can now be produced by plotting *case* on the X-axis and *dx2* and *ddev* on the Y-axis's. Additionally, the above saved and computed values can be used to produce the Figures displayed on text pages 600 and 601.

INSERT 14.7

```
✔ File
✔ New
✔ Syntax
* Type the following syntax.
compute rspi=zre_1/(sqrt(1-lev_1)).
compute dx2=rspi**2.
compute ddev=lev_1*(rspi**2)+(dev_1**2).
compute cook=rspi**2*lev_1/(5*(1-lev_1)**2).
exe.
✔ Run
✔ All
```

Inferences about Mean Response

On text page 603, the authors wish to calculate the 95% confidence interval for a new observation. SPSS's *Binary Logistic* procedure does not make the variance-covariance matrix

$[s^2\{b\}$, equation 14.51, text page 578] available to the user. Thus the information to generate this confidence interval is not readily available.

Prediction of a New Observation

The Disease Outbreak Example

Data File: appenc10.txt

We use the first 96 cases of appenc10.txt for this analysis (use *Select Cases* as in Insert 14.6). The authors wish to obtain the best cutoff point for predicting whether the outcome of a new observation will be "1" or "0". If 0.5 is used as the cutoff point, for example, the prediction rule would be: if $\hat{\pi}_h \geq 0.5$, predict the outcome will be "1"; otherwise predict it will be "0". Table 14.12, text page 605, displays the classification table based on text equation 14.95. That is, because 0.316 is the proportion of the diseased individuals in the sample, we use 0.316 as the starting cutoff point.

In the SPSS *Binary Logistic* dialogue window, ✔ Options and then enter the *classification cutoff*. The box allows for only 2 decimal precision and, therefore, will not accommodate 0.316. To remedy the problem, set the *Binary Logistic* dialogue box as needed and then ✔ Paste instead of OK. This will paste the current syntax into the Syntax Editor (*Binary Logistic* procedure in Insert 14.8). The /*criteria* statement will indicate *CUT(.5)*, the default cutoff value. Edit the line to read *CUT(.316)* and then execute the syntax. This will produce SPSS Output 14.17. We can produce text Table 14.12 (b) by changing the statement to read *CUT(.326)*.

SPSS Output 14.17

Classification Table[a]

			Predicted		
			Y		Percentage
Observed			0	1	Correct
Step 1	Y	0	49	18	73.1
		1	8	23	74.2
	Overall Percentage				73.5

a. The cut value is .316

There were 8+23=31 (false negatives + true positives) diseased individuals and 49+18=67 (true negatives + false positives) non-diseased individuals. The fitted regression function correctly predicted 23 of the 31 diseased individuals. Thus, the sensitivity is calculated as true positives / (false negatives + true positives) = 23 / 8 + 23 = 0.7419 as reported on text page 606. The regression function's specificity is defined as true negatives / (false positives + true negatives) = 48 / (18 + 48) = 0.727. This is not consistent with text page 607 where the specificity is reported as 47 / 67 = 0.7014. The difference comes from the numerator used to calculate specificity. Table

357

14.12 (a), text page 605, uses 47 and we determined the number of true negatives to be 49. We assumed that the regression model was to be based on using *age, status, and sector* as predictors. If we use only *age* and *sector* as suggested by the stepwise regression above, the classification table will match Table 14.12, text page 605.

To approximate Figure 14.17, text page 606, we first saved the predicted group membership (✔ Save ✔ Group Membership ✔ Continue) while in the *Binary Logistic* dialogue box and ✔ Paste instead of OK. (Again we use *age, status,* and *sector* as predictors.) In the Syntax Editor, edit the /*cut* statement to read *CUT(.316)* and execute the syntax as shown in Insert 14.8. This will save the predicted group membership (*pgr_1* by default) based on the cutoff point of 0.316. In Insert 14.8, we then request an ROC curve. We use the Chart Editor (steps not shown) to draw a vertical line (*reference line*) at 1-specificity on the *X*-axis and a horizontal line at sensitivity on the *Y*-axis. Results are shown in SPSS Output 14.18.

INSERT 14.8


```
LOGISTIC REGRESSION VAR=y
 /METHOD=ENTER age status sector
 /SAVE PGROUP
 /CRITERIA PIN(.05) POUT(.10) ITERATE(20) CUT(.316) .
exe.

* Type the following syntax.
roc pgr_1 by y (1)
 /plot = curve
 /print = coordinates.
exe.
 ✔ Run
 ✔ All
```

SPSS Output 14.18

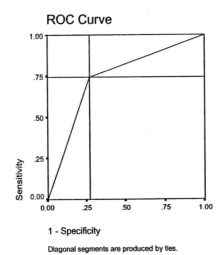

ROC Curve

1 - Specificity

Diagonal segments are produced by ties.

358

The Disease Outbreak Example – Validation Data Set

Data File: appenc10.txt

On text page 607, the researchers now wish to use the regression function based on the Model Building data set to classify observations in the Validation data set. After opening appenc10.text and renaming your variables *case, age, status, sector, y,* and *other*, the Data Editor should look like SPSS Output 14.19.

SPSS Output 14.19

	case	age	status	sector	y	other
1	1.0	33.0	1	1	0	1
2	2.0	35.0	1	1	0	1
3	3.0	6.0	1	1	0	0
4	4.0	60.0	1	1	0	1
5	5.0	18.0	3	1	1	0

appenc10.sav - SPSS Data Editor
File Edit View Data Transform Analyze Graphs Utilities Window Help
1 : case   1

In Inert 14.9, we first must compute *X1, X2, X3,* and *X4* so that these values will be contained in the Data Editor. We compute the estimated probabilities (*yhat*) for each case using the regression function (equation 14.46, text page 574) based on the model-building data set. We then use prediction rule 14.96 (text page 606) to establish group membership (*newgp*) as *newgp* = 1 if *yhat* ≥ 0.325 and *newgp* = 0 if *yhat* < 0.325. We also select on cases 99 to 196. Finally, we use the *crosstab* procedure to form a 2x2 table of *Y* values (diseased versus non-diseased) by *newgp* values (predicted diseased versus predicted non-diseased). Results are shown in SPSS Output 14.20 and are consistent with text page 608. The off-diagonal percentages indicate that 12 (46.2%) of the 26 diseased individuals were incorrectly classified (prediction error) as non-diseased. Twenty-eight (38.9%) of the 72 non-diseased individuals were incorrectly classified as diseased.

✔ File
✔ New
✔ Syntax
* Type the following syntax.
compute x1=age.
if (status=2) x2=1.
if (status ne 2)x2=0.
if (status=3) x3=1.
if (status ne 3)x3=0.
if (sector=2) x4=1.
if (sector ne 2)x4=0.
compute yhat=(1+exp(2.3129-(.02975*x1)-(.4088*x2)+(.30525*x3)-(1.5747*x4)))**-1.
if (yhat ge .325) newgp=1.
if (yhat lt .325) newgp=0.
exe.
✔ Run
✔ All

✔ Data
✔ Select Cases
✔ If Condition is Satisfied
✔ IF (under If Condition is Satisfied
* Type "case >= 99" (in the upper right hand corner box).
✔ Continue
✔ OK

✔ Analyze
✔ Descriptive Statistics
✔ Crosstabs
✔ Y
✔ ► (to the Row box)
✔ newgp
✔ ► (to the Column box)
✔ Cells
✔ Row
✔ Continue
✔ OK

SPSS Output 14.20

Y * NEWGP Crosstabulation

			NEWGP		Total
			.00	1.00	
Y	0	Count	44	28	72
		% within Y	61.1%	38.9%	100.0%
	1	Count	12	14	26
		% within Y	46.2%	53.8%	100.0%
Total		Count	56	42	98
		% within Y	57.1%	42.9%	100.0%

Polyotomous Logistic Regression for Nominal Response

Pregnancy Duration Example

Data File: ch14ta13.txt

On text page 609, a researcher studied the relationship between gestation period and several predictor variables. Gestation period was classified as less than 36 weeks (Y=1), 36 to 37 weeks (Y=2), and greater than 37 weeks (Y=3). After opening ch14ta13.txt and renaming the variables *case, Y, RC1, RC2, RC3, X1, X2, X3, X4,* and *X5*, the Data Editor should look like SPSS Output 14.21. In Insert 14.10, we request SPSS's *Multinomial Logistic* program. We specify *X1* as a covariate; *X2, X3, X4,* and X5 as factors; and *Y* as the dependent variable. Main effects only is the default model in *Multinomial Logistic*. Results are shown in SPSS Output 14.22. Notice that our results are not entirely consistent with Figure 14.18, text page 613. Our odds ratios are generally less than 1 where those in Figure 14.18 are generally greater than 1. Our constants also do not match Figure 14.18. By some hand calculations we could show that our findings are equivalent to those presented in the text. It is easier, however, to re-parameterize our model by executing the *recode* statement in Insert 14.10 and then re-running the *Multinomial Logistic* procedure. That is, we simply reverse the values of *X2, X3, X4,* and *X5* into *x2rev, x3rev, x4rev,* and *x5rev* for use in the *Multinomial Logistic* procedure. We repeat the first *Multinomial Logistic* procedure and use *x2rev, x3rev, x4rev,* and *x5rev* rather than *X2, X3, X4,* and *X5* as independent variables (factors) and *X1* as a covariate. This results in SPSS Output 14.23, which is entirely consistent with text Figure 14.18.

SPSS Output 14.21

	case	y	rc1	rc2	rc3	x1	x2	x3	x4	x5
1	1.0	1	1	0	0	150.0	0	0	0	1
2	2.0	1	1	0	0	124.0	1	0	0	0
3	3.0	1	1	0	0	128.0	0	0	0	1
4	4.0	1	1	0	0	128.0	1	0	0	1
5	5.0	1	1	0	0	133.0	0	0	1	1

361

INSERT 14.10

- ✔ Analyze
- ✔ Regression
- ✔ Multinomial Logistic
- ✔ Y
- ✔ ▶ (to the Dependent box)
- ✔ X1
- ✔ ▶ (to the Covariate box)
- ✔ X2 X3 X4 X5
- ✔ ▶ (to the Factor box)
- ✔ OK

- ✔ File
- ✔ New
- ✔ Syntax

* Type the following syntax.
recode
  x2 x3 x4 x5
  (1=0) (0=1) into x2rev x3rev x4rev x5rev .
execute .

- ✔ Run
- ✔ All

SPSS Output 14.22

Parameter Estimates

Y^a		B	Std. Error	Wald	df	Sig.	Exp(B)	95% Confidence Interval for Exp(B)	
								Lower Bound	Upper Bound
1	Intercept	14.987	3.322	20.357	1	.000			
	X1	-.065	.018	12.865	1	.000	.937	.904	.971
	[X2=0]	-2.957	.964	9.400	1	.002	.052	.008	.344
	[X2=1]	0^b	.	.	0	.	.	.	.
	[X3=0]	-2.060	.895	5.299	1	.021	.127	.022	.736
	[X3=1]	0^b	.	.	0	.	.	.	.
	[X4=0]	-2.043	.710	8.285	1	.004	.130	.032	.521
	[X4=1]	0^b	.	.	0	.	.	.	.
	[X5=0]	-2.452	.732	11.239	1	.001	.086	.021	.361
	[X5=1]	0^b	.	.	0	.	.	.	.
2	Intercept	12.057	2.852	17.866	1	.000			
	X1	-.046	.015	9.736	1	.002	.955	.927	.983
	[X2=0]	-2.913	.858	11.542	1	.001	.054	.010	.292
	[X2=1]	0^b	.	.	0	.	.	.	.
	[X3=0]	-1.888	.809	5.446	1	.020	.151	.031	.739
	[X3=1]	0^b	.	.	0	.	.	.	.
	[X4=0]	-1.067	.650	2.699	1	.100	.344	.096	1.229
	[X4=1]	0^b	.	.	0	.	.	.	.
	[X5=0]	-2.230	.668	11.143	1	.001	.107	.029	.398
	[X5=1]	0^b	.	.	0	.	.	.	.

a. The reference category is: 3.

b. This parameter is set to zero because it is redundant.

Parameter Estimates

Y[a]		B	Std. Error	Wald	df	Sig.	Exp(B)	95% Confidence Interval for Exp(B)	
								Lower Bound	Upper Bound
1	Intercept	5.475	2.272	5.809	1	.016			
	X1	-.065	.018	12.865	1	.000	.937	.904	.971
	[X2REV=.00]	2.957	.964	9.400	1	.002	19.241	2.906	127.408
	[X2REV=1.00]	0[b]	.	.	0	.	.	.	.
	[X3REV=.00]	2.060	.895	5.299	1	.021	7.843	1.358	45.304
	[X3REV=1.00]	0[b]	.	.	0	.	.	.	.
	[X4REV=.00]	2.043	.710	8.285	1	.004	7.713	1.919	31.001
	[X4REV=1.00]	0[b]	.	.	0	.	.	.	.
	[X5REV=.00]	2.452	.732	11.239	1	.001	11.616	2.769	48.720
	[X5REV=1.00]	0[b]	.	.	0	.	.	.	.
2	Intercept	3.958	1.941	4.158	1	.041			
	X1	-.046	.015	9.736	1	.002	.955	.927	.983
	[X2REV=.00]	2.913	.858	11.542	1	.001	18.420	3.430	98.912
	[X2REV=1.00]	0[b]	.	.	0	.	.	.	.
	[X3REV=.00]	1.888	.809	5.446	1	.020	6.603	1.353	32.227
	[X3REV=1.00]	0[b]	.	.	0	.	.	.	.
	[X4REV=.00]	1.067	.650	2.699	1	.100	2.907	.814	10.382
	[X4REV=1.00]	0[b]	.	.	0	.	.	.	.
	[X5REV=.00]	2.230	.668	11.143	1	.001	9.304	2.511	34.470
	[X5REV=1.00]	0[b]	.	.	0	.	.	.	.

a. The reference category is: 3.

b. This parameter is set to zero because it is redundant.

Polyotomous Logistic Regression for Ordinal Response

Pregnancy Duration Example

Data File: ch14ta13.txt

The steps to perform Polyotomous Logistic Regression for a nominal response completely generalize to the case where there is an ordinal response. Use Insert 14.10 and ✔ *Regression* ✔ *Ordinal* instead of ✔ *Regression* ✔ *Multinomial Logistic*. Results are not shown but are similar to Figure 14.19, text page 618.

Poisson Regression

Miller Lumber Company Example

Data File: ch14ta14.txt

The Miller Lumber Company conducted an in-store customer survey. The researcher counted the number of customers who visited the store from each nearby census tract. The researcher also collected and subsequently retained five (quantitative) predictor variables for use in the Poisson Regression. See Chapter 0, "Working with SPSS," for instructions on opening the Miller Lumber Company data file and renaming the *X1*, *X2*, *X3*, *X4*, *X5*, and *Y*. After opening ch14ta14.txt, we opened the Syntax Editor and executed *compute id=$casenum* (syntax not shown) to produce the *ID* variable. After this maneuver, the first 5 lines in the SPSS Data Editor should look like SPSS

Output 14.24. Note that Table 14.14, text page 622, appears to indicate that the variables are ordered in the data file as *X1*, *X2*, *X3*, *X4*, *X5,* and *Y*. By examining the Data Editor you will sometimes find that the data file was not structured as expected.

SPSS Output 14.24

Miller Lumber Company.sav - SPSS Data Editor

File Edit View Data Transform Analyze Graphs Utilities Window Help

1 : y   9

	y	x1	x2	x3	x4	x5	id
1	9.0	606.00	41393.00	3.0	3.04	6.32	1.00
2	6.0	641.00	23635.00	18.0	1.95	8.89	2.00
3	28.0	505.00	55475.00	27.0	6.54	2.05	3.00
4	11.0	866.00	64646.00	31.0	1.67	5.81	4.00
5	4.0	599.00	31972.00	7.0	.72	8.11	5.00

There are some complexities in performing Poisson regression in SPSS. The following eight paragraphs of verbatim explanation are used with permission from SPSS and can be found using AnswerNet at spss.com (Resolution number: 16655).

"The easiest way to handle Poisson regression models in current releases of SPSS (Release 6.1 and above) is to use the GENLOG procedure, which does general loglinear and logit modeling. The simplest type of Poisson model for our purposes is one in which the counts are modeled without denominators (i.e., we are modeling counts rather than rates), and all predictors are categorical. Rate models and quantitative predictors introduce further complexities into the process, but they can still be handled."

"If the simplest form of the model is being used, the variables in the data set consist simply of the predictors and a count variable. If rates are to be modeled, an additional variable representing the denominators is added. If quantitative predictors are to be used, a subject ID variable is also added, with a unique value for each case. The file preparation is completed by weighting the data by the count variable (in the dialog boxes, click on Data->Weight Cases, then click on Weight cases by and move the appropriate variable into the Frequency Variable box). The status bar at the bottom of the screen should have "Weight On" in one of the boxes."

"The Poisson option in the GENLOG procedure is the default for Statistics->Loglinear->General. If there are only categorical predictors, these are selected as factors. The count variable is not used (the data are already weighted by this variable). If a rate model is being fitted, the denominator variable is specified as the Cell Structure variable. The model is specified by clicking on the Model button, checking Custom and specifying the desired predictive model (or leaving it as full factorial, if that's the desired model). Continue out of the Model dialog and make any desired specifications in the Options or Save dialogs (printing of parameter estimates is specified in the Options dialog). When finished with these, click OK, or Paste if you want to see the GENLOG syntax that's been created."

"If quantitative predictors are used, the situation will generally be more complicated. The term quantitative predictor here is used to denote a predictor to be treated as is for modeling purposes. That is, you want to use that variable's values directly in the design matrix and produce a single parameter estimate, rather than creating a set of indicator variables to represent the unique levels or categories. Though technically not entirely accurate, the term continuous is commonly used for such a variable."

"The complications arise because GENLOG analyzes data on a cell by cell basis, with cells defined by the combinations of the factor variables. Cell covariates are handled by averaging all values in the same cell defined by the factors specified in the main dialog (on the GENLOG command). Thus, information about individual case values will be lost unless each case is treated as a cell. This is why we have to use the subject ID variable when there are cell covariates: to trick GENLOG into seeing each case as a cell. Note that if you have any quantitative/continuous variables, you must treat all predictors that way (as cell covariates). If you use any factors other than the subject ID variable, the table that GENLOG will construct internally will have as many cells as the product of the number of subjects and the number of levels of those factors, and you will not get the desired results."

"If there are only quantitative variables to be used, specify them as cell covariates and the subject ID variable as the only factor. Then go into the Model dialog box and specify a Custom model that includes the desired predictors and any interactions, but not the subject ID. After requesting any other specifications in other dialogs, click OK or Paste. Since you have a subject ID variable specified as a factor that is not being used in the model, a popup warning will tell you about this. Simply click OK."

"If there are any 0 counts that produce 0 marginals (as any 0 count will with quantitative predictors using the subject ID trick), these should be recoded to a small positive number (e.g., 1E-12). ML estimates will still be good to as many decimals as are printed on the output. 0 marginals will cause the procedure to refuse to run."

"As noted above, if there is a combination of quantitative/continuous and categorical predictors, all predictors must be treated as cell covariates. This means that for categorical factors, you must first dummy or effect code variables to represent the desired contrasts among their levels, and include these dummy variables as cell covariates"

In Insert 14.11, we produce the *ID* variable if not done above. To check for zero counts, we use *Frequencies* to obtain a frequency distribution of *Y*. For the Miller Lumber example we find that 3 of the counts (*Y* values) are zero. As instructed above, if *Y*=0 then we set *Y* = 0.00000001. We continue with the instruction as presented above to produce SPSS Output 14.25. The parameter estimates are consistent with Table 14.15, text page 622. Deviance Residuals can be displayed by checking Residuals under Display in the Options window of the *General Loglinear* window.

INSERT 14.11	INSERT 14.11 continued
✔ File	✔ Analyze
✔ New	✔ Loglinear
✔ Syntax	✔ General
compute id=$casenum.	✔ Poission (under Distribution of Cell Counts)
execute.	✔ x1
✔ Run	✔ ▶ (to Cell Covariates)
✔ All	✔ x2
	✔ ▶ (to Cell Covariates)
✔ Analysis	✔ x3
✔ Descriptive Statistics	✔ ▶ (to Cell Covariates)
✔ Frequencies	✔ x4
✔ y	✔ ▶ (to Cell Covariates)
✔ ▶	✔ x5
✔ OK	✔ ▶ (to Cell Covariates)
	✔ id
✔ File	✔ ▶ (Factor(s))
✔ New	✔ Model
✔ Syntax	✔ Custom
if (y=0) y=.00000001.	✔ ▼
exe.	✔ Main effects
✔ Run	✔ x1 x2 x3 x4 x5
✔ All	✔ ▶ (Under Build Terms)
	✔ Continue
✔ Data	✔ Options
✔ Weight Cases	* Uncheck all options under Display except **Estimates**.
✔ Weight Cases By	* Uncheck all options under Plot.
✔ y	✔ Continue
✔ ▶	✔ OK
✔ OK	* Your will get an error: "The following factors are".
	* not used in the Model: id".
	✔ OK